MW01017197

Studies in Military and Strategic History

General Editor: **Michael Dockrill**, Professor of Diplomatic History, King's College, London

Published titles include:

Nigel John Ashton
EISENHOWER, MACMILLAN AND THE PROBLEM OF NASSER
Anglo-American Relations and Arab Nationalism, 1955–59

Peter Bell
CHAMBERLAIN, GERMANY AND JAPAN, 1933–34

G. H. Bennett
BRITISH FOREIGN POLICY DURING THE CURZON PERIOD, 1919–24

David A. Charters
THE BRITISH ARMY AND JEWISH INSURGENCY IN PALESTINE, 1945–47

David Clayton
IMPERIALISM REVISITED
Political and Economic Relations between Britain and China, 1950–54

Michael J. Cohen and Martin Kolinsky (*editors*)
BRITAIN AND THE MIDDLE EAST IN THE 1930s: Security Problems, 1935–39

Paul Cornish
BRITISH MILITARY PLANNING FOR THE DEFENCE OF GERMANY, 1945–50

Michael Dockrill
BRITISH ESTABLISHMENT PERSPECTIVES ON FRANCE, 1936–40

Robert Frazier
ANGLO-AMERICAN RELATIONS WITH GREECE
The Coming of the Cold War, 1942–47

John P. S. Gearson
HAROLD MACMILLAN AND THE BERLIN WALL CRISIS, 1958–62

John Gooch
ARMY, STATE AND SOCIETY IN ITALY, 1870–1915

G. A. H. Gordon
BRITISH SEA POWER AND PROCUREMENT BETWEEN THE WARS
A Reappraisal of Rearmament

Stephen Hartley
THE IRISH QUESTION AS A PROBLEM IN BRITISH FOREIGN POLICY, 1914–18

Brian Holden Reid
J. F. C. FULLER: Military Thinker

Stewart Lone
JAPAN'S FIRST MODERN WAR
Army and Society in the Conflict with China, 1894–95

Thomas R. Mockaitis
BRITISH COUNTERINSURGENCY, 1919–60

T. R. Moreman
THE ARMY IN INDIA AND THE DEVELOPMENT OF FRONTIER WARFARE, 1849–1947

Kendrick Oliver
KENNEDY, MACMILLAN AND THE NUCLEAR TEST-BAN DEBATE, 1961–63

G. D. Sheffield
LEADERSHIP IN THE TRENCHES
Officer–Man Relations, Morale and Discipline in the British Army in the Era of the First World War

Simon Trew
BRITAIN, MIHAILOVIC AND THE CHETNIKS, 1941–42

Steven Weiss
ALLIES IN CONFLICT
Anglo-American Strategic Negotiations, 1938–44

Roger Woodhouse
BRITISH FOREIGN POLICY TOWARDS FRANCE, 1945–51

Studies in Military and Strategic History
Series Standing Order ISBN 0–333–71046–0
(*outside North America only*)

You can receive future titles in this series as they are published by placing a standing order. Please contact your bookseller or, in case of difficulty, write to us at the address below with your name and address, the title of the series and the ISBN quoted above.

Customer Services Department, Macmillan Distribution Ltd, Houndmills, Basingstoke, Hampshire RG21 6XS, England

Leadership in the Trenches

Officer–Man Relations, Morale and Discipline in the British Army in the Era of the First World War

G. D. Sheffield
Senior Lecturer
King's College, London
based at
Joint Services Command Staff College

Foreword by Peter Simkins

in association with
KING'S COLLEGE, LONDON

First published in Great Britain 2000 by
MACMILLAN PRESS LTD
Houndmills, Basingstoke, Hampshire RG21 6XS and London
Companies and representatives throughout the world

A catalogue record for this book is available from the British Library.

ISBN 0–333–65411–0

First published in the United States of America 2000 by
ST. MARTIN'S PRESS, INC.,
Scholarly and Reference Division,
175 Fifth Avenue, New York, N.Y. 10010

ISBN 0–312–22640–3

Library of Congress Cataloging-in-Publication Data
Sheffield, G. D.
Leadership in the trenches : officer–man relations, morale and discipline in the British Army in the era of the first World War / G.D. Sheffield ; foreword by Peter Simkins.
p. cm. — (Studies in military and strategic history)
Includes bibliographical references and index.
ISBN 0–312–22640–3 (cloth)
1. Great Britain. Army—Officers—History—20th century.
2. Leadership. 3. Morale. 4. Command of troops. 5. Military discipline—Great Britain. 6. World War, 1914–1918—Great Britain.
I. Title. II. Series.
UB415.G7S54 1999
355.1'23'094109041—dc21 99–30645
CIP

This book is printed on paper suitable for recycling and made from fully managed and sustained forest sources.

10 9 8 7 6 5 4 3 2 1
09 08 07 06 05 04 03 02 01 00

Printed and bound in Great Britain by
Antony Rowe Ltd, Chippenham, Wiltshire

To the memory of Dr John Pimlott (1948–97), Head of the Department of War Studies, RMA Sandhurst, 1994–97: a fine historian and teacher and a good friend, sorely missed

Contents

Foreword

The massive and unprecedented expansion of the British Army in the First World War not only necessitated a correspondingly huge influx of new officers but also placed extraordinary demands upon those granted commissions. Many who, in normal circumstances, might never have contemplated entering the army or seeking a commission, found themselves called upon to lead men in some of the most protracted and ferocious battles ever fought by British soldiers. Even in so-called 'quiet' sectors of the Western Front, the daily struggle against the elements – particularly in the severe winter of 1916–17 – and the steady toll of casualties from enemy action and sickness, required officers to summon up all their inner resources of mental and physical strength in order to maintain morale and the fighting efficiency of their units. Such qualities of leadership were needed in yet greater measure in the grim battles of attrition on the Somme in 1916 and at Arras and Ypres in 1917.

This, of course, was also true of the officers and men on the other side of No Man's Land. Whatever their background in civilian life, their previous military service or their individual experience and training, few, if any, were really prepared for the ordeal of modern industrialised warfare. Nevertheless, it was a very remarkable national achievement for Britain – in the midst of such a war – to create the country's first-ever mass army almost from scratch and for those same citizen-soldiers to withstand, match and then finally defeat the formidable Imperial German Army.

Central to that achievement was the performance of British officers from brigade, battalion and battery commanders downwards, especially in 1918 when mobility was restored to the battlefield and junior leaders were obliged, after three years of trench warfare, to adjust swiftly to more fluid operations in which constant reference to more senior officers was neither possible nor desirable. The devolution of command downwards and the simultaneous process of tactical decentralisation in the British Expeditionary Force (BEF) between 1916 and 1918 are important aspects of First World War studies on which a great deal of detailed scholarly research still remains to be carried out. However, it is already clear from the work of Dr Sheffield and others that this devolution and decentralisation *did* occur and, moreover – as

Dr Sheffield himself observes – that it took place against a background of growing technological and tactical sophistication which, in turn, further tested the adaptability of young officers to the limit.

Evidence of these trends can be found, for example, in the increasingly elaborate trench raids planned at divisional, brigade and battalion level in 1916 and 1917 or in the thorough briefing given to junior officers before the 18th Division's assault on Thiepval on 26 September 1916. It must be admitted that, at Arras in 1917, British junior leaders, accustomed as they were now becoming to well-planned and pre-rehearsed set-piece attacks, did not always exploit successes as vigorously as they should have done once the set-piece phase was over and they were confronted with the problems of semi-open warfare. On the other hand, I have long contended that the British official historian, Brigadier General Sir James Edmonds, is excessively and unjustly critical of the performance of British battalion, company and platoon commanders in 1918. During the defensive battles of March and April that year, when direct links to higher headquarters were often temporarily cut, front-line officers were continually forced to rely on their own initiative and judgement in coping with local tactical crises. The fact that the German offensives were first blunted and subsequently halted was at least partly due to innumerable actions by small groups of officers and men who, day after day, fought hard to defend villages, ridges, crossroads or woods for perhaps two or three hours at a time, then fell back a mile or more to a fresh position and made a similar stand. Equally, as I have written elsewhere, the BEF's feats in the victorious 'Hundred Days' offensive of August to November 1918, when it was able to overcome a wide variety of tactical obstacles and challenges in rapid succession – including big set-piece assaults, street-fighting, the capture of railway cuttings and embankments, and canal-crossings on improvised bridges and rafts – would quite simply have proved impossible had its standards of junior leadership actually been as low or mediocre as Edmonds sometimes implies. I share the view of Dr Paddy Griffith, as expressed in his book *Battle Tactics of the Western Front*, that more than a whiff of 'sour grapes' is discernible in the tendency shown by Edmonds, a Regular, to underrate the tactical skills and initiative of temporary officers and citizen-soldier NCOs. Nor should it be forgotten that many British infantry units in the spring and autumn of 1918 were largely composed of inexperienced 18½ to 19½-year-old conscripts.

If we are fully to understand the various factors which contributed to the BEF's ultimate success, officer–man relations, morale and discipline are obviously all key topics to examine, as major problems in any of these areas would quickly have manifested themselves in battlefield performance and undermined the BEF's ability to wage war against such a powerful enemy. Although many writers have touched upon these subjects, and a handful of British and Commonwealth historians have recently produced notable analyses of certain aspects of morale and discipline, it is nonetheless surprising that, until now, no detailed scholarly overview of all three areas has appeared in Britain. This masterly new study, which brilliantly draws together and evaluates an immense and daunting array of evidence, therefore makes an invaluable and authoritative contribution to our knowledge of the British Army in the First World War, confirming Dr Sheffield's place in the front rank of British military historians.

Peter Simkins
Honorary Professor in Modern History
University of Birmingham

Preface

The publication of this book marks the end of a project that dates back to 1981, when as an undergraduate at the University of Leeds I wrote an essay on the factors that maintained the morale of the British army on the Western Front. I remember being struck by the importance of the paternalism of the upper-class junior officer towards his working-class soldiers, and wrote something to the effect that this showed the British class system at its best – much to the disapproval of a friend of pronounced left-wing sympathies. My understanding of the British army in the Great War has deepened over the subsequent years of research, and in many ways my views on the subject – not least on its social composition – have changed. However in 1999 I believe that my research has demonstrated that my initial hunch on the importance of the paternalism of regimental officers was essentially correct.

I have been immensely fortunate to be near the centre of the British military historical profession during the writing of this book. As a member of the Sandhurst War Studies Department, Secretary of the British Commission for Military History and formerly a doctoral student at King's College London I have been able to tap into the expertise of many friends, colleagues and students. They cannot, of course, be expected to take the blame for any blemishes in this book. Those who deserve especial thanks include Professor Brian Bond (who supervised the original thesis), Dr Stephen Badsey, who read through the entire manuscript and made numerous helpful suggestions, Lloyd Clark, Peter Simkins, Keith Simpson, MP (who gave me access to his unique questionnaires on officers of the Great War), Professor Ian Beckett, Dr Ilana Bet-el, Julian Putkowski, the late Rae Russell, Bob Wyatt, Michael Orr, Chris McCarthy, John Lee and, not least, my colleagues in the Department of War Studies and elsewhere at RMA Sandhurst. Without the help of successive Sandhurst librarians, Andrew Orgill and Diane Hillier, and the librarian at Staff College, Pam Bendell, and their staffs, this book could not have been written. The staff of various archives gave me assistance far beyond the call of duty: at the Australian War Memorial, Bet Dracoulis, Ric Pelvin and their staff; at the Imperial War Museum, Rod Suddaby, Nigel Steel, Simon Robbins, and Sarah Patterson; at the Liddle Collection, University of Leeds, Peter H. Liddle and Dr Ian Whitehead; at the Essex Regiment Museum, Ian Hook.

I have left thanking some of the most important people of all to last. Alan and Mandy Bird, by forcing the Sheffield family to take a holiday with them, saved me from a summer of self-imposed martyrdom in 1992, and also uncomplainingly provided accommodation on my visits to Leeds. Without the financial support of my parents in the early 1980s, I would never have been in a position to pursue academic research. Above all, I wish to thank my wife, Vivienne, for her steadfast support for a project that must have seemed never-ending, and for coping with Jennie and James when I was distracted by work.

G.D. SHEFFIELD
The Royal Military Academy Sandhurst

List of Abbreviations Used in Text and Notes

2/Lt	Second Lieutenant
AA and QMG	Assistant Adjutant and Quartermaster-General
AAG	Assistant Adjutant-General
AG	Adjutant-General
AIF	Australian Imperial Force
AOC	Army Ordnance Corps
ANZAC, Anzac	Australian and New Zealand Army Corps
AP & SS	Army Printing and Stationery Services
APM	Assistant Provost Marshal
Appx	Appendix
A and Q	Adjutant-General's and Quartermaster-General's Branches
AO	Army Order
AOH	Australian Official History
AQ	*Army Quarterly*
Argylls	Argyll and Sutherland Highlanders
AR	*Army Review*
ARO	Army Routine Orders
ASC	Army Service Corps
Aus.	Australian
AWM	Australian War Memorial
BAR	*British Army Review*
Bde	Brigade
Beds	Bedfordshire
BEF	British Expeditionary Force
Berks	Berkshire
BJS	*British Journal of Sociology*
BLPES	British Library of Political and Economic Science
Brig. Gen.	Brigadier-General
BSM	Battery Sergeant-Major
Bucks	Buckinghamshire
Buffs	Buffs (East Kent Regiment)
Camerons	Queen's Own Cameron Highlanders
Can.	Canadian
Capt.	Captain
CEF	Canadian Expeditionary Force

CIGS	Chief of the Imperial General Staff
CJ	*Cavalry Journal*
CJH	*Canadian Journal of History*
CMP	Corps of Military Police
Cpl	Corporal
CO	Commanding Officer
Col	Colonel
CSM	Company Sergeant Major
CUP	Cambridge University Press
DAC	Divisional Ammunition Column
DCLI	Duke of Cornwall's Light Infantry
DERR	Duke of Edinburgh's Royal Regiment
Div.	Division
DLI	Durham Light Infantry
DPM	Deputy Provost-Marshal
DSO	Distinguished Service Order
Dvr	Driver
DWR	Duke of Wellington's Regiment
ed.	editor, edited
edn	edition
EHR	*English Historical Review*
E. Lancs	East Lancashire Regiment
E. Lothian	East Lothian
E. Surreys	East Surrey Regiment
E. Yorks	East Yorkshire Regiment
ERM	Essex Regiment Museum
FGCM	Field General Court Martial
FH	*French History*
FO	Foreign Office
FSR	Field Service Regulations
Fus	Fusiliers
Gen.	General
GHQ	General Headquarters
Glosters	Gloucestershire Regiment
GOC	General Officer Commanding
Gordons	Gordon Highlanders
Gnr	Gunner
Gren.	Grenadier
GSO	General Staff Officer
GRO	General Routine Orders
GW	*The Great War*

HAC	Honourable Artillery Company
Hants	Hampshire
HJ	*Historical Journal*
HLI	Highland Light Infantry
HM	His Majesty's
HMSO	His Majesty's Stationery Office
IGC	Inspector General of Communications
IHS	*Irish Historical Studies*
IWM	Imperial War Museum
JAWM	*Journal of the Australian War Memorial*
JCH	*Journal of Contemporary History*
J. Mil Hist.	*Journal of Military History*
JRA	*Journal of the Royal Artillery*
JRAMC	*Journal of the Royal Army Medical Corps*
JRUSI	*Journal of the Royal United Services Institution*
JSAHR	*Journal of the Society for Army Historical Research*
King's	King's (Liverpool) Regiment
King's Own	King's Own Royal Lancaster Regiment
KOSB	King's Own Scottish Borderers
KOYLI	King's Own Yorkshire Light Infantry
KRRC	King's Royal Rifle Corps
KRS Q	K.R. Simpson's Questionnaires
KSLI	King's Shropshire Light Infantry
L/Cpl	Lance Corporal
LC	Liddle Collection
LHCMA	Liddell Hart Centre for Military Archives
Lincs	Lincolnshire
Londons	London Regiment
LOOB	Left Out of Battle
Loyals	Loyal North Lancashire Regiment
Lt	Lieutenant
Lt Col	Lieutenant-Colonel
Lt Gen.	Lieutenant-General
LRB	5th Londons (London Rifle Brigade)
MA	*Military Affairs*
Maj.	Major
Maj. Gen.	Major-General
MC	Military Cross
MEF	Mediterranean Expeditionary Force
MGC	Machine Gun Corps
MG Coy.	Machine Gun Company

MP	Member of Parliament
MUP	Manchester University Press
NA	New Army (also known as Kitchener's Army)
NAM	National Army Museum
NCO	Non-Commissioned Officer
nd	not dated
NF	Northumberland Fusiliers
NLS	National Library of Scotland
Northants	Northamptonshire
Notts	Nottinghamshire
N. Staffs	North Staffordshire
NZ	New Zealand
NZEF	New Zealand Expeditionary Force
OBLI	Oxfordshire and Buckinghamshire Light Infantry
OC	Officer Commanding
OCA	Old Comrades' Association
OCB	Officer Cadet Battalion
OH	Official History
OTC	Officer Training Corps
OUP	Oxford University Press
P and P	*Past and Present*
PP	Parliamentary Papers
PRO	Public Record Office
Pte	Private
QMS	Quartermaster-Sergeant
QR	*Quarterly Review*
Queen's	The Queen's (Royal West Surrey Regiment)
R	Royal
RA	Royal Artillery
RAMC	Royal Army Medical Corps
Rifle Bde	Rifle Brigade
R. Berks	Royal Berkshire Regiment
RDF	Royal Dublin Fusiliers
RE	Royal Engineers
RFA	Royal Field Artillery
RFC	Royal Flying Corps
R. Fusiliers	Royal Fusiliers
RGA	Royal Garrison Artillery
RHA	Royal Horse Artillery
RIHM	*Revue Internationale d'Histoire Militaire*
R. Innis. Fus.	Royal Inniskilling Fusiliers

R. Irish Rif.	Royal Irish Rifles
RKP	Routledge & Kegan Paul
RMA	Royal Military Academy
RMASA	Royal Military Academy Sandhurst Archives
RMASL	Royal Military Academy Sandhurst Library
RMC	Royal Military College
RMCR	Royal Military College Record
RMO	Regimental Medical Officer
RMPA	Royal Military Police Archives
RND	Royal Naval Division
RO	Routine Orders
RQMS	Regimental Quartermaster-Sergeant
RSF	Royal Scots Fusiliers
RSM	Regimental Sergeant-Major
R. Sussex	Royal Sussex Regiment
R. Warwicks	Royal Warwickshire Regiment
RWF	Royal Welch Fusiliers
RWK	Royal West Kent Regiment
SCL	Staff College Library
Sgt	Sergeant
SLI	Somerset Light Infantry
Scot. Rif.	Scottish Rifles
ST!	Stand To!
Supp.	Supplementary
TF	Territorial Force
TLS	*The Times Literary Supplement*
TMB	Trench Mortar Battery
UP	University Press
USM	United Services Magazine
vol.	volume
VB	Volunteer Battalion
VC	Victoria Cross
WD	War Diary
WIH	*War in History*
WOL	War Office Library
W&S	*War & Society*
W. Yorks	West Yorkshire Regiment
YOC	Young Officer Companies
York and Lancs	York and Lancaster Regiment
Yorks	Yorkshire Regiment (Green Howards)

Acknowledgements

Crown copyright material appears with the kind permission of the Controller of Her Majesty's Stationery Office. Crown copyright material in the Haig papers is reproduced by permission of the Trustees of the National Library of Scotland. I am also grateful to the following for granting me permission to quote from material to which they hold the copyright: the Australian War Memorial; the British Library of Political and Economic Science, London School of Economics; Trustees of the Imperial War Museum; the Trustees of the Liddell Hart Centre for Military Archives, King's College London; Dr Peter Liddle, Keeper of the Liddle Collection, University of Leeds; National Army Museum; Keith Simpson MP.

I am also grateful to the following for permission to quote from material located at the Imperial War Museum, to which they hold the copyright.

Mrs M. Abraham (A.J. Abraham papers)
Mrs E.P. Adams (H.L. Adams papers)
Mrs J.E. Anderson (J.W. Riddell papers)
C.J. Ashdown (C.F. Ashdown papers)
Mrs S.M. Ashton (R. Cude papers)
Mrs S. Bond (W.P. Nevill)
Mrs D. Brooking (J.M. Thomas papers)
Mrs J. Bull (J.W. Mudd)
M. Calcraft (P.R. Hall papers)
Mrs M. Delgado (H.E. Politzer papers)
R.H. Earp (A. Surfleet papers)
M.R.D. Foot (R.C. Foot papers)
Mrs J.M. Francis (L. Parrington papers)
W.D. Fulton (J.O. Coop papers)
K. Griffiths (J. Griffiths papers)
G. Gower (J.W. Gower papers)
Dr R. Hamond (D. Hamond papers)
Mrs C.A. Hankey (W.B. Medlicott)
Mrs G. Hardie (M. Hardie papers)
Mrs J. Hewitt (C.J. Lodge Patch papers)
N. Lucas (G.D.A. Black papers)
R.G.S Johnston (R.W.F. Johnston papers)

P.P.H. Jones (P.H. Jones papers)
A.J. Maxse (Sir F.I. Maxse papers)
Mrs G. Millman (J. Williams papers)
Mrs D.K. Parker (R.E. Barnwell papers)
Mrs R. Pearson (R.S. Cockburn papers)
M. Pease (N.A. Pease papers)
Lord Robertson (W.R. Robertson papers)
Mrs M. Ryder (R.D. Fisher papers)
E. Tobitt (C.R. Tobitt papers)
Mrs M. Walker (G. Brown papers)
C.J. White (J. Woolin papers)
P. Williamson (H.S. Williamson papers)
A. Young (A. Young papers)

The IWM proved unable to trace the current whereabouts of the copyright holders of a number of collections I consulted. To anyone whose copyright I have unwittingly infringed I offer my sincere apologies.

Introduction

'An army, like any other human society, is an organism, whose well-being depends on the interplay of human relationships'.[1] I wrote this book in the belief that this assertion by a British temporary officer of the First World War was correct, and that a full-length study of one aspect of the 'interplay of human relationships', the relationship between officers and men in the British army of the First World War, remained to be written. Indeed, until fairly recently, historians' judgements on the subject of officer–man relationships in the British army in the 1902–22 period tended to be superficial and polemical. Writing in 1961, Brian Gardner, for instance, claimed that officers and men of the Edwardian army 'normally disliked, and often despised, each other'.[2] Gardner, who offered no evidence to support this statement, was a member of the 'lions led by donkeys' school of the 1960s,[3] and to depict the prewar army as a class-ridden, divided organisation suited his thesis that the Somme campaign was the consequence of British social and class failings. Other writers moved to the opposite extreme, painting a picture of prewar regiments as happy families, devoid of inter-rank tensions, consisting of loyal, contented soldiers and paternal, benign officers.[4]

The first serious modern analysis of officer–man relationships in the Great War was John Baynes' *Morale* (1967) in which he examined the relationship in 2/Scottish Rifles from the immediate prewar period to the aftermath of Neuve Chapelle in March 1915.[5] While my own study of a New Army battalion remains the only full-length work devoted to officer–man relations in the British army of 1914–18,[6] many works on the British army of the period make some reference to officer–man relations.[7] The debate has tended to focus on whether the experience of the officer differed significantly from that of the Other Rank, or if it is possible to speak of a common experience. Ian Beckett, for instance, stresses the considerable privileges enjoyed by officers but denied to Other Ranks, while Peter Liddle regards the idea that it was impossible to bridge the 'socio-military gap' between officer and Other Rank as misconceived.[8] John Bourne provides a short but important 'vindication of the traditional stereotype' of the officer as a leader and a maintainer of morale.[9]

Turning to Dominion forces, Isabella D. Losinger has produced a revisionist study of Canadian inter-rank relations in the Great War, questioning the notion of informality in the relationship.[10] By contrast, historians of the Australian Imperial Force have tended to repeat C.E.W. Bean's views on its supposedly informal discipline and officer–man relations.[11] Christopher Pugsley's work on the NZEF makes some brief, but interesting points about the nature of officer–man relations in New Zealand forces.[12]

Pugsley's *On the Fringe of Hell* is almost the only major study of British military discipline in the Great War, although A.J. Peacock has published a number of short articles on the subject, concentrating particularly on field punishment.[13] Three other aspects of discipline, namely military executions,[14] military police,[15] and mutiny,[16] have also received informed attention, and these works have much to tell us about the nature of the British army in the era of the Great War. J. Brent Wilson discussed the morale and discipline of the BEF in his excellent, although now rather dated, overview, and in recent years other valuable studies of particular aspects of morale have appeared.[17]

The bulk of this book was completed, as a doctoral thesis, in 1993. In revising it for publication, I have taken account of work that has appeared since then. Two recent books deserve particular attention. Joanna Bourke, as part of a wider study, has argued that wartime 'male bonding was limited and contingent on a huge range of factors. The struggle for comradeship with other men, if it was attempted, failed.' Those factors included rank, social class and 'political identity', military units, 'marital status, religion and ethnicity'. Her thesis represents a healthy correction to over-sentimental views on 'comradeship', but she somewhat exaggerates her case. Moreover she devotes little space to officer–man relationships and fails to differentiate sufficiently the experience of the front-line soldier from those of other servicemen, for example, unhelpfully including Black South African labourers under the heading of 'Fighting men'.[18] By contrast, Malcolm Smith has briefly referred to officer–man relationships as 'an opportunity for the more benevolent aspects of male solidarity to work through into a redefined paternalism'.[19]

Moving away from the traditional study of 'why men fight', 'power relationships' in the French 5th Infantry Division are the subject of Leonard V. Smith's recent study.[20] He uses the methods of Michel Foucault and others to demonstrate that ordinary soldiers could, and did, influence the actions of their superiors, even in battle. Notwithstanding some methodological and interpretative weaknesses,[21] Smith's thesis is both intellectually stimulating and generally persuasive. It takes a completely different approach from the present work but the two books have in common a challenge to the idea that the ordinary soldiers of the Great War were mere passive victims of the war and the military hierarchy. Some research on the BEF undertaken along the same lines as Smith's work would be a very valuable contribution to our knowledge of the Great War.

The lack of a major study of officer–man relations in the British army of the Great War means that a vital element is missing from our understanding of that army. My 1984 study amply confirmed Tony Ashworth's view that a military unit could be become a community, a substitute family for the soldier.[22] An examination of such a broad theme is fraught with dangers. As a TF officer remarked, 'There were marked but perhaps subtle differences' in the inter-rank relations existing in Territorial, Regular and New Army units.[23] I would go further, to say that no two units were exactly alike. To take but one example, the ethos of 2/4 and 2/6 Duke of Wellington's Regiment (DWR), two Territorial battalions of the same regiment, were very different.[24]

Ideally, to build up a coherent picture of officer–man relations, discipline and morale in the British army of the first two decades of the twentieth century, one would need an immense number of detailed case-studies of individual units. The aim of this book is to lay the foundations for future studies of individual units. It is to be hoped that the usefulness of such a *tour d'horizon* will outweigh the generalisations that will inevitably occur.

This book is based largely on published and unpublished writings of junior officers, NCOs and private soldiers; the latter being a hitherto somewhat underexploited source. By definition these writings reflect the opinions of individuals. How reliable are the postwar writings and memories of old soldiers? A former artillery officer wrote in 1964 that he found himself

> looking at the 1914/18 war in some astonishment that the events of fifty years ago can still be so vivid in an old man's memory ... it is a conscious effort to recall those World War I days. Yet when the effort is made, the name of a place, or of a soldier, will immediately bring back a lively recollection of an event, and of the circumstances that led to, and followed after, that event.[25]

It is, perhaps, not surprising that veterans of the Great War should be able to recall the most intense, exciting, traumatic and even, paradoxically, in many cases, happy period of their life with great clarity.[26]

War fiction that was based at least in part on the author's own experience, has also proved very useful as a source. A novel, whatever its literary merit, even if unreliable on matters of fact could nonetheless 'describe the sensations of military service' just as effectively as a memoir. Indeed the ability of the novelist to 'recreate soldiers' conversations' increases the usefulness of fiction for the historian.[27]

Several frequently used terms need explanation. Formally, private soldiers were referred to as 'Other Ranks'; informally, as 'rankers', and both terms are used here. For simplicity's sake, I have generally used the title of 'private' in place of some regimental ranks such as 'rifleman'. The term 'ranker-officer' is used to describe a former Other Rank, regardless of his social background, who received a commission. 'Regimental officers', as opposed to 'staff officers', were those who served with a unit.

When citing unpublished evidence it did not always prove possible to use conventional forms of notation. It is unclear, for example, to whom many rankers' letters were addressed. Generally, the following system has been used. Firstly, the type of evidence is mentioned (diary entry, unpublished account, letter). This is followed by the date or page number, as applicable, the collection from which the evidence is drawn, the reference number of the collection, and finally the archive. Unless otherwise stated, the individual named in the reference was the author of the letter, memoir or diary. As an individual may have held various ranks at different times, rank is omitted from the citation. However, an indication of their rank, or at least their hierarchical status at the time to which the evidence refers, is generally given in the body of the text and in the sources.

1
Officer–Man Relations and Discipline in the Regular Army, 1902–14

The social composition of the prewar Regular army had a major impact on the nature of the officer–man relationship. Broadly speaking, 'Kipling's army' recruited from the highest and lowest strata of British society. The social gulf between officers and Other Ranks was very wide.[1] Edmonds, the British official historian of the Great War, claimed that recruitment was aided by 'the compulsion of hunger'.[2] Various estimates of the proportion of unemployed men enlisting in the army ranged from 70 per cent in one area to more than 90 per cent for the country as a whole.[3] Skilled and unskilled labourers accounted for 24 and 44.5 per cent respectively of men joining the army in 1913 and a further 25.5 per cent of recruits came from other working-class occupations such as carmen and carters, outdoor porters, and domestic servants. Professional men/students and clerks constituted only 1 and 3 per cent respectively.[4] Thus the intake of recruits in a not untypical year came almost entirely from the working classes, with labourers, rather than artisans, predominating.

The typical Regular soldier of the period was of urban provenance[5] and his educational standard was low. On enlistment, recruits were graded according to their educational class. These ranged from A, 'men of good education', through to E, illiterates. In 1913 only 6 per cent to classed as A, 11 per cent were classed as E, while 58 per cent were men graded as C and D, men of moderate and inferior education.[6] When an officer in the RHA reported that W.J. Nicholson was 'well educated', Nicholson realised 'that he meant [well educated] by army standards' and was not particularly flattered.[7]

The health of the Other Ranks was equally unimpressive.[8] An MP believed that the majority of men who enlisted in London were 'poor, half-starved, unintelligent boys'.[9] The rejection on medical grounds of

40 per cent of those offering for enlistment during the Boer War had caused a great deal of national soul-searching, but little result: in 1910, 52.2 per cent of potential recruits failed the medical.[10] Others scraped through the medical by a subterfuge.[11] British medical officers were resigned to recruits being 'probably underfed'.[12] In short, the ranks of the British army of the Edwardian era were filled with ill-educated men of indifferent health, from poor, urban backgrounds.

Most officers, by contrast, came from 'traditional sources of supply': families with military connections; the gentry and peerage; and to some degree, the professions and clergy.[13] British officers did not form a distinct caste. Regimental messes were open to men from landowning *parvenu* families, or with a family background in 'trade'.[14] The ethos of the officer class remained that of the landed interest, however. There are several reasons for this. The military leadership was overwhelmingly rural in background.[15] Thus even 'middle-class' officers were usually of rural rather than urban/industrial provenance, and therefore likely to be influenced by the traditions of the landed, rather than the commercial, classes. Moreover, for a son to be commissioned into a smart regiment was tangible evidence of the social arrival of a family, and the son would be keen to conform and to be accepted as an officer and a gentlemen. To deviate from acceptable codes of behaviour was to risk ostracism or worse. As a contemporary writer observed, those who worried about the commissioning of officers who had 'nothing to recommend them but the riches which their parents had acquired in trade' could rest assured: 'we can safely leave these young men to the tender mercies of their brother officers'.[16] Unfortunates deemed 'unsuitable' by their peers were hounded out of the army, sometimes being subjected to a mock court-martial.[17]

The British officer class was educationally homogeneous. An education at a public school, especially a Clarendon school, was an almost essential rite of passage for the aspirant officer.[18] By 1913, the majority of officers also passed through Sandhurst or Woolwich. The common educational background of the majority of the officer class also helped to ensure that 'country-house' values permeated the officers' mess. (These values, and the effects they had on officer–man relations, are discussed below.)

In 1899 a 'genuine' ranker, as opposed to a gentleman-ranker, was described as a man who joins the army 'without money or scrip, without influential or sympathetic friends'.[19] In earlier times, the commissioning of such men had been relatively common, particularly in wartime, but by 1913 only about 2 per cent of Regular officers were commissioned from the ranks each year. In addition there were a number of ex-ranker quartermasters and riding masters, but these were 'dead end' promotions given to senior NCOs and WOs nearing the end of their careers.[20] Low rates of pay deterred other soldiers from contemplating a commission.[21] Although ranker-officers were not necessarily treated unkindly,[22] the privations endured by impecunious officers ensured that the impact of the ex-ranker on the prewar British officer class was a minor one.

The British army was a collection of individual regiments and corps, each fiercely independent, with its own traditions and customs. There was an unofficial 'league table' of exclusivity, with some regiments demanding a large private income of their officers. Indeed, by 1912 there was a shortage of officers in the cavalry of the line, and the War Office was forced to ask that Sandhurst cadets be acquainted with the fact that it was possible in some regiments to survive on an income of £300–400.[23] In spite of these qualifications, the British officer class shared common values, and can be treated as a body that shared a 'collective mentality'.[24]

Modern scholarship has thus confirmed the essential accuracy of J.F.C. Fuller's view of the army of the period, that 'Recruited from the bottom of Society, it was led from the top'.[25] The inter-rank relationship was, however, rather more subtle than traditional views, discussed in the Introduction above, would allow. The structure of the army was rigidly hierarchical, but there was scope for informal relations. However, the relationship of the officer to the soldier can be properly understood only if the position of the Non-Commissioned Officer (NCO), who played a key role in the enforcement of discipline, is first examined.

Sergeants were the non-commissioned equivalents of the junior officer, wrote an observer in 1914, but were actually more important, since the officer was not infrequently absent from parades, while the sergeant was 'at all times responsible for the efficiency of his men'.[26] If a private wished to speak to an officer, an NCO had to be present throughout the conversation.[27] The NCO might have been on friendly terms with his men before his promotion, but once he had gained a stripe, all social contact between them had to cease. Sergeants had their own mess and billets, to 'emphasise their separateness from the rank and file' and to improve discipline, but the unfortunate lance-corporal had to sleep in the same room and share the same meals as his section.[28] A prewar NCO stated that the newly promoted lance-corporal could be the 'unhappiest man in the army. He is immediately isolated from his old companions, and his new friends jealously watch him for faults.'[29]

This policy of segregation seems to have been effective. In the army of the 1870s it was said that 'The sergeant might belong to a different race', and an NCO could be arrested for calling a private by his Christian name. Similarly, forty or so years later, a corporal of 2/Rifle Bde was reprimanded for overfamiliarity with the men.[30] In the 2/Cheshires corporals were not allowed to mix with lance-corporals, although this does not seem to have been a general practice in other regiments.[31] An exception to this general rule of segregation, if a ranker's memoirs are to be believed, occurred in 19th Hussars.[32] A natural result of the segregation of junior NCOs from the men was that many men were reluctant to become lance-corporals.

However, relations between NCOs and privates varied from unit to unit. In 2/R. Bde NCOs were generally detested by privates. An NCO of this unit felt that the army would have been more efficient if a closer relationship could have developed between NCOs and men, allowing men to feel that they could go to their sergeants for advice and help. Instead,

> The whole trouble about the pre-war army was fear, Privates for Cpls & Cpls for Sgts (*sic*) and so on. There was no trust among the troops, and as a result it was impossible to get the best out of the men, although they were excellent....[33]

One historian used this evidence to imply that this situation was common to the whole army,[34] but NCO–man relations seem to have been unusually poor in 2/R. Bde. By contrast, a ranker's account of life in the 20th Hussars mentions one NCO who abused his authority, but does not give the impression that the privates lived in 'fear' of their superiors.[35] Indeed, some young soldiers admired their NCOs.[36] The role of the NCO was not merely that of coercion. He also looked after the welfare of his men and to some extent played the role of a father-figure, but most of all he was the vital link between commissioned officers and Other Ranks. Inevitably, since the smooth running of the unit was largely dependent on co-operation between officers and NCOs, relations were less distant, although far from intimate.[37]

Officer–man relations followed the pattern of the ideal 'country-house' relationship between the landlord and tenant, with loyalty and deference (considered in a later chapter) being given in exchange for paternalism and leadership. Paternalism was a set of widely held social attitudes rather than a coherent social theory. Paternalists of the Victorian era believed society 'should be authoritarian, hierarchic, organic, and pluralistic'. A belief in a society that was hierarchical was a central pillar of the paternalist's *Weltanschauung*. In an egalitarian society, the poor would lack incentive to work, and the affluent would not possess 'the wherewithal ... to rule, develop the arts of government and do charitable work'. Society was organic, in that every individual had his place, his responsibilities, 'his reciprocal obligations, and his strong ties of dependency'. Finally, society consisted of a number of different hierarchies, each contained within the greater hierarchy.[38]

The core of the paternalist's creed was *noblesse oblige*, the belief that privilege entailed responsibility, in this case ruling, guiding and helping. The first two responsibilities involved keeping order, punishing anyone who posed a threat to the stability of society, and preventing turmoil through spiritual, moral and political guidance of those in the lower reaches of the hierarchy. The third consisted of helping the poor, whether it was by building cheap housing or dispensing food in time of want. By the early twentieth century, paternal owners of industrial factories accepted that they had to provide continuous work, good conditions and materials. Such men were perceived by their workforce as being 'fair'.[39]

The attitudes of the Edwardian officer reflected these concepts of paternalism. In the wake of the Curragh Incident in 1914 a retired Guards officer, Lord St Audries, made a classic statement about the nature of discipline and the officer–man relationship in the British army:

> Discipline ... is not kept up by fear of punishment, by threats, or by bullying. Discipline is kept up, partly no doubt by training, but a great deal more by tradition, by *esprit de corps*, by the confidence, respect, yes and the affection, which exists between officers and men....[40]

St Audries's belief that the regiment formed a community, in which officers and men were bound together in friendly common interest was widely held by Regular officers.[41] Another common belief was that Other Ranks preferred to be led by gentlemen.[42]

The officer's concept of inter-rank relations had evolved over the years. From the mid-nineteenth century onwards, officers' attitudes to their men gradually changed. This reflected the evolution of the social philosophy of the landed interest during this period. The image of the Christian gentleman, a model of sobriety and propriety, replaced that of the port-sodden rake of the Regency period. In part, pragmatism accounted for this change in officer–man relations. Some senior officers believed that recruiting would suffer unless they instituted a more enlightened regime.[43] One subaltern had a more immediate application for his practical paternalism:

> [After a route march] we inspect all the men's feet to see that they haven't any blisters etc.... If you left a man alone he would never wash his feet.... But it is most important, as if your men can't march, they are no use.[44]

A similar message of pragmatism and duty emerges from a book by a general, in which two pictures are displayed side by side. One depicts a man standing beside a horse, asking the groom '"Well, Jim, has he fed all right?"' The second shows a group of soldiers seated around a fire. An officer is asking them '"Dinners all right, men?"' Underneath both pictures appears the words 'Noblesse Oblige'. The original watercolours have, since the 1920s, hung in the officers' mess at Sandhurst.[45]

Philanthropy was as at least as important as pragmatism in the refashioning of officers' attitudes. *Noblesse oblige* was not an idea that was new to the army. What was new was that by Edwardian times the officer class had, almost without exception, accepted the notion that privilege entailed responsibility. In the 1870s, such responsibilities were not so widely accepted. 'Old Paddy', the CO of a Fusilier battalion, was renowned for his paternalism, and was 'loved' by the men 'like a father', but one of his subalterns 'used to talk to the men as if they had been dogs'.[46] Thirty or forty years later, such behaviour on the part of a subaltern would have been considered unofficerlike. When a subaltern joined 2/Argylls in 1908, his company commander impressed on him that he had to 'get to know the character and personal problems of each soldier' in his half company.[47] Similarly, upon joining 4/Rifle Bde eight years earlier, a young officer was taken aside by a senior colleague and told that his 'first and most important duty ... [was] the care and welfare of the men' he commanded.[48]

Officers tended to regard their men as children: their lives needed to be closely supervised, for, left to their own devices, they were untrustworthy. An officer of the Leinsters wrote of the 'naiveté, almost of childishness' of his men.[49] It was not only Irish soldiers who were regarded in this light, and this curious mixture of respect and mistrust formed the basis of the average Regular officer's attitude towards the Other Ranks, and the disciplinary system reflected this. Officers' opportunities for contact with their men in peacetime were limited by the factors already discussed and the fact that they enjoyed long periods of leave, from four to six months a year.[50] Could there, then, be any commonality of experience in the peacetime army?

In the sense that officers were not exempt from the disciplinary process, there was. Officers could be disciplined for misbehaviour and pressurised to conform by their peers. Officers of the 2/Essex were

> enjoined to set an example to those whom they are placed in authority by preserving at all times a gentlemanly bearing, both in language and demeanour.[51]

What is more important, young officers had a taste, albeit brief, of life at the bottom of the military pile. It has been argued that Sandhurst cadets had only a limited amount of instruction in the tasks of the ordinary soldier, in contrast to cadets at West Point and Kingston where the cadet 'carried out all the functions' of an NCO.[52] In fact the system of cadet government used at Sandhurst meant that at least some Gentlemen Cadets took on the responsibilities of NCOs, which included a number of unpleasant and tedious tasks.[53] Sandhurst was a hierarchical society, with junior cadets in much the same position as privates in a battalion. Until 1913, cadet under-officers and NCOs could award punishments to fellow cadets.[54] This situation, and the inevitable bullying[55] was, of course, not dissimilar to life in public schools.

Virtually all Regular officers shared one experience with rankers. One former cadet commented that

> the Sandhurst man ... has in some respects to undergo what the private soldier undergoes ... by day he is the simple recruit, hectored and crimed for the slightest fault, savagely drilled ... in all ways harshly and meticulously schooled.

This individual considered this an advantage, although it was not appreciated as such at the time.[56] There was no teaching on leadership: one officer recalled that he had little about 'man management' at Sandhurst, while another recalled that it had the reputation of 'turning out a very good private soldier'.[57]

Once they reached their regiments, newly commissioned subalterns had to 'pass off the square' – that is, undergo training in drill under a senior NCO alongside private soldiers. Highland subalterns had to undergo the even greater ordeal of passing off the square in Highland dancing.[58] A young officer would find himself 'less than dust ... learn[ing] his drill as an ordinary "rooky"'[59]. Indeed, during the three months that J.A. Halstead (l/Loyals) spent on the square, he felt that he 'learned something of the private soldier'.[60] A Territorial officer recalled the humiliation of making blunders in front of social inferiors, but thought it an important stage in gaining the respect of his men.[61]

One should not exaggerate the extent to which the young officer became familiar with the lot of the ranker. The experience was of relatively short duration, and led directly to a privileged lifestyle. But it did give officers a glimpse into the life of the rank and file, and ensured that they had first-hand experience of the methods favoured by the army for instilling discipline.[62]

The use of small mobile columns by the British during the Boer War seems to have caused a diminution in differences between the ranks. Officers and men shared the hardships of campaigning, and leadership became less a matter of 'the formal dictates of rank', and came to rely more on the 'informal, personal qualities of individual officers'.[63] Certainly, some senior officers were perturbed by the decline of formal discipline which was tacitly condoned or actively encouraged by regimental officers.[64] Some officers carried into the peacetime army some first-hand experience of the life of the ranker on campaign.

Some officers certainly believed that they enjoyed close relationships with their men, one writing of the family atmosphere in the 16th Lancers: at 'stables' an 'opportunity was provided for the most intimate relationship to be established between officer and men'.[65] It remains to be seen to how far Other Ranks shared this rosy view of officer–man relations.

The ranker's experience of military discipline coloured his view of the officer–man relationship. One believed that the intention was 'to destroy all vestige of individualism in us and to remould the messy remains into an unshaken loyalty and devotion to the regiment'.[66] The informal hierarchy of the barrack room reinforced formal discipline.[67] Other recruits were not overly worried by discipline[68] possibly because they were prepared for the rigours of army life by their experience as a civilian.

For the experienced soldier, discipline was irksome rather than savage, except for those who chose to fight the system. An NCO of 21st

Hussars learned from bitter experience that military law was based on the principle of 'Heads I win, tails you lose'.[69] The standing orders of the Irish Guards (a regiment that had an enlightened attitude to officer–man relations) presupposed rankers lacked the most basic standards of cleanliness, thrift and honesty.[70] While there was some merit in this approach, given the impoverished background of many soldiers, the army treated the NCOs of the Corps of Military Police (CMP) in much the same way. Yet CMP NCOs were very different from the average private, having vastly more individual responsibility, and being characterised by a high level of self-discipline.[71] The situation varied from regiment to regiment. One Life Guardsman thought that the Footguards were engaged in 'manufacturing crime' by deliberately looking for minor faults in dress.[72] The importance attached to correct dress is indicated by the issue of passes to Territorials temporarily stationed in Aldershot, which stated that they were entitled to walk out in service, rather than full, dress.[73]

Some rankers recognised and resented the basic premise underpinning military discipline. Pte Grainger (9th Lancers) confided in his diary that as a soldier 'I did not belong to myself ... I am only a small nut in the Great Indian War Machine.... I am a number, still retaining my name, but that being of secondary importance.' He believed that 9th Lancers placed more emphasis on 'bull' than some other regiments.[74] Other, no less articulate men did not object to the disciplinary system. Nicholson of the RHA wrote that the Regular soldiers were 'properly, though (*sic*) not harshly, disciplined'.[75] Hawke of 2/Cheshires believed that strict discipline 'did not completely deaden the intellect' and discipline allowed the BEF to survive in 1914.[76]

Deference aside, there were good reasons why men accepted the system without complaint. Regular battalions offered 'a nexus of relationships, familiar faces and existing loyalties'.[77] The peacetime army offered some men a more comfortable and fulfilling existence than civilian life, freeing them from many of the worries of everyday living. Army food, monotonous and inadequate as it might have been, at least was served regularly, which had novelty value for some recruits.[78] There were other benefits, ranging from the issuing of good, waterproof boots, to the provision of facilities and time for leisure, whereas many working-class recruits would have had little time or money for leisure.[79] When stationed in India, the private could live well; he could hire servants to shave him in bed, and to carry out his fatigues. A more subtle benefit was that the meanest British soldier exercised real power over Indians.[80]

For every disgruntled soldier like Pte Grainger, it is likely that there were several more like H.J. Coombes (RWK), who enjoyed the outdoor life of the army, which gave him 'a feeling of well being and fitness' or T.A. Silver (E. Surreys) for whom the red uniform brought a sense of self-esteem.[81] Probably even more common was the soldier who sought a quiet life by avoiding trouble. Riddell believed that after passing through the recruit stage, army life was tough for dirty, dim or unathletic soldiers, but was easy and tolerable provided one was proficient at 'spit and polish', possessed the Third Class Educational Certificate, or was a good sportsman.[82] The prospect of a comfortable and trouble-free existence was a powerful inducement for men to internalise discipline and to conform.

The implications of the Curragh Incident for military discipline alarmed a number of officers. As one subaltern wrote:

> how are you to preserve discipline after this, how are you to use your Army to keep law and order against strikers when once (*sic*) the officers have successfully resisted an attempt to use them to enforce a law which they do not approve?[83]

Indeed, one source claimed 'hundreds' of men were prepared to come 'out from the ranks' if given a lead by the Labour Party.[84] At the very least discipline would have been placed under severe strain. In the words of one officer, 'It's all very unfortunate & terribly bad for discipline.'[85]

Did the rank and file see their relationship with their officers as 'petty and heavy handed interference in the private lives of vulnerable people?'[86] Certainly, that disgruntled 9th Lancer, Pte Grainger, denounced his officers as 'blue Blooded Bacon dryers, cheese mongers, or pork butchers that are in command and have money' and said that soldiers sometimes had 'the greatest contempt and hatred for those in position and command, whether civilian or military'.[87] Conversely, other men were fulsome in their praise of their officers.[88] Perhaps more typical than either type was Pte Fanton (1/Cheshires). In the course of his unpublished memoir of army life, he mentions officers only once, and then in connection with criticism of 'bull'.[89]

Officers played a very marginal role in the life of the average private. There were several areas in which the ranks did come together. One was on the sports-field, where officers and men played in the same teams. This served the invaluable function of allowing feelings to be relieved without endangering discipline, for the private could hurl abuse at an officer from the touch-line with impunity. Another meeting-place was the regimental dramatic troupe.[90] A third was the freemasons' lodge, which men of all ranks could join, apparently without damaging discipline.[91]

Horace Wyndham, who served as a gentleman ranker in a Fusilier regiment from 1890 to 1897, stated that officers saw little of their men, and did not know them individually. This was in part due to the nature of short service and the constant movement of personnel. He claimed that regimental officers were forced to spend hours before an inspection committing soldiers' details to memory so that they might impress a visiting general with their 'active and intelligent interest … in the affairs of the rank-and-file'.[92] Little had changed 15 years later. John Lucy (2/R. Irish Rifles), writing of his service just before the Great War, likewise argued that that the officers' fond belief that they understood the men was misguided. In reality the officer 'was not very much in touch' with the men.[93]

Wyndham believed that one of the principal obstacles to establishing closer relations was the attitude of the soldier himself, who, 'like the average schoolboy' was 'most comfortable when he is furthest removed from those in authority over him'. Thus the private preferred to sing in the wet canteen rather in the presence of officers who, meaning well, had organised a concert. Indeed, Wyndham implied that the officer class sometimes carried this aspect of paternalism to excess. What the private desired from his officers, argued Wyndham, was not close friendship but tact and 'an intelligent appreciation of their work and the conditions under which it is performed'.[94] Frank Richards (2/RWF) commented upon the occasion when a lonely officer struck up a conversation with him: 'it shows how hard pressed he must have felt …'.[95]

Interestingly, in view of the alleged preference of the ordinary ranker for gentleman amateur officers, Wyndham claimed that men greatly preferred officers who were knowledgeable about their work. Soldiers appreciated paternal officers who, for instance, ensured that the men ate properly cooked food at the end of a day on manoeuvres. Soldiers disliked selfish officers, and the small number who made a fetish of 'spit and polish'. Above all, Wyndham stressed the need for officers to display man-management skills: to give the odd word of praise on parade or to visit a sick man in hospital.[96]

Wyndham, Lucy and Richards came to broadly similar conclusions about the nature of the officer–man relationship. The former wrote that relations needed 'no great alteration' provided that the 'officer is tactful'. Respect for officers, Lucy said, was high, and in time of war the gallantry of the officer 'won the greatest devotion, and very often the affection, of the men'. Richards believed that the social code of the army was based around 'mutual trust in military matters and matters of sport, but no social contact'.[97] Although some evidence suggests that Other Ranks were influenced by *esprit de corps* and regimental pride[98], neither the view that the prewar army was one big happy family, nor that which sees the inter-rank relationship as one of mutual antipathy, is entirely correct. Generally, officers and men did respect each other, but relations were far from intimate. In the words of a private of 4th Dragoon Guards, 'There was an enormous gulf' between officers and men.[99]

The officer–man relationship could be very complex. Shortly before the Great War, some men of 20th Hussars refused to parade because the cookhouse had closed before they could eat. A deputation of 'old soldiers' conferred with Maj. Cooke, the squadron commander. Cooke defused the situation by buying food for the men with his own money, and the men then went on parade. Cooke backed up this compromise with a visible show of force. The army soon gained its revenge by criming the protesters on various charges.

This incident, technically a mutiny but effectively a strike, is most instructive. Cooke displayed an impeccable grasp of the paternalistic relationship by acting to rectify a genuine grievance. He also displayed considerable managerial skill. Had he attempted to act in a doctrinaire fashion, applying the full weight of military law, this relatively trivial incident could have escalated into something more serious. By his subsequent actions Cooke reasserted his authority. As one ranker who was involved noted, the men's grievances might have been real but 'orders given in the Army simply had to be obeyed, and objections made afterwards'.[100] A similar lesson emerged from a recruits' 'round robin' complaining of a bullying NCO which was sent to the commander at an RFC base in 1913. The recruits were arrested and received a 'severe choking off' – but the NCO was posted away almost immediately.[101]

Both officers and men recognised that their relationship was governed by certain unwritten rules. Provided both sides observed those rules, the relationship, although devoid of intimacy, was nonetheless effective.

2
The Prewar Army: the Auxiliary Forces and Debates on Discipline

Two distinct 'strands' of discipline and officer–man relations co-existed in the prewar British army. The previous chapter discussed the Regular strand but here, using a similar approach, the focus is on the Auxiliary forces, which consisted of Yeomanry (cavalry) and Volunteer (from 1908 Territorial) infantry and artillery units. In 1908 the Yeomanry merged into the newly created Territorial Force (TF). Although technically disbanded in 1908, many Volunteer units simply changed their name, and there was much continuity between the old Volunteer Force and the TF.[1]

In 1928 it was claimed that:

> A Yeomanry regiment may be said to be an expression of the best of the county on horseback ... its ranks manned from the home-steads and farms whose tenure has often been held for successive generations, and officered ... from the great houses'.[2]

As far as the ranks were concerned this ideal may have been generally realised in earlier times, but by the twentieth century the social composition of the Yeomanry was undergoing significant change. By the turn of the century most Yeomanry regiments were enlisting increasing numbers of urban recruits. This reflected the increasing urbanisation of British society.[3]

There were two broad categories of Auxiliary infantry and artillery units: the 'class corps', which like the London Yeomanry regiments were recruited from men of some social standing, and the rest. Although the former have received more attention, the Edwardian Volunteer and Territorial Forces recruited the bulk of their members from the working classes. Although the numerous Volunteer Rifle

Corps raised in 1859 had had a strongly middle-class character, by 1904, much to the disgust of some venerable Volunteers,[4] some 70 per cent of the ranks of the Volunteers were drawn from the working class. Artisans made up 40.3 per cent, 9.2 per cent were clerks and only 1.6 per cent were professional men.[5] One battalion, 2nd V[olunteer] B[attalion] East Surreys, roughly matched this national profile. This unit, recruited partly in the London suburbs, in 1904 comprised a 'few gentlemen', a 'fair proportion of clerks, some small tradesmen, a large proportion of artisans', and in the 'country companies ... labourers and men in country pursuits'. The unit contained very few casual labourers, and the service of those who did join was generally short. The average earnings of the rank and file were estimated at 35s per week, some 15s above the poverty line, which was in itself a further indication of the predominance of artisans within this corps.[6]

It is possible that the next few years saw a slight decline in the social standard of the TF infantry. In 1911 it was noted that increasing numbers of labourers, and fewer tradesmen and clerks, were joining the TF, and the intelligence of the average Territorial recruit 'differs but little, if anything, from that of the Regular recruit'.[7] Like their Regular counterparts, many recruits to auxiliary units suffered from the effects of poverty. The 17th to 20th battalions of the London Regiment recruited 'mainly from artisans who are badly fed and nourished and who are frequently small, of poor physique, and with little stamina'.[8] 'Slum battalions' were also found elsewhere; the general standard of physique of 7/DWR, recruited in Yorkshire, was described as 'miserable'. In general, battalions which included large numbers of mill-hands were noted for the poor physique of their soldiers.[9]

In general, however, in contrast to the Regular army, the Volunteers and TF recruited from the artisan rather than the labourer.[10] The London Regiment offers many examples of these artisan dominated corps, 6/Londons having companies sponsored by the South Metropolitan Gas Co., Amalgamated Press and Associated Newspapers.[11] In some units, such as the Cambridgeshire Battalion, and 5/DCLI middle-class and working-class men served together.[12]

Infantry 'class corps' such the London Scottish (14/Londons), 6/Manchesters, and 5/Scottish Rifles recruited very largely from white-collar workers: it was said that London class corps contained 'men of higher intelligence and education, and finer physique than most of the urban corps of the country'.[13] Some units charged entrance fees, and all were concerned to exclude proletarians from their ranks,[14] while 60 per cent of the Other Ranks of 7/Welsh (Cyclists) were 'young professionals and businessmen of the same social standing of (*sic*) the officers'.[15] A member of the London Rifle Brigade (LRB) said that a strong motivation for 'young men in the banks, insurance offices, the Civil Service, and the City' to join class corps was to enjoy social activities and sporting facilities.[16]

In 1899 an officer claimed that half of the Yeomanry's officers 'consist of retired cavalry officers, landed proprietors, and enthusiastic fox-hunters'.[17] This assessment was broadly accurate. Service in the local regiment was a favourite pastime of many pillars of the Edwardian social and political elite; the Oxfordshire Hussars included the Duke of Marlborough and Winston Churchill among their officers. Auxiliary infantry officers were generally of a lower social class than their Regular and Yeomanry counterparts.[18] In 1904 just over 6 per cent of Volunteer officers were 'Gentlemen of independent means', while 65 per cent were either 'Professional men' or 'Men in business on their own account'. That a proportion of officers were of lower or lower middle-class origin is suggested by the 21.3 per cent classified as 'employees', although this term is too imprecise to easily ascertain their exact social status.[19]

It is unlikely that there was a high incidence of working-class officers in the Auxiliary Forces. In 1904, 59 per cent of the Volunteer officers had attended public schools and universities.[20] Other evidence also suggests substantial numbers of middle-class TF officers. A 1911 report said that that Territorial infantry officers were 'almost all business men', and the officers' mess of 1/8th Londons (Post Office Rifles) certainly contained many men of middle-class occupations.[21]

The middle-class domination of the auxiliary officer corps resulted in part from the fact that the upper and upper-middle classes found service in the TF, except for the Yeomanry, unattractive. *Pace* the alleged impact of the public schools in promoting militarism, in the period 1908–12 only 4 per cent of former members of cadet corps took a Territorial commission, and less than 2 per cent took a commission in the Special Reserve. A report by Col. L. Banon, the Assistant Adjutant-General, in 1912 concluded that this situation arose from a variety of factors ranging from family hostility to the TF (either because of attachment to the old militia and Volunteers or a desire to introduce conscription), to the decadence of the young which had eroded the military spirit. The high standards required of an officer were a deterrent, as was the expense involved, and many suitable candidates for commissions preferred to serve in the ranks of class corps. Banon did not, however, question the basic patriotism of British youth.[22]

Many observers regarded the officer corps of the Volunteers as socially second-rate.[23] According to a wartime TF officer, pedantic and officious men who held relatively lowly positions in civil life 'formed the great majority [of officers] in the city regiments'.[24] Although one Regular believed that successful businessmen were 'accustomed to think for themselves' and were well-placed to judge character, and in 1908 the CIGS stated that middle-class auxiliary officers were 'enthusiastic' if 'touchy', many Regular officers were concerned about the low social status of auxiliary officers.[25]

Col. Banon's report of 1912 expressed the view that auxiliary units should be 'officered by young men of good social standing', whom he defined as being 'the sons of the gentry and professional classes', on the grounds of their paternalism and 'their hereditary aptitude for command'. He doubted whether members of *nouveau-riche* families possessed these qualities. However, he saw that eventually the Auxiliary Forces would draw officers from a wider social background, and believed that some could 'acquire the ideas, manners and standard of the professional classes'.[26]

The social status of non-Yeomanry auxiliary officers thus gave cause for official concern. However, Banon's pragmatic conclusion reflects the spirit that underlay the decision to commission Regular NCOs on mobilisation, and anticipated the discarding by the wartime army of the dogma that only men with a public school background could possess leadership qualities.

The disciplinary regime of Auxiliary units reflected their social composition. An important difference between the disciplinary structures of Auxiliary and Regular regiments lay in the role of the NCO. The foreman class provided many NCOs for the TF,[27] and some contemporaries saw this as being beneficial to discipline. The 'foreman, small manager or head employee' would allegedly make a good NCO, 'for the same qualities that have got him on in civil life get him on in the regiment and give him a sense of discipline and command';[28] they were 'ready-made' NCOs 'accustomed to take responsibility and to take charge of men'.[29] Kipling, writing of a TF battalion in 1914, claimed that the officers knew their men 'intimately' in civilian life, and the relationships built up between sergeants and privates who normally knew each other in the role of foreman and employee enabled the unit to work 'with something of the precision of a big business'.[30] A related point was made by an officer of the West Kent Yeomanry:

> Frequently each troop was a happy family from the same town or district, and if a Private received an order from a Sergeant with the reply 'All right, Ginger,' it was not a term of disrespect but merely the natural manner of answering a friend, who has, we will say, married your sister.[31]

Others had a less rosy view of the grafting of an industrial hierarchy on to a military one. Some men became NCOs in units recruited largely from a single workplace because of their hierarchical status rather than their military ability. Among the factors cited in 1909 as tending to undermine the position of the NCO were, besides the dread of losing face by the admittance of ignorance of an NCO's powers, the fear of upsetting subordinates who were socially superior, or who 'held higher positions in business'.[32]

Many Regular officers identified the NCO as the weakest link in the disciplinary chain of the TF. In 1909 a report stated that the NCO had 'little authority over his men', did not understand his role in 'the maintenance of discipline' and 'the men will readily obey' a Regular NCO attached to the unit, 'but do not attach much importance to the orders of the Territorial non-commissioned officer'.[33] In 1911 Col. Fanshawe, the Regular commander of a TF brigade, stated that Territorials were enthusiastic, but he feared the time when the novelty of soldiering palled and the NCOs had to drive the men on. Territorial NCOs and officers gave orders 'rather as if they were asking a favour' and the men responded 'often after some delay.... This leads to slowness and indecision.'[34] Things were little different in the Yeomanry.[35]

The leadership of auxiliary units possessed little effective coercive power. A fine or dismissal from the unit (which carried some social disgrace) was, in practice, the only formal sanction available.[36] Volunteers could resign after giving 14 days' notice. Territorials had to complete their four-year engagement, but could get a free discharge under certain conditions.[37] By the end of the nineteenth century, the discipline of the Volunteers had improved,[38] but still fell far short of Regular standards. To take one example: in 1896, a Volunteer NCO was reduced to the rank of private 'for writing a disgusting letter to the Adjutant'.[39] An attempt to impose Regular-style discipline would have led to men leaving the Auxiliary forces. Thus although Volunteers were subject to the Mutiny Act while brigaded with Regulars, company commanders of 4th VB E. Surreys did not have 'the courage to hold an Orderly Room in camp'.[40] In short, the discipline of auxiliary units was a very different nature from that of the Regulars. As, under conditions of peace, Auxiliaries, unlike Regulars, were under military discipline for only a few hours a week, it could hardly be otherwise.

Given the absence of the tools of formal discipline, in auxiliary units discipline was largely reliant on the social authority of the officer. Not surprisingly, considering the social profile of Yeomanry officers, the closest approach to the inter-rank relationship in the Regular army was found in the Yeomanry. The ideal of officer–man relationships in the Yeomanry was that they should be the 'happiest and most cordial possible', based on 'a true feeling of comradeship and mutual confidence'.[41] Yeomanry officers had their share of paternalism; in March 1907, E.E. Fiennes, MP, stated that many Yeomanry units, at considerable expense to their officers, bought tents and 'complete camp equipment for the messing of the men'.[42]

Yeomanry officers generally mixed paternalism with tact. It is significant that a prewar member of R. Wilts. Yeomanry complained bitterly about the arrival of a Regular officer who ignored the easy-going traditions of the Yeomanry and instituted Regular practices:

> we are not a blithering pack of fools! This is an instance of the lack of imagination of a Regular Commander, who all his life has commanded Regular Army men, and who now cannot see the difference in the mode of treatment desirable with a totally different style of man that he has under him in the Territorials, and Yeomanry at that....[43]

Underlying the distaste of many Regular officers for middle-class auxiliary officers was the belief that, unlike Regular and Yeomanry officers, Volunteer and later Territorial officers lacked the social authority to command. An ungentlemanly Volunteer officer, argued one writer in 1905, was 'apt to play the Jack-in-Office'.[44] Certainly, such units were heavily reliant on the individual qualities of the officer. An officer of the 4th VB E. Surreys declared that officers could secure obedience 'only by their personal magnetism in handling their men'.[45] Tactful handling of their men was essential. The discipline of one class corps, the LRB, was described as 'exceptional' but 'incomprehensible to a critical outsider', and newly arrived Regular adjutants in particular found it difficult to comprehend that discipline could be maintained 'without the administration of military law'.[46] Officers and men were not divided by a social chasm – in some units officers were often promoted from the ranks[47] – and it was perfectly possible for a private and an officer to be friends or workmates in civilian life. Peer-group pressure was an effective way of dealing with misdemeanours. By the exercise 'of good sense and tact on both sides', discipline was maintained 'without familiarity' while on duty.[48] Off duty, at social events 'every member of the LRB, whatever his rank, met on a basis of comradeship; on parade Army discipline and routine took over again'.[49] In short, both an informal officer–man relationship and good discipline were maintained by the men's self-discipline, and by peer-group pressure and sense of duty: as a London Scottish officer explained, 'Esprit de corps is the foundation of all real discipline.'[50]

An informal disciplinary regime was not the exclusive preserve of the class corps. The 6/W. Yorks recruited mainly from the mills and warehouses of Bradford, but also had a company raised from the old boys of Bradford Grammar School.[51] The 7/Manchesters recruited from the suburbs of the city, and serving in the ranks were both unskilled labourers such as carters and packers and skilled men such as builders and joiners. Other soldiers had solidly middle-class occupations such as clerks and draughtsmen.[52]

Neither battalion set much store by formal discipline. The 6/W. Yorks, for instance, marched off parade in January 1914, 45 minutes late, only 80 strong and 'Even this was considered a good attendance!'[53] The ethos of the Territorial was very different from that of the Regular. The former took pride in their civilian, non-military – or even anti-military – attitudes. Gerald Hurst, an officer of 7/Manchesters wrote that they were

> almost arrogantly civilian.... The social traditions of the middle-class urban population, from which the Territorials were drawn, had never fostered the military spirit, nor the power to recognise and understand that spirit in others.[54]

Similarly, most members of the 6/W. Yorks enlisted 'as a relief from the monotony of civil life, as an outlet for high spirits, and as a means of spending a healthy holiday with good comrades' rather than for militaristic reasons.[55]

Both units enjoyed good inter-rank relations and high *esprit de corps*. Regular modes of officer–man relations and discipline were simply inappropriate. A Regular brigadier described the discipline of 6/W. Yorks as being that of 'good will',[56] while Hurst wrote of the 'comradeship' which produced an 'easy relationship between officers and men ... [which] was the despair of the more crusted Regular martinet', a form of discipline which was maintained without requiring the 'banishment of individuality and of the exercise of intellect from Regimental life'.[57]

Its historian claimed that in 6/W. Yorks men initially obeyed orders 'simply because of a mutual confidence and respect between officers and men, similar to that in a workshop or any small society'. While the 'first bond was personal', discipline, when it developed, was used to 'strengthen and regularize relationships already existing';[58] for officers and men came from the same city. Both Bradford and Manchester had a strong tradition of civic pride, which was reflected in battalion *esprit de corps*. Manchester's Territorials had the advantage of possessing one of the few brigadiers to be appointed from a TF unit, Noel Lee, allowing Hurst to claim that 'all ranks, from Brigadier General to private, came from one neighbourhood, and viewed life from much the same angle'.[59] The 6/Lancashire Fusiliers offers another example of how the importance of the person increased when an officer began to take a direct personal interest in the welfare of the men, which included the provision of extra sporting facilities.[60]

In sum, although the social composition of class corps was rather different from that of the 7/Manchesters, many, although not all[61] TF units battalions adopted, from a mixture, one suspects, of pragmatism and genuine pride in the nature of the unit, a style of discipline and officer–man relations that was very different from that of a Regular unit.

Many Regulars understood the importance of handling their Auxiliary soldiers with care. The commander of North Midland Mounted Brigade went on record that Regular NCOs posted to TF units as Permanent Staff should undergo a six-month probationary period to judge their skill as an 'instructor and a disciplinarian', and, significantly, their 'tact and zeal in the performance of his duties'.[62] Most Regular officers and NCOs posted to TF units managed to adapt to their peculiar disciplinary system and officer–man relationship, although one wonders how a Regular officer would have adjusted to the state of affairs in the 9/Manchesters, where, among other things, an officer told a soldier to 'b——r off'.[63]

Territorial discipline coloured many Regular officers' views of the military efficiency of the TF.[64] One of the kinder views was that of Lt Gen. Sir Arthur Paget, GOC Eastern Command, who in 1909 likened the relationship of the Regulars and TF to that of a 'strong, well grown man in the prime of life and a young, immature but healthy child who ... should some day develop into maturity'.[65] Two years later Paget recorded his belief that after four months' training, Territorial infantry would still be 50 per cent less effective than Regulars, and he believed that TF artillery was even worse off.[66] That the TF was intended for home defence only was scant comfort, given the prevalent fears of invasion.

The seriousness of Territorial training varied from unit to unit; much depended upon the character and inclinations of individual officers and men. The Yeomanry seems to have been a byword for inefficiency, while one soldier of 6/Essex said that he took training seriously, because the TF was his hobby, but the unit as a whole did not.[67] In 6/Londons, a new CO organised a drive for efficiency, while Pte S. Blagg of the South Notts Yeomanry welcomed the replacement of a lenient commander by a stricter officer because it brought a sense of purpose to their training.[68] The persistent absenteeism from which most auxiliary units suffered was not wholly attributable to indiscipline. Poorer men could not always afford to attend all of the annual camp, and many employers were reluctant to give Territorials time off from work.

In sum it seems that much of the criticism of the standard of Auxiliary training was justified and some of the problems resulted from indiscipline. However, to have imposed tighter discipline would have been counterproductive. The ultimate test of Territorial efficiency came on the Western Front in 1914–15 where TF units performed surprisingly well. Their discipline, although very different from that of Regular units, was sufficient to withstand the strains of industrialised, attritional warfare.

In common with other European armies, many British officers believed that the offensive could succeed in the face of the increased destructiveness of modern weaponry if the morale of the assaulting troops was sufficiently high.[69] According to Tim Travers, 'many' officers conflated this view with that of another 'vague camp' which emphasised increased discipline, producing a demand for a well-trained and highly motivated soldier with high degree of self-discipline.[70] The products of the Edwardian army fulfilled the first criterion, but not the second, for Regulars 'were highly trained and well disciplined, but initiative in the ranks was discouraged and had been drilled out of them'.[71] This dichotomy between desired results and the realities of army life fuelled a lively debate.

A prominent contributor to the debate was the foremost British military intellectual of the day, Col G.F.R. Henderson, Professor of Military Art and History at Staff College from 1892 to 1899.[72]

Henderson argued that the ranks of the armies of the American Civil War contained men, many of whom were of high intelligence, who did not take easily to formal discipline. They would willingly follow men who proved themselves or whom they liked, but proved resistant to 'military etiquette' in such matters as familiarity between officers and men. They were 'thinking bayonets', in contrast to the armies of Europe (although Henderson was usually content to leave this point implicit).[73]

Henderson was careful to point out the shortcomings of the lack of discipline in the armies, but he stressed that successful generals recognised the nature of their armies and adapted their methods of command and leadership accordingly.[74] Thus Stonewall Jackson recognised that 'his citizen soldiers were utterly unfamiliar with the forms and customs of military life' and thus with his troops 'tact, rather than the strict enforcement of the regulations, was the key-note of command', although offenders were harshly punished.[75]

Henderson saw the handling of Civil War armies as directly relevant to the Volunteer movement, of which he was a supporter. In 1894, he suggested that one day Britain might raise a large army composed 'at least in part' of temporary civilian volunteers, and the

> habits and prejudices of civil life will have to be considered in their discipline and instruction, and officers will have to recognise that troops without the traditions, instincts, and training of regular soldiers, require a handling different from that which they have been accustomed to employ.

An understanding that conventional discipline was inappropriate for citizen volunteers was, he argued, 'one of the most important lessons to be learned from the American War by English soldiers'.[76]

Henderson was more far-sighted than most in his suggestion that this might have some relevance to a future mass volunteer army. Indeed, in a discussion of 'The tactical methods of handling partially trained troops' at a General Staff Conference in 1909, J.E. Edmonds, one of Henderson's former students, denied that the greater intelligence of wartime volunteers would 'make a very great difference, when we actually get into its zone of fire' and thus dense formations would be needed to harness the energies of troops that lacked the lengthy training of the Regulars.[77]

In the 1890s Henderson held largely conventional views about the discipline of Regular troops.[78] As a result of the initially poor performance of the army in the Boer War he paid more attention to this subject. In an article written in 1903, published in a widely read volume in 1906, Henderson drew a useful comparison between 'mechanical' and 'intelligent' discipline. He compared the latter to

> a pack of well trained hounds, running in no order, but without a straggler, each making good use of his instinct, and following the same object with the same relentless perseverance.

Henderson believed that the army of 1899 relied on 'mechanical' discipline. but under the conditions of modern warfare this was inadequate. Instead the soldier had to use his initiative.[79] Henderson did not take his argument to its logical conclusion by attacking the nature of the disciplinary regime in the army, although he did admit that 'Monotony and routine' were part of soldiering, but were 'certain, if unrelieved, to deaden ambition and to contract the intellect'. He also criticised the lack of educational opportunities for officers and of training facilities.[80]

The views expressed in Henderson's 1903 article represent the development of previous ideas. It is possible, if he had lived beyond that year, that his ideas would have developed further as a result of the debate on discipline conducted within the army in the decade before the outbreak of the First World War. As it was, Henderson's views were important because he was the most influential military intellectual of the period. Furthermore, in 1914 many officers faced just the situation that Henderson had predicted twenty years earlier, commanding untrained but enthusiastic civilian volunteers, unused to military discipline.

Other Regular soldiers were thinking along similar lines and some exceeded Henderson in their radicalism. Another prominent thinker, Sir J.F. Maurice, called for training that would produce thinking, individualistic 'light infantry'.[81] At a much more junior level, Lt E.L. Spiers (11th Hussars) became dissatisfied with the 'insufferably dull routine' of training, and tested some new schemes on his troop. Spiers deliberately strove to make these interesting for his men, and was careful to take time explain them.[82] In a lecture in 1906 General Sir Richard Harrison argued for a higher level of man-management skills. Other Ranks should not be treated as if they were stupid and the lot of the men should be improved, by the provision of amusements, more interesting training, and by preventing the men from being 'humbugged about'.[83]

The nature of the officer–man relationship itself received little attention in the Edwardian army. Discussions on this subject tended to focus on the practical side of paternalism.[84] Harrison's lecture, while touching on relevant matters, did not go into details on the subject of the officer–man relationship, while Henderson subscribed to a romanticised image of the relationship, particularly regarding the role of the NCO.[85] Even Maurice subscribed to a rosy view, which was not wholly erroneous, yet strangely at odds with the perspective of the Other Ranks.[86]

Perhaps it was a futile exercise to attempt to inculcate initiative among the ranks without altering the basis of the relationship between the officers and men. It was difficult in the extreme to graft a system of devolved command on to a rigidly hierarchical army whose Other Ranks were regarded and treated as little more than cogs in a machine, or, to change the analogy, as children who had to be spoon-fed by their officers. While some officers did criticise discipline and training, few publicly challenged existing views on an equally vital and directly related topic, the nature of the officer–man relationship.

Despite the paucity of public debate on the inter-rank relationship, some officers' attitudes to their men were undergoing change on the eve of the Great War. This was, in part, a pragmatic response to recruiting problems.[87] In early 1914, some officers identified the treatment of recruits as a factor that affected recruiting and tactful handling of recruits was urged, without, it must be said, much discernible effect.[88]

There is also evidence of a greater willingness to trust soldiers with responsibility. Most strikingly, in 1909–10 the Army Council accepted, however reluctantly, the need to overcome the shortage of officers by commissioning selected NCOs upon mobilisation.[89] Early in 1914 the army planned to commission as many as 50 NCOs on the outbreak of war although French, the CIGS, opposed large-scale promotions from the ranks on the pattern of the French army, since this would upset the 'exceptionally happy' officer–man relationship.[90]

This shift in opinion is also evident in the pages of the *Soldier's Small Book*, issued to every ranker. The 1905 edition contained three-and-a-half pages headed 'OBEDIENCE IS THE FIRST DUTY OF THE SOLDIER'. This section set out in graphic detail the various punishments, including death, for disobedience. It is perhaps significant that the 1909 edition omitted most of this material.[91] Changes in attitude also occurred at regimental level. Lt Col the Hon. G.H. Morris (1/Irish Guards), who had a reputation as a radical, in a 'revolutionary step', set up a weekly consultative meeting with the other ranks on welfare and similar issues, along the line of the Indian *durbar* (see below).[92] The regimental historian of the King's Own discerned a considerable change in attitudes in this period. The extension of the franchise had given some rankers the vote, and on polling day the regiment provided four motor-cars to take soldiers to the polling station. In 1913 each company of the first battalion sent a representative to a messing committee, chaired by the sergeant-major, while four years earlier well behaved men earned the privilege of walking out in civilian dress.[93] These concessions were trivial enough in themselves, but do appear to mark an increasing acceptance that the ranker was capable of behaving responsibly.

On occasions, the officer–man relationship involved a measure of consultation and even democracy. In 1903 the War Office proposed to extend the length of service in the Foot Guards. The regimental adjutant of the Scots Guards ordered that officers should obtain 'as far as possible the individual or prevailing opinion' on this subject. The commander of 3/Scots Guards outlined the scheme to senior NCOs, told them to ascertain the opinion of the men in the following week and report back. The commander spoke to 40 men in person.[94] A similar process took place in 2/Coldstream Guards in 1903 when Col Ivor Maxse asked the men whether they were prepared to forgo two days' pay in order to have the use of a miniature rifle range. The men were given 15 minutes in which 'they could regard themselves as a republic and talk the matter over'. At the end of that time a vote was taken, which produced a unanimous 'yes'.[95] One may question whether the other ranks were truly free to express their own opinions, and suggest that the Guards may have been a special case, but the mere fact that officers were prepared to go through the motions of consultation indicates that attitudes towards discipline and the officer–man relationship were somewhat more complex than has sometimes been assumed.

This chapter concludes with an examination of one officer's views on discipline and officer–man relations, and the way in which he put his ideas into practice at regimental level. At the time of his death in 1916, Brigadier-General Philip Howell (GSO1, II Corps) was widely regarded as one of the rising stars of the army.[96] He served with the Guides in India and entered Quetta Staff College in 1905. Howell then took a series of staff posts and in 1913 he transferred into the 4th Hussars as major and second in command.[97] Although his close, if not uncritical, association with Haig would appear to mark him down as a conventional officer,[98] Howell was far from being the stereotypical cavalryman (although some cavalry officers argued that they had to be more flexible than the infantry). He was a keen student of his profession and had political views of a liberal character.[99] As an Indian army officer he was a relative outsider to the British military establishment. This position allowed him to take a reasonably detached view of the British army, and to come to some radical conclusions about officer–man relations and discipline.

There are indications that Howell was thinking along radical lines on these topics as early as 1908,[100] but his appointment as second-in-command of the 4th Hussars in 1913 was a turning-point. Howell brought to the 4th Hussars experience of the *silladar* system used in many Indian cavalry regiments where rankers, who tended to be high-caste, were in the position of a contractor, rather than a mere hireling.[101] Such units made effective use of self-discipline. In the Guides, a *durbar* or 'open court' was held twice a week, at which the men could air grievances and raise matters relating to welfare.[102]

Howell painted a grim picture of life in the 4th Hussars:

> Shortage of strength means more frequent fatigues ... excessive youth [of soldiers] means much elementary work, more boredom & more mistakes: boredom & mistakes lead to punishments: & punishments to desertions & unpopularity of recruiting (*sic*) – and so round and round we go.

Howell considered that modified discipline and officer–man relations would produce a more contented soldiery, who would work more efficiently, and this would lead to a more professional army. For example, he regarded the military obsession for cleaning as a symptom of a concern for 'outward appearances', which, in his wife's words, 'torment the soldier without ... increasing his efficiency'.[103]

Howell was no less critical of the frequent use of punishments in the regiment, making clear that he would look with favour on the sergeant who produced the shortest, rather than the longest, list of defaulters. In a similar vein, he would attempt to seek the root cause of indiscipline, rather than simply punish it. On one occasion, tired of punishing him, Howell wrote to the parents of one young private, a persistent defaulter, and then gave him home leave: the soldier returned a changed man.[104] Howell came to see that reforms were necessary to get the best from the human material. He attempted to make his subordinates lead rather than drive the men, and to improve training techniques, believing that men could only learn when in a receptive frame of mind. But, Howell realised, when faced with 'cursing, swearing and noise[.] A man becomes either frightened or surly.' Howell, supported by his like-minded CO, Lt Col Ian Hogg, does seem to have brought a more enlightened disciplinary system to the regiment, although some of his ideas did not go down well with the men – a suggestion-box remained empty, a reminder of the innate conservatism (and desire for self-preservation) of the Regular ranker.[105]

Howell's aims and methods discussed thus far would have won the approval of many thinking officers. However, at least as early as 1908, Howell had come to believe in commissioning men of natural authority, regardless of class. Education and training, Howell believed, could supply the army with suitable leaders.[106] The Curragh Incident of 1914 led to calls, mainly from the political Left, for a 'democratic' army: this notion appealed to Howell. In a letter to Ramsay MacDonald, he argued that the interests of army officers and Labour leaders coincided. A 'sound system' of promotion of rankers would aid political impartiality and military efficiency: 'A stratum of rankers of the right sort would soon break down prejudices & make itself felt.' The main problem Howell foresaw was not opposition to the scheme but the low quality of the ordinary ranker, which Howell blamed on the lack of career prospects.[107]

In April 1914 the possibility of democratising the officer class appeared remote. Within twelve months, under the pressure of war, the army had begun the wholesale commissioning of men drawn from the lower reaches of British society, vindicating Howell's predictions. Army officers and military authorities pragmatically accepted as officers men from a far humbler social background than the average prewar officer and gave them appropriate training. Howell's fears concerning the poor quality of the available material proved largely unfounded because considerable numbers of high-calibre wartime volunteers and conscripts, who would never have joined the prewar army, provided the army with an excellent source of officers.

This chapter and the last argued that two distinct strands of discipline and officer–man relations co-existed in the prewar British army. The Regular version was characterised by a rigidly hierarchical approach, reliance on 'imposed' discipline, and distant although mutually respectful relations between officers and men. Auxiliary discipline was, by contrast, much looser, with greater emphasis being placed on self-discipline, and inter-rank relations were characterised by informality. Both of these strands contributed to the disciplinary system and officer–man relationship of the army of 1914–18. The paternalistic ethos of the prewar Regular officer infused the wartime officer class and was, it will be argued, a crucial factor in maintaining the morale of the British soldier in the First World War.[108] The disciplinary structure of the prewar Regular army was used, for the most part, in the mass army of 1914–18, to some extent inhibiting the development of the independently minded soldier called for by many of the protagonists in the prewar debate on discipline. However, the relationship between the officer and soldier in the wartime army resembled that of the Auxiliary forces. Thus there was much continuity between the officer–man relationship and the disciplinary system in the British army of the Great War and its Edwardian predecessor. While little was new, most wartime units took elements from both the Regular and Auxiliary traditions, in varying proportions depending on the unit, to create a style of inter-rank relations and discipline that showed traits inherited from both parents.

3
The British Officer Corps, 1914–18

The British officer corps of the Great War was an heterogeneous body consisting of seven varieties of regimental officers. First, there were Regular officers with peacetime service; second, officers who were granted permanent commissions during the war years; third, civilians who were granted Temporary commissions valid for the duration of the war only; fourth, Territorial officers commissioned in peacetime. In addition, there were 'prewar' ranker-officers, men who had served in the ranks of the Regular army before the war and 'prewar Territorial' ranker-officers, men who had served as privates or NCOs in TF units before the war. The final category of officer was the 'wartime Temporary' ranker-officers. The latter term does not necessarily indicate that an individual was of the class that had dominated the ranks of the prewar Regular army, for large numbers of middle- and even upper-class men served in the ranks of the army in the first few months of the war.

On 17 September 1914, the Secretary of State for War, Lord Kitchener, announced that he was looking to volunteers and Regular NCOs to provide officers needed for the newly expanded army.[1] Larger numbers of officers drawn from 'non-traditional' sources were commissioned during 1914–18 than in any previous war. This was, in part, a consequence of the very high casualties sustained by junior officers, and the disproportionate heavy losses inflicted on the social elite.[2] It was also a reflection of the sheer size of the army raised between 1914 and 1918. Approximately 5 704 000 men served in the army during the course of the war. The total strength of the army grew from 733 514 on 1 August 1914 to an estimated 3 563 466 on Armistice Day, 11 November 1918.[3]

On 4 August 1914 there were 28 060 officers in the British army, of which 12 738 were Regulars. On 11 November 1918, the army possessed 164 255 officers. (See Table 3.1.) From the beginning of the war up to 1 December 1918, 247 061 commissions were granted to combatant officers, chaplains, and the RAMC.[4] While the average number of commissions granted per annum between 1908 and 1913 had been a mere 649,[5] in 1910 the Adjutant-General had asserted that 'We are coming to the end of our tether as regards candidates from the limited class which has hitherto supplied the commissioned ranks' of the Regular army.[6] Thus it is not surprising that the War Office had to reach far beyond its traditional sources of supply of officers to provide enough leaders for the mass army of 1914–18. It was not the original intention of the military authorities to broaden the social base of the officer class to any great extent. The usual insistence that potential officers possess OTC Certificates A (from a public school OTC) or B (from a university OTC), and the recruiting policies of many of the raisers of units in 1914–15 suggest that it was hoped that officers could be provided for the enlarged wartime army with the minimum disturbance to the social *status quo*. The pressure of war quickly forced the abandonment of this policy, although not without some official misgivings.[7] In the long term, the policy of awarding Temporary commissions insulated the British officer corps from major social change.

Table 3.1 Officers serving in the British army at the beginning and end of the Great War

	4 August 1914	*11 November 1918*
Regular and New Army:	12 738	74 200
Special Reserve:	2 557	28 000
Territorial:	9 563	60 055
Reserve of Officers:	3 202	2 000
Total	28 060	164 255

NB: New Army figures apply to 1918 only. All figures are approximate.
Source: *SME*, p. 234.

It was possible, although not easy, to convert from a Temporary to a Regular commission; 1 109 Temporary officers were among the total of 16 544 men who were granted Regular commissions during the war, although not all were commissioned into their original regiment.[8] Sandhurst and Woolwich ran abbreviated courses for Regular cadets fresh from civilian life.[9] The parsimonious distribution of Regular commissions ensured that the impact of the influx of lower-class men into the British officer class was short-lived. The vast majority left the army in 1918–19, so the postwar British officer class more closely resembled that of 1913 than that of 1918.

However, in the short term, a revolutionary change occurred in the social composition of the wartime British army. This can be illustrated by the figures for demobilisation of officers. The dispersal certificates of 144 075 of the officers who had been demobilised by 12 May 1920 were analysed by the War Office. There are some anomalies in the data, and the occupational categories employed are annoyingly broad. However, these statistics do give a clear picture of the social status of the officer corps in 1918. The three largest single categories of officers' occupations are 'commercial and clerical' (group 37) 'students and teachers' (group 43) and 'professional men' (group 42). These groups, which cover broadly middle-class occupations, account for 59.5 per cent of the total (85 889 men).

The 'commercial and clerical' category might contain men of a lower social status than the other two groups. Even omitting this category, the not unimpressive total of 58 706 men, or 36.5 per cent, remains. (See Table 3.2.) Furthermore, when expressed as a percentage of all men, both officers and Other Ranks, demobilised from the army from these occupational groups, other significant patterns emerge. About 44 per cent of those in the 'professional men' group, and 38 per cent of all 'students and teachers' were demobilised as officers. By comparison, only 8 per cent of commercial and clerical workers and a mere 0.2 per cent of general labourers demobilised in this period served as officers.[10]

Table 3.2 Principal occupations of officers demobilised, 11 November 1918 to 12 May 1920, as expressed as a percentage of those returns analysed

Commercial and clerical 38 572 (27 per cent)
Students and teachers 25 577 (18 per cent)
Professional men 21 740 (15 per cent)
Engineering 11 389 (8 per cent)
(144 075 returns analysed out of 161 103)

NB: All percentages have been rounded up to the nearest 0.5 per cent.
Source: *SME*, p. 707.

As J.M. Winter has suggested, the raw data suggests a 'pattern of middle class domination of the officer corps' with only the body of engineering workers providing an exception.[11] However, by comparison with the prewar structure of the officer class, some significant variations can be noted. The term 'middle-class' is a somewhat elastic one and undoubtedly includes men who would not have been thought suitable for a commission in the old Regular army. In the course of the war some men who were on the social borderline between the lower-middle and upper-working classes came to be regarded as suitable to take commissions. In October 1916, a general commented approvingly upon the type of man to be found in 2/5 Buffs, a 'Bank Clerks Battalion', whom he considered to be officer material. They were, of course, likely to have been very few former bank clerks serving as officers in the prewar Regular army.[12]

It will be recalled that in 1913 only about 2 per cent of those commissioned had passed through the ranks. If this may be taken as a benchmark against which to judge working-class penetration of the prewar officer corps, the demobilisation figures reveal that the officer class had been considerably democratised by 1918. There are difficulties in interpreting the occupational groups, but numbers of working-class men may be discerned, such as 'carters' (148 officers), 'dock and wharf labourers' (184), 'seamen and fishermen' (638), 'leather tanners' (99) 'coal and shale miners' (1 016) and 'warehousemen and porters' (266). These figures may include some of the foreman/overseer type who perhaps shade into the lower reaches of the middle classes.

A rough-and-ready guide to the social status of each category is to compare the numbers of men who became officers with those who remained in the ranks. From our sample, 21 740 professional men became officers, as against 26 988 who remained in the ranks. By contrast, 7 495 workers in agriculture became officers, while 301 770 served in the ranks. One hundred and fifty-seven coachbuilders and woodworkers were demobilised as officers, and 34 222 left the army as Other Ranks. In all, 54 584 officers, or 39 per cent of the sample, were demobilised from the 41 remaining industrial groups (i.e. not commercial and clerical, students and teachers, and 'professionals' which, with the reservations noted above, may be classified as broadly 'working-class' or at least lower-middle-class occupations). However imprecise the figures, it is clear that in 1918 many men were serving as army officers who would have been effectively excluded on educational and social grounds from obtaining a Regular or even a Territorial commission before the war. Many of these men, to judge from the occupational categories, were drawn from the artisan class, with the engineering industry providing the largest proportion of officers (11 389 officers, 359 948 Other Ranks).[13]

These bare statistics are supported by a wide variety of other evidence. R.C. Sheriff used 'Trotter', a working-class ranker-officer archetype in his play *Journey's End*,[14] for instance, while a middle-class ranker complained that by 1917 any 'Tom, Dick or Harry' could be put forward for a commission.[15] More specifically, on joining a Special Reserve unit, 3/SLI, in October 1917, a public-school officer wrote that 'about sixty-five percent' of the officers were 'gentlemen', which, he opined, was about as many as could be expected.[16] The original officers of 22/R. Fusiliers (Kensington), a New Army battalion raised in 1914, were mostly educated at public schools and universities, but many officers who joined the battalion on the Western Front came from humbler backgrounds.[17]

While it was true that a middle-class man was far more likely to receive a commission than a member of the working classes, and the latter were grossly under-represented in officers' messes given the numbers of them serving in the army, a limited meritocracy emerged in the British army during the Great War. The official claim that 'no barrier' existed to prevent talented rankers gaining commissions was an exaggeration, but perhaps a pardonable one.[18]

By the end of 1914, 4 270 officers of the BEF had become casualties, of whom 1 278 had died.[19] As early as mid-September 1914, 593 officers had been sent to the BEF as battle-casualty replacements, a rate which caused Kitchener some concern.[20] Kitchener had to balance the requirements of the BEF and other units on active service against those of the newly raised Territorial and New Army units. The latter units, in training in the UK, necessitated the retention of a number of officers who could have been used to supplement the reservists and Special Reserve officers being sent as reinforcements to France, much to the disgust of Henry Wilson, who complained that Kitchener's phantom forces were robbing the BEF of much-needed soldiers and officers.[21] The military authorities were thus compelled to look elsewhere for officers to fill the gaps in the battalions of the BEF.

Men serving in the ranks of Territorial class corps in France, who combined the advantages of reasonably high social status with experience of active service, were an obvious source of officers. Class corps began to be viewed with envious eyes by Regular commanding officers. Indeed, it had been proposed by the War Office at the time of the creation of the TF that the Artists' Rifles (1/28 Londons) should become an officers' training corps, but in the event, it was agreed that

10 per cent of its rank-and-file would be made available for commissions on mobilisation.[22]

Training for Territorials commissioned in the field was initially scanty. In November 1914, some men of the Artists' were commissioned and returned to the trenches as officers in Regular battalions within 24 hours, wearing a second-lieutenant's 'pip' on a private's tunic. By the beginning of December 1914 potential officers (or 'Commissionairs') were being given crash courses in officership, the Artists' having effectively been turned into an officer training unit.[23] The training given was both theoretical and practical. Potential officers received lectures and instruction and spent 48 hours in the trenches, before being posted to a Regular battalion.[24] This system, rough and ready as it may have been, seems to have been regarded as a success. In November 1914 a divisional commander told Kitchener that the subalterns commissioned from the ranks of the Artists' were 'first class ... and the cry is "give us more of them"'.[25]

The ranks of Territorial class corps continued to be a fertile recruiting ground for officers during 1915. In eight months on the Western Front, 100 men of the Queen's Westminster Rifles (1/16 Londons) were commissioned or transferred to cadet schools.[26] In the spring of 1915, Regular commanding officers were 'only too eager' to take LRB veterans as officers, but the battalion CO was apparently reluctant to release his men for fear of his entire unit evaporating, some 95 per cent of Other Ranks being suitable to take commissions.[27] Similarly, not all the rankers of class corps were eager to see their units transformed into officer training units.[28]

Reserve and Special Reserve officers, and those cadets hastily commissioned from Sandhurst and Woolwich in the autumn of 1914 were mostly sent as reinforcements for the BEF, and were therefore unavailable for the New Armies. Kitchener went to extraordinary lengths to provide officers for his volunteer army.[29] He kept back three officers and a proportion of NCOs from each BEF battalion; detained 500 Indian army officers on leave in Britain; and tapped the reservoir of 'dugouts' (retired officers). Kitchener also ordered that the commander of every unit in the country, including Territorials, should be asked to put forward the names of likely WOs and NCOs for commissions. By methods such as these Kitchener hoped to bring about a situation in which newly raised battalions and units of the BEF 'have the same proportion, as far as possible, of experienced and inexperienced officers'.[30]

Another potential source of officers was men who had served in the

Officers Training Corps. An announcement appeared in newspapers on 10 September 1914 that '2000 Junior Officers (unmarried) are immediately required' to serve with the Regular army until the conclusion of the war. They had to be cadets or ex-cadets of a university OTC, or members of a university aged between 17 and 30. Even at this stage, the War Office was prepared to accept men without OTC experience, providing they possessed the necessary educational and social qualifications. 'Other young men of good general education' were advised to apply, in person, to the commander of their nearest army depot.[31] Depot commanders would be able to judge, from seeing would-be officers in the flesh, whether they would be acceptable in an officers' mess.

Formed in 1908, the OTC consisted of Junior and Senior divisions, the first consisting of public school, the latter of university corps.[32] In 1907, just prior to the formation of the OTC, there existed eight cadet battalions, three Rifle Volunteer Battalions and 152 cadet corps or companies.[33] These units were of doubtful military value. Great attention was paid to shooting and drill, but very little to tactical training.

In 1907 a War Office committee recommended that the OTC should be established on the basis of existing corps, as a means of overcoming the shortage of officers for the army.[34] By 1914 OTCs existed at 79 per cent of all public schools, which amounted to about 60 per cent of all junior OTCs.[35] The OTC was unpopular and underfunded in peacetime, and relied so heavily on the goodwill of the part-time officers of the OTC that one claimed that the country had 'solved the riddle of getting something for nothing'.[36] The success of the scheme may be gauged from the fact that between August 1914 and March 1915 commissions were awarded to approximately 20 577 (and possibly as many as 27 000) men serving, or who had served, in the ranks of the OTC. This was exclusive of the 6 322 men who had been gazetted from the OTC between 1908 and the outbreak of war, or who had served in school cadet corps prior to the formation of the OTC in 1908.[37] In all, the OTC produced some '100,000 of the 230,000 officer candidates who were found to lead Britain's soldiers in battle'.[38]

Peter Simkins has pointed out that the use of the OTC as a source for officers ensured 'that there was no sudden and radical change in the social composition of the officer corps in the first year of the war'.[39] However, the use of the OTC did occasion a more modest social change in the social profile of the British officer. No less than 41 per cent of the officers who served in the Boer War were drawn from the 'ten great public schools', with Eton alone providing 11 per cent.[40] While impressive numbers of commissions were granted to alumni of the OTCs of such schools between August 1914 and March 1915 (350, 253, 506 and 403 from Eton, Harrow, Marlborough and Wellington respectively), just 10 per cent of OTC commissions went to products of Clarendon schools.

Even omitting the 4697 commissions given to Senior Division candidates from the total, Clarendon commissions amounted to only 13 per cent. In absolute terms more commissions were granted to the more numerous 'lesser' schools, although taken individually, such schools tended to have greater numbers of former pupils serving in the ranks: Dollar Institution provided 35 officers and 200 Other Ranks, and Wellingborough Grammar School 42 and 44 respectively.[41] Even allowing for the fact that the available figures are incomplete, given the strong correlation between the prewar officer class and the great public schools,[42] there is a strong suggestion that the typical officer of the New Armies was socially inferior to his prewar regular counterpart.

The term 'public school' had been defined widely when the OTC was created. Of the schools which possessed an OTC in 1914, 43 were not actually listed in the public schools yearbook of 1908, and 19 of those that were did not possess an OTC.[43] St Dunstan's College, Catford, did have an OTC, but its pupils' fathers tended to have relatively low-status occupations. Clerks abounded, along with accountants, a 'grocer and provisions merchant', an engraver and a 'retail fishmonger and poulter'.[44] Similarly, the boys of Royal Grammar School, Newcastle-upon-Tyne, tended to go on to study locally at Armstrong College rather than to attend the ancient universities, and made careers in such solidly middle-class occupations as 'colliery manager', 'assistant surveyor of shipping' and 'architect'.[45]

In 1908, two criteria were employed when deciding which cadet corps were to be allowed to convert to OTC. Firstly, corps had to 'show an enrolled strength of not less than 30 cadets' over the age of 13 years. Secondly, it had to be militarily efficient. Efficiency was gauged by such factors as the availability of training facilities and qualified officer instructors, and the Army Council reserved the right to disband units which failed to maintain required standards.[46] Social considerations therefore appear to have played a relatively minor role in deciding whether individual units were to be allowed to be upgraded to OTC status. One commentator suggested that by using military effectiveness as the primary criterion for conversion, the military authorities armed themselves with means of checking the abuse of the system by schools who wished to enhance their social prestige by acquiring an OTC.[47]

The implied division between schools of similar types was resented by some. In 1918 the Public Secondary Schools Cadet Association called for 'The sweeping away altogether of the invidious and anomalous distinction at present existing between OTC schools and Cadet Corps schools'.[48] Even in official circles, men who had served in a cadet unit were generally assumed to be of low social status.[49] The familiar story of R.C. Sherriff's unsuccessful attempt to gain a commission in 1914 should be viewed against this background. Sherriff, who was educated at Kingston Grammar School, an institution founded in 1567, was refused a commission because it was not listed as a 'recognised public [school]'. Sherriff's rejection is perfectly explicable on military, rather than social grounds, for his school did not have an OTC or even a cadet corps, while the Wykehamist who preceded him in the queue, and was accepted, had presumably obtained his certificate A from his school's OTC.[50] Many young men educated at grammar schools little different from Kingston received commissions in this period, on the strength of their OTC certificates. The War Office's policy on officer recruitment was not entirely consistent in the early months of the war. However, it is clear that the desire for military experience in potential officers played a greater role, and snobbery a lesser role, in its thinking in this period than has sometimes been allowed.

In March 1915 12290 men possessing OTC certificates were reported as serving as Other Ranks in the army.[51] Large numbers subsequently applied for commissions. Others did not, but were commissioned anyway. P.G. Heath, who served in the OTC at Malvern, enlisted in 23/Londons on the outbreak of war, only to be told one day in the autumn of 1914 that he was now an officer, as a relative had applied for a commission on his behalf.[52]

Some newly raised battalions were especially rich in potential officers. The four battalions of the 'Universities and Public Schools Brigade' (18-21/RF) had provided 50 officers for the New Army as early as 23 September 1914, and eventually provided more than 7000 before their disbandment in early 1916.[53] The potential of the rankers of the Public Schools Battalion (16/Middlesex) was soon recognised, and 108 men from the battalion were commissioned by mid-October 1914. This led to increasingly desperate, but unsuccessful, attempts by its CO to stem the flow of other ranks leaving the battalion.[54] Less socially prestigious units, such as the four northern Pals battalions of 93rd Brigade,were also rich sources of officers.[55]

Given the widespread belief that officer-like qualities were mainly to be found among the social elite, and the shortage of officers for both the peacetime and wartime army, it is strange that the War Office should authorise the raising of units such as the 16/Middlesex. It is stranger still that having sanctioned these units to be raised, they were not converted into officer training units along the lines of the Artists' Rifles. Instead, large numbers of potential officers were killed serving as privates, as an RAMC officer lamented in late 1917.[56] The loss of such men was the penalty paid by the British army for operating a *laissez-faire*, voluntary system of recruitment at the beginning of the war. However, those who lamented such losses overestimated the importance of the loss of such 'officer material', for , as will be seen, their lower middle-class and working-class replacements were usually just as effective as junior officers.

In the summer of 1915, a rumour circulated among 1/Coldstream Guards that no more privates would be considered for commissions.[57] In reality, within a few months the army adopted a new system of officer-recruitment that ensured that the vast majority of new officers came from the ranks. This new system formally recognised the trend of commissioning from the ranks which had become apparent from at least early 1915 and was, of course, a reversal of prewar policy on officer recruitment.

As early as 16 September 1914, commanders of New Army divisions and brigades were given discretion to nominate men for commissions, who then joined their units in anticipation of their confirmation by the War Office.[58] A War Office circular of December 1914 stated that candidates could be accepted even if they were not trained to command men in the field, if they were 'otherwise suitable in all respects to hold temporary commissions'. On this occasion, no guidance was given on the criteria of suitability.[59]

In fact, commanding officers would have known exactly what criteria to apply to potential officers. Regular officers shared what has been described, in another context, as 'unspoken assumptions'.[60] In this case, it was assumed that potential officers had to be gentlemen, or at least possess enough social skills so that they would not disgrace themselves in an officers' mess or be ridiculed by the Other Ranks. These unspoken assumptions were not simply the product of snobbery, but also reflected the deeply held belief that unless an officer was a gentleman he could not be an effective leader, and his men would not follow him willingly. A question posed in a set of instructions issued to commanding officers in France in late 1914 struck at the heart of the matter. If considering recommending a ranker for a commission, the commanding officer had to ask himself whether he would be prepared to 'accept him as an officer in his own unit'.[61] This concern for social standards was reflected by a man commissioned from the ranks of the Wiltshire Yeomanry in 1915, who asserted, tongue-in cheek, that provided the candidate did not pick his nose while being interviewed and swore that he hunted he was sure to be commissioned. It is interesting that he did not refer to military prowess as a criterion of officership.[62] This man was commissioned from a home service unit. In France, individuals had the opportunity to prove their leadership qualities on the battlefield.

In April 1915 a more detailed set of instructions was issued concerning candidates for commissions. Rankers had to possess 'adequate military knowledge', 'a public school education or its equivalent' and had to be under 27 years of age, except in 'exceptional circumstances'.[63] These criteria were to hold broadly true for the rest of the war. Stated more simply, the first criterion was considered fulfilled if the soldier demonstrated qualities of leadership, or at least had the potential to develop them. The possession of a public school education implied far more than educational achievement or even social status; it also suggested that a man was imbued with leadership qualities and *noblesse oblige*, and that he shared a certain set of social values. The provision of references from public figures such as JPs or clergymen acted as a further filter to prevent socially unacceptable men from becoming officers. Thus Pte Percy Copson supplied two character references, including one from the former mayor of Northampton, with his application.[64]

In practice, the phrase 'a public school education or its equivalent' was capable of considerable elasticity in its interpretation. One brigade commander would satisfy himself of the educational qualifications of potential officers by asking them to spell the word 'Mediterranean'.[65] In late 1917 a letter from the headmaster of Diamond School, Sunderland, sufficed to satisfy the army; it simply stated that '[Cpl William Allen]

> left when in the Senior Class having done well in the various forms, and I consider that he is well qualified educationally to hold a commission in HM forces.[66]

This reliance on the subjective system of personal recommendations, rather than a formal, structured system for admitting men to officer training, was an extension of normal peacetime military practice, whereby patronage and personal connections played an important role in the career structures of British officers.[67] It could be a laborious process to obtain recommendation for officer training. In August 1915 a gunner of 48th Divisional artillery recorded that he obtained permission from his sergeant 'to apply for a commission'. The sergeant took him to see his section commander, with whom, in the would-be officer's words:

> I had a talk over it (*sic*), and finally, I got his permission to be taken to the Colonel commanding 2nd battery ... [who had] no say in the matter beyond recommending me to the Brigade Commander.... He alone signs the forms finally – after seeing me – and he alone recommends me to General – to be seen in turn by him, the latter being the Commander R.A. for the Division....[68]

Thus this man underwent a number of informal as well as formal interviews. In the infantry, prospective officers were usually interviewed by their battalion and by their brigade commander, and often had a preliminary discussion with their platoon or company commander. This system was generally adopted after February 1916.[69]

Likely men could be recommended for commissions even if they had not applied to become officers. By the spring of 1917, battalions had to furnish the names of five NCOs or privates suitable for commission every month.[70] Pte G. Hall (13/Yorks and Lancs) was informed by his company officer that Hall had been recommended for a commission, and Hall's diffident protests that he was 'only a working lad' were overridden.[71] Although some soldiers saw the system as arbitrary, in general it worked well.[72] However, before they could be sent to the Western Front to command men, potential officers had to be trained. That is the subject of the next chapter.

4
British Military Leadership, 1914–18: Influences and Training

Before examining British regimental military leadership in the Great War, it is essential to establish a theoretical framework. One authority states that leadership is:

> the phenomenon that occurs when the influence of A (the leader) causes B (the group) to perform C (goal-directed behavior) when B would not have performed C had it not been for the influence of A.[1]

This definition is useful, but incomplete. In 1927, F.C. Bartlett identified three broad categories of leaders:

(a) institutional leaders, 'who maintain their position by virtue of the established social prestige attaching to their office';
(b) dominant leaders, 'who maintain their position by virtue of their personal capacity to impress and dominate their followers';
(c) persuasive leaders, 'who maintain their position mainly by virtue of their personal capacity to express and persuade their followers'.[2]

Bartlett's work suggests that leadership involves more than the imposition of one individual's will on a reluctant group. In Correlli Barnett's words,

> Leadership is a process by which a single aim and unified action are imparted to the herd. Not surprisingly it is most in evidence in times or circumstances of danger or challenge. Leadership is not imposed like authority. It is actually welcomed and wanted by the led.[3]

A combination of Henderson's and Barnett's assessments of leadership produces the following definition:

> Leadership is the phenomenon that occurs when the influence of A (the leader) causes B (the group) to perform C (goal-directed behaviour) when B would not have performed C had it not been for the influence of A; the influence of A being generally welcomed by B.

Officers have five sources of influence over their men, influence being defined as the process by which A modifies the behaviour and attitudes of B, and power being defined as that which enables A to influence B.[4] These are: 'resource' power, the ability to give rewards; 'physical' or 'coercive' power; 'institutional' power, that is, power which flows from the acceptance of a leader's hierarchical status; 'personal' power, that is power which flows from the identification of the soldier with his officer, leading to the creation of ties of trust and admiration; 'expert' power, the belief that the leader is professionally competent, and has the necessary expertise and information that makes him well-equipped to lead the unit.[5]

The ideal leader is one who relies mainly on personal and expert power. A poor leader is one who relies mainly on institutional and coercive power. Used wisely, the officer's influence can be instrumental in creating and sustaining morale, for certain factors that tended to undermine the ordinary soldiers' morale have emerged from sociological and psychological studies of men in combat:

(1) Military service involves a loss of freedom, and better-educated men feel status deprivation more strongly than ill-educated men.
(2) Military service involves suffering personal discomfort, and often conveys the impression that the individual is a mere 'cog in the machine', powerless and impotent in the face of an impersonal and arbitrary coercive authority.
(3) Military life involves isolation from accustomed sources of affection.
(4) Individual rewards are subject to both poverty and uncertainty in the army, and 'job satisfaction' is often denied to the soldier.
(5) Military service involves continual uncertainty, insecurity and inadequate cognitive orientation.[6]

Good officer–man relations and enlightened leadership can alleviate many of these factors. The leader has two main functions. First, he helps to create and sustain unit cohesion. This is a broadly managerial function. The officer has to ensure that his men have sufficient supplies of food, drink and other essentials. He also has to exercise man-management, to prevent the alienation of the soldier, perhaps defending him against the unreasonable demands of higher authority. The officer also has to detect and correct behaviour that deviated from group norms. Ultimate success is judged by the achievement of cohesion, when the formal military unit becomes for its members a substitute for family, the core of their social and emotional lives. Second, the leader has to mould the cohesive group so that their goals are congruent with those of the greater organisation, the army; in short, he has to lead the group into battle.[7] Cohesion is a two-edged weapon; survival of the group, rather than the accomplishment of the aims of the army (usually, combat missions) can become the object of some military primary groups.[8] The regimental officer is the vital link between the primary group and the army.

In the era of the Great War soldiers rarely questioned the notion that there was a direct connection between the ethos of public schools and the ability to lead men in battle. Elite schools emphasised the development of 'character', the virtues of 'physical and moral courage, loyalty and co-operation and the ability both to command and obey'.[9] Col C. Bonham-Carter commented that the 'gift of leadership' was almost a race inheritance among public schoolboys'.[10] A prominent industrialist commented that a public school and university education produced men who were loyal to the organisation, and were able 'to get on with their subordinates by just treatment mixed with kindness' and treated their superiors 'with deference that does not deteriorate into subservience'.[11] It is not surprising that such members of the establishment should praise the public schoolboy. However, confirmation of their views comes from the pen of an individual who rebelled against the establishment: Richard Aldington.[12]

Aldington attended a public school, served in France in the ranks of 11/Leicesters and was commissioned into 9/R Sussex. In his novel *Death of a Hero* (1929) Aldington draws a pen-portrait of 'Lieutenant Evans', who was described as a typical public schoolboy, 'ignorant', 'inhibited' but '"decent" and good humoured'. He 'accepted and obeyed every English middle-class prejudice and taboo' and was totally convinced of his superiority to the working classes. Evans was xenophobic, sexually repressed, and philistine in his cultural tastes. In sum, Evans was depicted as a living example of the values that Aldington despised as obsolete. Evans was also, as Aldington made clear, an excellent officer:

> [Evans] was honest, he was kindly, he was conscientious, he could obey orders and command obedience in others, he took pains to look after his men. He could be implicitly relied upon to lead a hopeless attack and to maintain a desperate defence to the very end. There were thousands and tens of thousands like him.[13]

Aldington believed that public schools produced ideal subalterns for the British army on the Western Front. Aldington was essentially correct. Furthermore, non-public school officers soaked up public school values through education, training and socialisation. As a result, the lower-middle-class and working-class officers of the latter part of the war were imbued with much the same values as their public school-educated predecessors. The public school ethos, described by a headmaster in 1913 as 'something essentially English' consisting of 'a sense of honour, of *esprit de corps*, and ... a spirit of self-sacrifice'[14] therefore influenced far more officers than had ever attended public schools.

Judging by the problems experienced by the Edwardian army in attracting officers, public schools might appear to have been inefficient instruments of militarisation.[15] However, the inculcation of values of officership by overt means, such as near-compulsory service in the OTC, was less effective than other, less tangible factors such as athleticism, the classics, and the concept of chivalric gentlemanliness and self-sacrifice, all of which were very much a part of the public school ethos. Here it is not intended to duplicate the work of other writers on the public schools, but rather to discuss some aspects of their work relevant to the officer–man relationship in the Great War.

In the nineteenth and early twentieth centuries the idea that sport was a useful training for war was widely accepted.[16] Games allegedly produced the 'character' necessary for leadership. Sport channelled aggression into cooperation, taught self- and corporate discipline, made boys physically fit and created team spirit. By teaching boys to take rapid decisions on the playing field, games prepared them to take similar decisions on the battlefield. On the sports field, as on the battlefield, boys learned to take risks and to disregard their personal safety. Team games, declared Haig in 1919, required

> decision and character on the part of the leaders, discipline and unselfishness among the led, and initiative and self-sacrifice on the part of all.

The 'inspiration' of games, Haig claimed, 'has brought us through this war, as it has carried us through the battles of the past'.[17]

In recent years writers have questioned the correlation between the cult of athleticism and military leadership. Geoffrey Best is generally critical of the 'games ethic' but he recognises that games were useful as a means of inculcating loyalty to the group, be it the team, school, regiment or Empire.[18] Peter Parker is dismissive of the whole concept, suggesting that obsessive devotion to the public school, or the house, detracted from loyalty to nation. Moreover, Parker argued that a games player might be respected by his peers, but not necessarily by strangers in the army, and the cult of the athlete encouraged boys to work for individual, rather than group goals.[19]

Parker's criticisms are overstated. J.D. Burns, an Australian subaltern, argued in his school magazine that loyalty to the school was the first stage in the development of a more mature code of loyalty. Later, Burns argued, horizons widen and 'patriotism for the Empire will succeed, though not replace, loyalty to the school'.[20] Parker's second argument ignores that the public schoolboy, on joining the army, would find himself in an organisation that, like his school, placed a high degree of emphasis on sport, and in which sporting prowess was greatly prized. The individual would very soon have an opportunity to win the respect of his fellow officers on the sports field. He would also be well placed to win the respect of his soldiers, for interest in sport was by no means confined to the social elite. Watching and playing sport, especially Association football, was one of the principal forms of working-class leisure in the early twentieth century.[21] On the Western Front sport was important as a recreational activity for the British working man in uniform. A Regular officer noted in his diary in August 1914 that 'The men spent the evening in their usual manner, kicking a football about', while in June 1915 a popular paper devoted a double-page spread to a picture of a typical scene of men 'resting' behind the lines by playing football.[22]

In the later nineteenth century, former public schoolboys codified sport. They converted football from a rough-and-tumble into a disciplined activity conducted according to carefully defined rules. Thus the public school games ethic, in a diluted form, influenced the working classes.[23] It is not surprising that in the autumn of 1914, the football stadium proved to be fertile ground for the recruiting officer.[24] A soldier thought it essential to bring his officer news of the 1915 FA Cup Final during a tense moment in the second battle of

Ypres. This is suggestive of both lower-class enthusiasm for sport, and that a mutual interest could bring officers and men together.[25]

Burns refuted the idea that sport encouraged the boy to strive to achieve renown for himself, rather than to seek honour for the team. He argued that the object of the athlete in training was not self-glorification but the 'setting of an example', this being the 'duty of each individual in the school'.[26] Other evidence would suggest that this was a common belief.[27]

Parker also attacks the notion that the cricket or rugby pitch was a valuable training ground for subalterns. He argues that the taking of decisions on the sports-field was very different from battlefield leadership, in which the officer had to take 'snap decisions' that place men's lives at risk. Moreover, the discipline of the sports-field could not be compared with that of the battlefield. To some extent, Parker is supported in these criticisms by Best.[28] Superficially, these criticisms seem irrefutable; there was indeed a world of difference between football and war. However, these criticisms are exaggerated, not least because leadership on the battlefield occupied a relatively small proportion of the military leader's time.

Stuart Sillars has argued that the familiar sporting imagery used in Ernest Raymond's immensely popular 1922 novel *Tell England* provided psychological consolation for the bereaved by helping them to make sense of their loss.[29] Similarly, sport helped young officers to avoid the stress caused by 'role ambiguity'[30] on joining his unit by helping the newcomer to make sense of that situation by relating it to a familiar experience. The young officer, faced by his platoon for the first time, was able to fall back on his experience in a sporting team. One extremely effective leader of men, the product of a public school, was 'convinced' that games developed 'the ability to deal with and handle men in later years'.[31]

Regimental officers had to train men as well as lead them into battle. Throughout the war, drafts arrived at front-line units having received only the sketchiest of training at home depots.[32] Changing tactical situations also called for the officers to retrain their men, especially during the re-emergence of mobile warfare in 1918. An officer, wounded in 1917, returned to France in October 1918 to be 'astonished' to find how profoundly the nature of warfare had changed.[33] Writing of the fighting on the Marne in July 1918, the commander of 62nd Division confessed that 'these Operations were practically our first experience of open fighting'.[34] Mobile warfare exposed any lack of appropriate training and made hasty improvisation necessary. A typical formation, 17th Division, reported that in August 1918 that its troops were suffering from

> lack of tactical knowledge, which resulted in formations not being adapted to suit [the] ground, and lack of power to apply the principle of fire covering movement, also lack of intelligent patrolling.[35]

Junior officers were often thrown back on their own resources and imaginations to training their men for a type of warfare for which they were themselves ill-prepared.[36] They naturally drew upon their own sporting experiences for inspiration, encouraged by official enthusiasm for sport as a training aid. One official manual stated that 'there is a close analogy between cricket and an exercise of this kind'; the 'players play the game under the agreed laws and, under the orders of their Captain ... the platoon commander', the umpires adhere strictly to their allotted role, and the spectators 'keep away from their pitch'.[37] To maintain interest and foster *esprit de corps*, another official pamphlet recommended competitions 'in Lewis and Machine Gun Practice, Revolver Shooting and Musketry, Bayonet Fighting, Assault Training, Wiring, Yukon Pack Carrying, etc.'[38] By treating training as a form of sport, officers tried to make training comprehensible to their fellow sports-enthusiasts in the ranks.

The Great War belief that sport was essential training for war was an extension of a much older belief that hunting developed courage, *coup d'oeil* and leadership.[39] One 'fire-eating' gunner officer described combat in terms of hunting, paying tribute to his 'stout-hearted men' who were 'good partners ... in the game of killing the Hun'.[40]

During periods of trench warfare, patrols into No Man's Land and trench-raids gave subalterns their main opportunities to show initiative. Many contemporaries noted the similarities between raiding and sport. There was an obvious relationship between stalking and carrying out stealthy operations;[41] and cricket can be seen as a 'tamed' version of stalking. The language used to report in an account of a raid carried out by 5th and 7th Canadian battalions is revealing. The writer asserted that the men practised for the attack 'with the same relish as if training for a football match', and mentions that an officer commanding an attacking party was 'a footballer'.[42] On occasions, raids appear to have become another form of sporting contest between rival units.[43]

When viewed against this background, the provision of footballs by

Capt. W.P. Nevill for his company of 8/E. Surreys to kick into action on 1 July 1916 becomes comprehensible. Far from being an act of public school bravado or the ludicrous action of a man obsessed with sport[44] it can be seen as a shrewd psychological stroke. Nevill intended the footballs to distract his men from the terrors of their baptism of fire.[45] It is arguable that the use of football on this and other occasions[46] also helped soldiers to make sense of a new and terrible experience by presenting it in terms which they were immediately able to comprehend; a common interest in sport 'bound together' a band of individuals about to undergo a tremendous ordeal.[47] Far from being a ludicrous anachronism, the public school games ethic played an important role in the war on the Western Front.

Mark Girouard has argued that the nineteenth century saw the resurgence of the medieval knightly code of chivalry, or at least a reinterpretation of it. Chivalry, a code of conduct that stressed honour, bravery, loyalty, courtesy, generosity, mercy and self-sacrifice, had by 1914 metamorphosed into gentlemanliness. From the mid-nineteenth century, this code was transmitted via the reformed public schools, and thus came to be the dominant ethos of the upper-middle and upper classes: the 'Victorians selected the qualities which they admired in chivalry and remodelled games in the light of them'.[48]

Great War subalterns gained from sport a sense of responsibility towards other members of their team; in the army, this readily translated into paternalistic responsibility for the well-being of their men. Capt. Oates of 6th (Inniskilling) Dragoons, who died on the polar expedition of 1912, and the male passengers on the *Titanic* who, having given up their places in the lifeboats to women and children, calmly awaited death, were held up as examples of gentlemen who had sacrificed their lives for the good of the larger group.[49] Christ's dictum that 'Greater love hath no man than this, that a man lay down his life for his friends' (John 15:13) dovetailed neatly with chivalric ideals. The result was a view of life summarised by the Latin tag *Dulce et decorum est pro patria mori*: it is sweet and right to die for one's country.

Chivalric influences, ingested via the public school, pulpit and sports-field, left the young men who served as subalterns in the trenches with little doubt as to the standards expected of gentlemen placed in command of men who were fighting for their country. In the church of St Michael the Archangel, Lyme Regis, Dorset, there is a memorial window to 25-year-old Capt. G.H. Bickley of the Devons. The stained glass depicts various scenes, including a knight in armour kneeling in vigil, and another knight receiving the Holy Grail from an angel. In sharp and ironic contrast to the mystical imagery of his memorial, Bickley was killed near Ypres on 4 October 1917 while attached to 236th Machine Gun Company. This was a unit created to make use of advanced technology, fighting in a battle that has come to symbolise twentieth-century techniques of industrialised warfare.[50] As late as 1973 a public school officer could write about the Great War in chivalric terms.[51] An equally powerful testimony is that of Wilfred Owen's poem 'Dulce et Decorum est' which is a denunciation of 'The old Lie' that underpinned the chivalric code of self-sacrifice.[52] The bitterness of this poem can only be understood if placed in the context of the *Weltanschauung* of the average public school-educated subaltern.

The classics dominated the public school curriculum. Not surprisingly, the Greek heroic tradition had an enormous impact on the public school ethos. Even in 1935, a classical scholar who had grown to manhood in the nineteenth century could write that 'There are worse ways of educating a boy than to familiarise his mind from childhood with great tales of splendid deeds and heroic men.'[53] Indeed, there were distinct similarities between Edwardian Britain and Homeric Greece, particularly in the great respect given in both societies to military and political heroes. The generation of 1914 was 'profoundly Homeric', and 'in Homer, the hero was a warrior'.[54]

The Dardanelles campaign, fought almost in sight of the remains of Troy, brought forth a particularly fine crop of Homeric allusions in soldiers' writings.[55] The Western Front also inspired references to ancient Greece. Pte Stephen Graham (Scots Guards) returned to Ypres shortly after the war, and in describing a monument in Polygon Wood referred to Thermopylae; an allusion that he took for granted his educated readership would understand.[56] While men who read the classics for pleasure were probably in a minority, even non-intellectual former public schoolboys were influenced by classical and heroic traditions, so deeply were they imbedded in the British public school ethos.[57]

The classics influenced the officer–man relationship in three major ways. Homer, Caesar and the like dealt with issues of war rather than peace, reinforcing the notion that leadership in war was a natural part of the gentleman's duty. Moreover, because the notion of chivalry was extended back into the classical era (Plato was once described as 'one of the greatest of gentlemen'), it gave support to the concept of the leader as a gentleman, with all that implied for paternalism.[58] Finally, it reinforced the belief that the leaders of society should be warriors, who physically led their men into battle. Hector, Achilles, or for that matter Alexander the Great, were not generals who calmly directed battles from a safe distance in the rear. They were warriors who fought hand-to-hand with their enemies.[59] A further, subsidiary influence of one classical tradition, the Spartan, was that the harsh regime at public schools, of which it was once claimed that it would have 'produce[d] an immediate revolution if applied to the masses', was intended to toughen future leaders.[60]

The success of Raymond's novel *Tell England,* which was made into an immensely popular film in 1930, demonstrates the continuing popular influence of the classical tradition after the war. Set partly at a public school, and partly on Gallipoli, *Tell England* is a deliberate echo of the epitaph of the Spartans who died at Thermopylae. The final frame of the film shows a gravestone inscribed with the epitaph:

> Tell England, ye who pass this monument
> We died for her, and here we rest content.

Thus a successful novel and a feature film explicitly linked traditions of classical and modern warfare. To continue Sillars' thesis, it is possible that a reason for the success of the film was that, by reaffirming heroic values, it offered psychological consolation to the bereaved at a time when the whole notion of glory and self-sacrifice in war was under attack from the 'disillusioned' school of war writers.[61]

One further influence, that of popular writing, warrants examination. The approved reading of middle- and upper-class boys tended to purvey militaristic and paternalistic values. Magazines such as *Young England* and the *Boy's Own Paper* (*BOP*) published articles designed to teach affluent readers their responsibilities towards the poor. In an article of 1906, for instance, a clerical writer stressed that *noblesse oblige* was a sacred duty:

> Christ taught that the strong should support the weak, and all should care for those who cannot help themselves.

In 1915 the *BOP* carried a typical article that featured the work of the Shaftesbury Society and the Ragged School Union. The author was explicit in his propagandistic aim in writing:

> [If] every bright Lad of Advantage would be a Big Brother to every Lad of Disadvantage, the occupation of the Ragged School Union would be gone. It would have no raison d'etre.[62]

In terms of sales, G.A. Henty was the most important boy's writer of the late nineteenth century, selling about 25 million books by 1914. His historical adventure stories, which often dealt with war, offered role-models for young military leaders. The protagonist was usually a boy of 15 or 16 who learns about the responsibilities of leadership during his adventures.[63] Henty was merely the most successful writer in the genre. His many rivals and imitators also propagated the militaristic elements of the public school ethos. An anonymous story, 'A Boy's First Fight' depicts the battle of Waterloo through the eyes of a 17-year-old subaltern. This story features a paternalistic quartermaster who attends to the material needs of the rankers, and courageous, phlegmatic and self-sacrificing officers and NCOs who offer leadership at the crisis of the battle. The rankers demonstrate their 'affectionate regard' for their wounded colonel. Heroism and dedication to duty enables the youthful subaltern and indeed the British army to overcome all difficulties and defeat the French.[64]

The one writer who was possibly even more influential than Henty in shaping the ethos of the officers of the Great War was Rudyard Kipling. It was largely through his work that the general public gained their impression of both the private soldier and the subaltern, although Kipling's military tales were based on his time in India in the 1880s, and army life had changed somewhat in the intervening years. He was also an advocate of the militarisation of British society. One story, published in the aftermath of the Boer War, depicts the entire British nation in arms.[65]

Kipling's subalterns were characterised by a moral code emphasising stoicism, self-denial, obedience, loyalty to the Regiment and to their men, and adventurousness.[66] In 1895 Kipling created 'Bobby Wicks', the ideal junior officer, a boy/man whose devotion to his men was so complete that he died of cholera after nursing a private. By giving prominence to private soldiers as characters in his stories, Kipling added another dimension to the portrait of the perfect officer. Kipling's Tommies eloquently admired officers who were liberally endowed with public school values. The soldier nursed by Bobby Wicks assaults a fellow private who failed to show sufficient sorrow at Wicks's death, uttering the immortal words 'Hangel! Bloomin' Hangel! That's wot 'e is!'[67] Kipling's message was clear. Applying the public school ethos to military leadership was effective. Paternal, courageous, self-sacrificing officers earned the loyalty and love of their men.

In *Puck of Pook's Hill* (1906) Kipling's ideals reached their apotheosis.

The centurion Parnesius, who doggedly defended Hadrian's Wall against the assaults of barbarian 'Wing Hats', exemplified devotion to duty and self sacrifice. The Parnesius section of the book depicts junior officers continuing to do their duty when their superiors have failed to do theirs. By holding their position when surrounded, by fighting on with no thought of surrender, they saved the province of Britain from the invaders.[68]

It is notoriously difficult to establish the degree of influence that ideas have on the conduct and opinions of individuals. In the case of the extent of the influence of Kipling's views on qualities of officership, the career and opinions of one man, Charles Carrington, offers a valuable guide. Carrington was an upper-middle-class boy of 17 when the war broke out in 1914. After a brief spell in the ranks in England, he was commissioned, and served for most of the war in 1/5 R. Warwicks. After the war he wrote a number of books, including his war memoirs and a life of Kipling. In the latter book, and in letters and conversation with the present author, Carrington evaluated the influence that Kipling had on his concept of officership.

Puck of Pook's Hill was published when Carrington was at the impressionable age of 9. Kipling's son, John, the model for 'Dan' in *Puck*, was Carrington's exact contemporary. Nine years later John, and thousands of boys like him, were serving as subalterns on the Western Front. In 1955 Carrington claimed that nothing else in the Kipling canon was 'more effective in moulding the thought of a generation' than the Parnesius stories. Certainly, the story was strangely prophetic. It took but a short leap of the imagination to see the Wall as the Western Front, and the Wing Hats as the Germans; and Kipling's Roman officers spoke, thought and acted very much like Edwardian subalterns. The Roman stories, Carrington believed, 'strengthened the nerve of many a young soldier in the dark days of 1915 and 1941'.[69]

In retrospect, Carrington saw Kipling's fictional subalterns, and the real junior officers of 1914–18 as part of a long English tradition of the boy-officer that included the Black Prince at Crécy, Chaucer's squire, and boy-ensigns such as James Wolfe at Dettingen.[70] Thus Kipling was tapping into a long tradition when he created his exemplary subalterns. Kipling, Carrington argued, had 'moulded a whole generation of young Englishmen' in the form of Bobby Wicks:

> They rose up in their thousands, in 1914, and sacrificed themselves, in the image that Kipling had created.[71]

John Kipling, killed in 1915 while serving as a subaltern in the Irish Guards, was among their number.

In 1940, George Orwell, who was at Eton during the Great War, argued that from their earliest days, the majority of the middle-classes 'are trained for war ... not technically but morally'.[72] It is difficult to disagree with this conclusion. The average Edwardian public schoolboy would have had to have been strong-willed indeed to resist the range of militaristic cultural influences that have been summarised under the heading of the 'public school ethos'.

The elements of the public school ethos relevant to military leadership were passed on to officer cadets who had not attended a public school. It is important to note that such men did not begin officer training totally ignorant of the public school ethos, for in the next chapter it will be argued that it also affected the thinking of lower-middle and even working-class youths.

In the autumn of 1914, training for junior officers was usually conducted within the unit, with subalterns learning the rudiments of drill at the same time as privates. Some units and formations made their own arrangements. In 16th (Irish) Division the divisional commander interviewed candidates for commissions. Successful applicants who lacked military experience were told to enlist in 7/Leinsters, which had a candidates' company. If they proved themselves fit to become officers, their names were forwarded for commissioning. In November 1914, this company contained 50 men. They attended lectures given to officers and had certain privileges, but remained in the ranks.[73] One hundred and sixty-one cadets graduated from this company to become officers in 16th Division between November 1914 and December 1915.[74]

W.P. Nevill of 8/E. Surreys was luckier than many temporary subalterns in that he went on a junior officers' course at the Staff College in November 1914. His description of a typical day's work conveys something of the intensity of the course. Parade was at 6.55 a.m., which was followed by bayonet-fighting, and PT before breakfast. Then came musketry instruction, lectures on military topography and tactics, and then more drill. Then there was a pause, followed by another lecture on 'organisation' at 6.00 p.m. Nevill spent the evening copying up rough notes. In early 1916 Nevill was sent on a course at Third Army Infantry School that was very similar to the earlier course.[75]

From January 1915, selected NCOs and men went on four-week courses organised by OTCs and by units such as the Artists' Rifles and the Inns of Court. If they passed their course, they went for further training at a Young Officers' Company (YOC) attached to a reserve brigade, before being sent on to their units. Some men from civilian life also went to a YOC where they experienced their first taste of 'fundamental army discipline'.[76] Later in the year, in an attempt to ensure a uniform standard of instruction, YOCs were grouped together in threes.[77]

Army Council Instruction 357 of 14 February 1916 created a number of Officer Cadet Battalions (OCB). This system of officer training endured until the end of the war. The majority of the 107 929 temporary officers commissioned from February 1916 until the end of the war first served in the ranks and then passed the four months' OCB course. In addition, Sandhurst and Woolwich continued to commission small numbers of Regular officers. It remained possible to take a commission without having first served in the ranks of an ordinary unit, or at least in an OTC, for some 'specially qualified young men' fresh from civilian life were sent to the Inns of Court or Artists' Rifles OTC for two months' training in the ranks, before going on to an OCB for cadet training.[78] Of the 100 cadets of D Company, No. 14 OCB in September 1914, 51 had previously served with Inns of Court OTC. This disproportionate number is probably explained by the fact that this OCB was affiliated with the Inns of Court OTC and co-located in Berkhamstead.[79] The institution of the OCB system virtually ended the usefulness of the university OTCs, as the majority of their members were eligible for training at an OCB.[80]

Under the old system, it had proved extremely difficult to deprive 'duds' of their temporary commissions.[81] With the OCB system, unsuitable men could be returned to their units, since cadets were only commissioned upon successfully completing the course. Undoubtedly, some men applied to become officers simply to gain a few months' respite from the trenches. A ranker of 8/Queens with 'no great ambition' to be an officer was recommended for a commission in October 1918 and decided to accept it for the sake of a spell in England.[82] In mid-1915, the Artists' OTC was unpopular because the course, which was only five days long, was held in France, not England.[83] But many cadets were genuinely enthusiastic about officer training. At No. 4 OCB in 1916 a group of cadets wrote to the War Office requesting that an inefficient dugout officer be removed from the staff.[84] In November 1916 the hardbitten combat veterans of No. 9 OCB howled down an instructor when he attempted to deliver a lecture that they considered insulted their intelligence.[85] One instructor at Inns of Court OCB noted that cadets could be difficult audiences, inclined to heckle during lectures.[86] In general, officer cadets were very well motivated: a 'very decent collection', as one instructor described them.[87]

The content of the various officer training courses remained fairly constant throughout the war. Potential officers had an enormous number of subjects to master.[88] Drill, tactics, interior economy and military law were the staple fare, supplemented by lectures on recent military history; an officer who had served as a brigade-major at Mons gave a lecture on the battle at the Artists' OTC at Bailleul in March 1915, which much impressed his audience.[89] In September 1916 a Cambridge don lectured at GHQ Cadet School on Balkan history and the causes of the war.[90] The most obvious difference between an OCB course and that of prewar Sandhurst or Woolwich lay in the fact that officer cadets were not assumed to be leaders. Wartime OCBs treated leadership as another subject to be taught, along with drill and tactics.

One of the primary motives of the course was to teach the officer cadets to think like officers. This involved lifting the mental processes 'to a different plane of vision ...'.[91] Basic training conditioned rankers to obey orders in an unquestioning fashion. As officers, the same men had to issue orders and think for themselves. At the OCB a rather more genteel version of basic training took place, the aim being to wean cadets from the ranker's mentality and to instil them with a version of the public school ethos. The change in posture from deferential ranker to officer and leader was summarised by an NCO instructor at an OCB at Keble College, Oxford, in July 1917, when he told his cadets that

> Soldiering is more than 'arf swank. You've got to learn to walk out as if the bloody street belonged to you.[92]

Socialisation into the officer's role was an important part of the process; in 1915 the CO of No. 19 OCB announced that he was going to turn them into 'officers *and gentlemen*'.[93] Many OCBs were located in Oxford and Cambridge colleges and other universities such as Queen's Belfast and Bristol. One cadet wrote in late 1915 from Pembroke College, Cambridge that officer training involved hard work but 'I love this life – it is what I have always longed for – and it will be with feelings of regret that I leave it behind.'[94] Another described himself as privileged to be housed in Trinity.[95] At a time when there were very few students at Cambridge University, the members of the three OCBs housed in the colleges provided a semblance of normal undergraduate life, playing sport, indulging in amateur dramatics, and producing magazines.[96] Billeted in gracious surroundings very different from their peacetime environment, living a relatively carefree life far removed from the drudgery of the office or factory, many lower-class cadets proved particularly responsive to the need to play the role that was demanded of them.

Part of the process of socialisation was to teach the cadets to behave in a gentlemanly fashion. Officer cadets ate in a series of messes, and in some OCBs the custom was for a member of the Directing Staff to sit at the mess table with the cadets, looking out for standards of behaviour and etiquette.[97] Capt. A.L. Bonham Carter lectured to Inns of Court OTC on 'Military Etiquette'. He gave such advice as:

(a) Keep any 'lady (?)' friends out of sight
(b) If you find it necessary to get drunk – go home in a cab and hide yourself
(c) bar unauthorised dress (socks (fancy) with shoes ... etc.)[98]

Similarly, the commanding officer of No. 5 OCB, based at Trinity College, Cambridge, lectured the cadets on correct behaviour and on one occasion publicly complained that cadets had 'neglected to flush the WCs after use', and that a cadet 'had been seen in the college precincts with his arm round a girl's waist'. 'Neither', he warned, 'must occur again!'[99] In this particular OCB such warnings seem to have been taken very much to heart, for a few months later the diary of 'Samuel Pepys, Cadet', a feature of the OCB magazine, noted that 'it is not seemly for a cadet to take the arm of a wench in the street'.[100]

The purpose of instruction of this kind was not simply to ensure that an officer did not disgrace himself and his uniform by slurping the soup in the mess, or by committing some other *faux pas*. Nor was it entirely a snobbish reaction to the perceived social shortcomings of 'temporary gentlemen'. Rather, it was part of the army's pragmatic response to the shortage of officers from the traditional officer-providing classes. It was an attempt to manufacture passable imitations of gentlemanly officers by providing, *via* the medium of an intensive course, the kind of social training that upper-class young men received in homes, at public schools and universities. The entire process was rooted in the belief that the only effective officers were those who possessed certain qualities – in this book described as 'public school values'.

By 1916 the well of genuine public schoolboys had largely run dry. Seven years earlier, during the debate on the shortage of officers,

General Sir Ian Hamilton had looked to the ranks of sports-playing, charismatic junior NCOs to supply additional officers for the army in wartime.[101] Likewise, during the war the army took rankers who already demonstrated a measure of leadership potential and tried to turn them into officers by exposing them to the public school ethos in a concentrated form. Teaching potential officers how to behave at the dinner table was not directly relevant to making them leaders of men in the same way as, for instance, instruction in tactics, but nonetheless it was considered vital, for it was an article of faith that soldiers would only follow gentlemanly officers.

The prewar habit of treating potential officers in a similar fashion to recruits to the ranks persisted at some establishments. The basic training at the RHA barracks at St John's Wood emphasised the more unpleasant side of soldiering. Officer candidates, who included students, barristers and dons among their ranks, were subjected to cleaning and polishing, mucking out stables with bare hands, PT, drill, riding, some basic instruction in gunnery, and much shouting by NCOs. As with prewar Woolwich, the army made no concessions to the social class of the recruits, or the fact that they were shortly to hold the King's commission.[102] One man who passed through St John's Wood and then went on to Royal Artillery Cadet School at Exeter described the change, significantly, as 'just like being transferred from a lunatic-asylum to a well-run public school'. At Exeter, unlike St John's Wood, cadets were not treated as 'half-witted shirkers'.[103]

Men who passed through other basic training establishments noted that their treatment was very different once they graduated to an OCB. While discipline was, initially at least, strict,[104] and the removal of rank badges reduced all cadets, whether sergeant-major or private, to the same level, they were treated as gentlemen and potential officers. In sharp contrast to the RE signal service training centre, where he had previously trained, one cadet found that at the OCB at Berkhamstead he was encouraged to think of himself as an officer, and was 'treated as such'.[105] Harsh treatment during basic training ensured that many non-ranker officers commissioned after January 1916 had, like their Sandhurst-trained predecessors, an insight into the life of the ordinary soldier, however brief, that no doubt served to emphasise just how privileged officers were in comparison to the rank and file.

Like the modern British army, the army of 1915–18 had a functional approach to training leaders, believing that the job of the OCB was to 'bring out' inherent leadership abilities that had been previously detected in an individual. Not surprisingly, given the prevailing ethos, instructors saw team games as an essential tool for training leaders. Accounts of training at officer cadet units are littered with references to sport.[106] In 1917, No. 4 OCB devoted two afternoons a week to sport. Naturally, the authorities encouraged extra-curricular sport. The cadets of No. 4 OCB (Hertford College, Oxford), took full advantage of their hosts' sporting facilities by establishing a rowing club.[107] One instructor recalled that the conduct of cadets on the football and rugby fields was regarded as a useful guide to their qualities as leaders: 'Those who played rough but not dirty, and had quick reactions, were the sort needed' and much of the staff's leisure time was taken up with playing sport with the cadets.[108] Cadets seemed to have recognised the importance of sport in their training. One, who trained with No. 10 OCB, noted that his fellow cadets generally believed that good sportsmen were sure to pass their final examination.[109] Since team games encourage team-work and cooperation, while leaving room for individual acts of initiative and leadership, this emphasis on sport was not misplaced.

Cadets also had the opportunity to learn to be leaders on training exercises. By 1916, fairly sophisticated schemes were in use at GHQ Cadet School. On one exercise a cadet commanded troops and on the following day, his peers and instructor discussed his scheme and leadership at a plenary session.[110] Under the critical gaze of his peers, a cadet's limitations as a commander and leader were cruelly exposed. But opportunities to command 'real' soldiers were limited. On active service, rankers might be less amenable than fellow cadets.[111]

Lectures on paternalism supplemented practical training in leadership. A standard lecture given to drafts of the Artists' Rifles before their dispatch to camp informed them that upon eventually reaching their regiment as officers

> your first job is to get to know your men, look after them, study their interests and show you are one of them, taking a share in their pleasures and interests as well as their work. If you do this you will find that when the time comes they will follow you to hell....[112]

Such advice seems to have been fairly standard. Lectures at other units advised officers to be the soldier's 'friend as well as officer' and to '[c]onsider [the] man's point of view', and informed cadets that '[a]n officer succeeds in so far as he lives up to his men's expectations of him'.[113] A plethora of publications, both official and unofficial, reinforced such advice. (See Appendix 4.)

The British army thus responded in a pragmatic and eminently practical way to the shortage of public school officers. The officer-training system institutionalised the ethos of the prewar officer class and ensured that it passed on to cadets who in many cases did not have a public school background. The OCB was a mixture of boot camp and public school or Oxbridge college. Besides training in military skills, cadets were taught how to behave in a gentlemanly fashion, how to think like leaders rather than followers, and, perhaps most important, to behave in a paternalistic fashion towards their men. The teaching of leadership was a new departure for the British army. A typical question that appeared on a final examination paper in December 1916 asked the candidate to imagine that his unit has been relieved from the trenches and that his men arrive, cold and wet, at their billets at midnight. The question asked 'Before going to your own quarters what will be your duties as platoon commander, with regard to your own platoon?'[114] A question of this nature would have had no place on an examination paper at prewar Sandhurst, but in a wartime context, such a question was designed to teach the paternalism and *noblesse oblige* which was the hallmark of the Regular officer.

Many regarded the wartime system of officer-training as a success. An officer who passed out from No. 17 OCB at Kinmel Park in 1917 described the training, which aimed to give cadets 'a full regard of responsibilities and all that it (*sic*) meant' as 'excellent'.[115] Another cadet, who had previously seen active service in the ranks, approved of the advice he received on how to look after his men as 'extremely practical'.[116]

The success of the OCB system was due in great part to the enthusiasm of the cadets themselves. Many ex-rankers would, of course, have had personal experience of the importance of paternal officers to the well-being of soldiers in the trenches. The wish to avoid being sent back to their units, or in the case of men fresh from civilian life, being drafted as a private, was an obvious incentive to do well, for by no means everyone passed the course.

Conversely, many cadets seemed to have internalised the demands of the course, that is, adopted them as part of their personal outlook, achieving a high level of commitment to the concept of officership that they were taught at OCBs.[117] The writer C.S. Lewis commented not unsympathetically on his fellow cadets at Keble College, Oxford, in June 1917 that they were 'mostly jolly good chaps' but 'their own naive conceptions of how gentlemen behave among themselves lead them into an impossible politeness that is really very pathetic'.[118]

These lower-class cadets were attempting to adopt the mannerisms of the Regular officer when off duty, not merely in front of their instructors. One cadet put the matter in a nutshell: 'We were all pretty keen and did our best and, provided that you were not lazy and kept alert, you could get by.'[119] Robert Graves, who served as an instructor and adopted many of the attitudes of a Regular officer, went as far as to claim that the OCB system 'saved the army in France from becoming a mere rabble'.[120]

Graves was essentially correct. The products of OCBs had an important role in maintaining the morale of their men during the latter years of the war, and displayed as strong a commitment to the Regular officers' paternalistic concept of leadership as had the public school temporary subalterns of 1914–16. OCBs aimed to produce officers who used personal and expert power to influence their men, who identified with their charges and placed their well-being above their own. Such leaders were far removed from the image of the British officer of the Great War depicted in popular films, books and television programmes.[121] Subsequent chapters will examine the practical application of paternalism.

5

Officer–Man Relations: the Disciplinary and Social Context

The hierarchical structure of the wartime army strongly resembled that of the prewar Regular army. A temporary officer believed that 'Between officers and men there was a great gulf fixed'.[1] An American who enlisted in 1914 had to come to terms with 'the class distinctions of British army life':

> The officer class and the ranker class are east and west, and never the twain shall meet, except in their respective places upon the parade-ground.[2]

It is somewhat misleading to refer to the wartime army as being class-based. Men were allotted places in a rigid hierarchy by virtue of their rank, not their social class. On receiving the King's commission, individuals were entitled to all the privileges of an officer: superior food and accommodation, deference from the ranks, even absolution from having their hair closely cropped,[3] regardless of their social background. Likewise, possession of a public school education gave the private no official privileges.

Inevitably the ranker's perspective on army life differed appreciably from that of the officer. Pte H.S. Williamson (8/KRRC) wrote in 1918 that 'Probably no one, except those who have been in the Army can fully appreciate the huge difference between the rankers & the officers (*sic*) point of view'; earlier, he had written that 'Even in the trenches, everything is done to smooth things for officers ...'. As a middle-class artist, whose brother served as an officer, Williamson was well placed to appreciate the variety of experience.[4] The most basic difference between the experience of the officer and that of the soldier lay in their treatment by the army itself.

With the brief exception of basic training, officers never experienced the way in which the army treated Other Ranks, unless they had themselves previously served in the ranks. Out of the trenches, rankers' lives were often characterised by boredom and monotony, and by the feeling of being a mere pawn, unable to influence their fate, moved hither and thither at the whim of faceless military bureaucrats. Ordinary soldiers tended to see the army as an impersonal, arbitrary coercive system. Typical complaints concerned moving men who had just established themselves in comfortable billets, because of a sudden change of plan ('How like the Army!' commented a victim of one such move) and the army's habit of withdrawing blankets in May and issuing them in October, regardless of the actual temperature.[5] An NCO of 22/Manchesters attended a specialised course in range-finding. After rejoining his unit, he never used a range-finder again.[6]

The military authorities also had it within their power to impose petty humiliations and restrictions on non-commissioned ranks ranging from the withdrawal of any shred of personal privacy by ordering a 'prick-inspection', to restricting certain shops and estaminets for the use of officers only, by way of depriving NCOs of acting rank by the summary award of a commanding officer.[7] On one occasion, an officer publicly humiliated a sergeant by having his stripes ripped off in front of the rest of the unit.[8] Soldiers used various imagery to describe their situation. T.P. Marks, a schoolmaster who served in the ranks of 1/Glosters, wrote that the soldier was 'no longer an entity, but a cog waiting to be set in motion'.[9] Another private reported that it was a common saying among Other Ranks that from the moment the soldier disembarked at Le Havre he was 'a slave'.[10]

As Christopher Duffy has remarked, 'It is notoriously difficult to evaluate the severity of the discipline of a military institution. If our attention is drawn by punishments of a spectacular and barbaric nature, we can easily overlook the small currency of blows and torments which have been meted out in every army known to history.'[11] In comparison with the discipline of the army before the abolition of flogging in 1881, the disciplinary code of the army of the Great War pales into insignificance. Likewise, the flogging of Indian soldiers during the 1914–18 war highlights the relative mildness of the treatment of their British counterparts.[12]

Nevertheless, British army discipline of the Great War was severe enough; '[w]artime courts martial found 92 per cent of men guilty as charged'.[13] One crude and not entirely satisfactory method of assessing the severity of discipline is to count the number of executions carried out. Three hundred and fifty-one British soldiers were executed, all but 37 for 'military' offences such as desertion, although this only represents 11.23 per cent of all death-sentences passed. By comparison, the army executed only 37 men between 1865 and 1898, and only one out of the four men executed during the Second Boer War suffered the penalty for desertion (as a comparison, between 1901 and 1924, 340 civilian criminals were executed in Britain).[14] The German army of the Great War, which was considerably larger than its British counterpart, executed only 48 men, British sneers at 'Prussianism' notwithstanding.[15] Indeed, German officers noted the 'iron discipline, maintained by a severe code of punishments' which was in the 'very blood' of the British prisoners they examined. They also commented on resilience of British morale, and the respect that Other Ranks had for their officers.[16]

There is some evidence to suggest that the effectiveness of the punishments available during the Boer War caused some dissatisfaction.[17] In 1904, a military committee even called for the reintroduction of corporal punishment on active service, although this recommendation was not accepted.[18] Although a firm judgement must await comparative study on military punishment in the two wars, the impression remains that the disciplinary regime of 1914–18 was harsher than that of 1899–1902.[19]

During the Great War, Field Punishment No. 1 was popularly known as 'crucifixion', from the practice of tying men to objects such as posts and wagon wheels for set periods. The milder Field Punishment No. 2, noted Pte Surfleet (13/E. Yorks), consisted of

> full pack drill; not ordinary drilling but with a police-[provost] sergeant standing by, shouting 'Right turn, left turn, about turn ...' one after the other, all done at something near the 'double'.[20]

Despite improvements in conditions in the late nineteenth century, brutal punishment regimes were the order at military prisons and field punishment centres.[21] One man who served a sentence at Gosport military prison in 1916 recalled the monotonous and pointless cleaning tasks, and the blows dealt out casually by the warders.[22]

Men punished by death, penal servitude or Field Punishment No. 1 constituted only a small minority of British rankers. Yet these punishments had an impact on British soldiers out of all proportion to the number of men directly affected. Severe punishments acted as a salutary reminder of the power that the army had over their lives, and that the military authorities on occasions regarded the enforcement and maintenance of military discipline as more important than justice. At one level, this could take the form of the imposition of collective (or 'vicarious') punishments. One example occurred in 1915, when the men of 2/RWK were refused permission to drink from their water bottles on a march through the Mesopotamian desert until a thief gave himself up.[23] Major-General Childs, the Director of Personnel Services, denounced vicarious punishments as 'hopelessly illogical', and in 1919 the Army Council officially discouraged their use. Other senior commanders, including Haig, were in favour of them.[24]

One of the factors taken into account by senior commanders when deciding whether to confirm a death sentence was the state of discipline within the prisoner's unit and whether 'an example was necessary'.[25] It is now generally recognised that at least some men were executed without having had a trial that would have been regarded as fair by the standards of the Edwardian civilian judicial system, and some men were actually suffering from psychiatric wounds that would have merited treatment, not punishment, in the war of 1939–45.[26] All of this served to remind ordinary soldiers that they were very much at the mercy of their hierarchical superiors. An RFA driver, who received a relatively mild punishment from an officer who 'was the prosecutor, judge, jury and jailer' noted that this 'seemed to me a curious combination of power in one man' (although, as noted in the next chapter, by awarding punishment at regimental level, a company commander could protect a man from the harsher punishments inflicted by courts martial).[27]

Although one officer claimed that many soldiers preferred field punishment to a route march,[28] the reaction of most rankers seems to have been horror and disgust. One NCO who witnessed 'crucifixion' wrote that it 'breaks a man's spirit';[29] another described it as 'disgusting and humiliating', especially with French civilians looking on;[30] yet another wrote that it made him

> sick with resentment. Of all the Army forms of punishment, and others are pretty rotten, I do not know of any more likely to embitter a man for ever.[31]

Pte Surfleet saw an artilleryman being punished:

> He was stretched out, cruciform-fashion, his arms and legs wide apart, secured to the wheel. His head lolled forward as he shook it

> to drive away the flies. I don't think I have ever seen anything which so disgusted me in my life and I know the feelings amongst our boys was (*sic*) very close to mutiny at such inhuman treatment.... I'd like to see the devils who devised it having an hour or two lashed up like that.[32]

Rankers' views on 'shirking' were complex. According to a recent writer, it was not a 'shameful private vice' but a 'companionable activity'.[33] However, rankers' attitudes towards the death-penalty for cowardice and desertion were rather more ambiguous, for as the padre of 12/HLI pointed out, there was a certain logic in making men who had failed in their duty as soldiers pay the ultimate penalty, given the death of many of their comrades who had not failed.[34] Many opposed such executions. A prewar Regular, Cpl A. Roberts, recorded in his diary that a sense of depression descended on 1/KRRC after an execution.[35] Two other Regular rankers, Cpl J. Lucy (2/R. Irish Rifles) and Pte F.M. Packham (2/R. Sussex) implied that executions did not act as a deterrent but instead depressed morale.[36] The practice of reading out sentences of death from routine orders to troops on parade shocked many soldiers and prompted some educated, middle-class volunteers to criticise the death-penalty's value as a deterrent for cowardice. Others resented it as casting an unnecessary slur on their motives for becoming soldiers, as another example of military mind failing to adjust to the fact that many units contained a very different type of man from the prewar Regular.[37]

Discipline cannot, of course, be measured purely by punishments. Regular officers' views on the purpose of discipline were often sharply at variance with the opinions of many rankers on the same subject. A.A. Hanbury-Sparrow (R. Berks), a Regular officer, argued that discipline was not 'meaningless, wooden obedience' but:

> the vehicle by which the superior will permeated the subconsciousness of the troops ... close-order drill and rifle exercises were ceremonies by which the superior will made its presence felt.

The individual ego, which 'subconsciously was continually in revolt' against discipline, had to be suppressed. Thus the cleaning of brass buttons was important because it was 'a daily disciplinary exercise'. Discipline was intended to promote unit cohesion and military efficiency by producing obedient men who took a pride in developing soldierly skills, and who did not give way to fear in battle.[38]

By contrast, a middle-class Scots Guards private, Stephen Graham, accepted the importance of discipline in battle but argued that for many, the greatest ordeal was not the battlefield but the training ground.[39] An LRB ranker believed that upon the 'calm self-confidence' produced by drill rested 'one's chances of survival'. However, he resented the 'arbitrary restrictions on the individual [which went] beyond the requirements of safety and good discipline'.[40] These two soldiers, among others, recognised that discipline was essential in battle, but did not agree with the army's methods of producing discipline, seeing it as designed to crush the temperament, destroy individuality, humiliate the soldier, and produce a sort of military robot.

The process of instilling discipline began in basic training. On reaching the depot, Pte Graham was told that 'They try and (*sic*) break you at the beginning and take all your pride out of you.'[41] The discipline of the Guards was perhaps exceptionally severe, but was probably not much worse than that of many other training units in 1916–18.[42] At the beginning of the war discipline was less strict in many newly raised units, and in some cases remained so for some time, but the honeymoon period experienced by some volunteers in 1914 was mostly short-lived. As one former ranker wrote:

> With so much improvisation, a few months passed before those who had enlisted in newly created battalions discovered exactly what it was they had let themselves in for. They had engaged to serve ... in a complex organisation incidentally designed to enforce the will of each and every superior on those in the lowest rank of all ... in order to carry out what they had conceived, for the most part romantically and generously as a patriotic duty, the young civilians were compelled to undergo a preliminary process ... [involving] an almost total surrender of personal liberty and an immediate, unconsidered obedience to orders.[43]

Similar methods of training continued at base camps in France, of which the 'Bullring' at Etaples was the most notorious.

Many rankers believed that there was a deliberate policy to 'crush the individual temperament ... as a means of enforcing his compliance with military discipline'[44] and indeed one 'fire-eating' officer candidly admitted what many Other Ranks suspected: that the purpose of training was to break down the natural humanity of the recruit and turn him into a ruthless killer.[45] While there was undoubtedly something in these views, the army's disciplinary regime is best seen as the result of preconceptions and traditions, and the inability, or unwillingness, of Regular officers and NCOs to adjust to the influx of a more educated type of man into the army, as much as the product of a deliberate policy.

Nonetheless, ordinary soldiers willingly believed in a conspiracy theory. One conscript believed that the purpose of training was to make the life of the soldier so miserable that he ceased to worry whether he lived or died: 'A happy soldier does not want to die and this looked like the start of our conditioning'.[46] One important result of this was that when soldiers finally arrived at their units, they were more inclined to look favourably upon the officers and NCOs who were responsible for the relatively mild disciplinary regime operated at unit level.

However, front-line units had their fair share of the irksome aspects of discipline. High command and Regular officers in general laid great emphasis on the external aspects of discipline. A particular criterion for the state of discipline in a unit was saluting. The very first routine order issued by the Inspector-General of Communications on arrival in France in August 1914 concerned the importance of paying 'strict attention' to the saluting of Allied officers.[47] A routine order later that year stressed that there was to be no relaxation in saluting in the field, 'except when active operations are actually in progress'.[48] Commanders at various levels issued reminders about the importance of saluting, for as a III Corps routine order stated, a failure of men to salute officers indicated 'a want of discipline [which] reflects seriously on the unit ...'.[49] Other official criteria for discipline included smartness of appearance and drill, cleanliness of billets and trenches, and good march discipline. While many, if not most, officers and men accepted the necessity of such things, some thought that excessive insistence on them was unnecessary and even counterproductive.[50]

Middle-class, educated soldiers in particular resented 'bullshit' and status deprivation, yet for most of the war, in most units, the army subjected them to the same training and disciplinary regime as the prewar soldier. Too often, any spark of independence was ground down – as a soldier remarked, a private 'could rarely walk twenty yards' without coming across someone with the power to "tick him off"'.[51] Arguably, the army failed to take full advantage of the talents of educated Other Ranks. The BEF made no general attempt to develop *Auftragstaktik* (directive control), a command philosophy that involved the delegation of authority and the cultivation of initiative among the lower ranks.[52] Historians have perhaps overestimated the lack of initiative of the British soldier, and some units did cultivate a form of *Auftragstaktik*, especially in the last two years of the war. But Frederic Manning, a novelist who served in the ranks of 7/KSLI, could write:

> Regular officers as a rule didn't understand the new armies, they had the model of the old professional army always in their mind's eye.... The majority ... [with] brilliant exceptions, did not understand that the kind of discipline they wished to apply to these improvised armies was only a brake on their impetus.[53]

This passage refers to 1916, and was in this period all too typical of the approach of the old army towards the new.

In view of the harsh nature of military discipline, from which officers were largely exempt, why did relationships between regimental officers and their men remain, for the most part, cordial? The nature of Edwardian society offers some clues. The British armed forces of the Great War did not represent 'a cross-section of the nation'. Rather, a variety of social, medical, economic and political factors led to non-manual workers and professionals being over-represented in the army. From 1914 to 1916 the enlistment rate for manual workers was approximately 30 per cent, but 40 per cent for non-manual workers.[54] Nonetheless, since the overwhelming majority of soldiers were examples of the 'British working man in uniform'[55] it is necessary to place the attitudes of working-class soldiers to military discipline and authority in the context of their experience as civilians.

Contemporary observers sometimes suggested that working-class soldiers of the 1914–18 war had natural discipline: 'However much they girded with coarse and biting irony at discipline, they were really very well disciplined.'[56] A number of factors, including strict parental discipline, prepared the working-class male for the realities of military life.[57] The school and the workplace were perhaps of even greater importance than the home in this respect.

Some wartime officers were horrified at the inadequacies of the elementary school system as revealed by their contact with products of that system.[58] Their comments echoed the emotive opinions of contemporary critics[59] and a modern historian: 'British elementary education served only to turn ... sickly and filthy children into robots able to read and write and count and obey.'[60] Nevertheless, in some ways the system was a resounding success. In 1908 a writer claimed that it produced 'the habits of obedience and regularity'[61] and some who experienced such schools came to appreciate the virtues of stern discipline.[62] Corporal punishment was ubiquitous.[63] Writing in 1911 a former school inspector claimed that the schoolchild had his spirit broken by 'severity and constraint'; this reduced the child 'to a state of mental and moral serfdom'; once this was achieved, 'the time has come for the system of education through mechanical obedience to be applied to him in all its rigours'.[64] The parallels with army basic training are striking.

The use of military-style drill in elementary schools was widespread. One of its primary purposes was to create discipline in schoolchildren. Many supervising instructors in schools were ex-soldiers. In 1903 an official committee recommended that physical training should be compulsory for school-leavers of 14 years and older to prepare the male population for future military service 'with very little supplementary discipline'.[65] Although there were some modifications from 1904 onwards, physical education remained influenced by military recruit-training down to 1914.

Uniformed youth organisations such as the Boys' Brigade, Church Lads Brigade, and the Boy Scouts acted as agents of social control, carrying inculcation of social discipline and orderliness a step further: 'Fear and self-interest' were high among the motives of the middle-class originators of such organisations.[66] Some organisations were overtly militaristic, such as military cadet corps, or the Boys' Brigade, which drilled with dummy rifles. Others, notably the Boy Scouts, the largest youth organisation, placed less emphasis on militarism.[67] Some historians have rightly questioned the extent to which 'popular militarism' prepared men for combat, but the overall effect was to produce a working class that, like the middle class, was trained for war, in Orwell's phrase 'not technically but morally'. Perhaps 41 per cent of all young Edwardian males may have joined a uniformed organisation at some stage.[68] In sum, the generation of young men that filled the ranks of the British army of the First World War carried into the army first-hand experience of military-style drill, which inculcated working-class youth with 'habits of sharp obedience, cleanliness and smartness'.[69]

Writing of his previous job in a department store, one Regular soldier stated that 'Discipline at Lewis's was the equal of any I've experienced since in the Army.'[70] Men drawn from regular employment made good soldiers, believed one staff officer, in part because they were 'accustomed to the discipline of the workshop', and were a distinct improvement on the quality of prewar recruits, many of who mhad been unemployed.[71] Working-class ideas about work patterns had changed significantly by 1914. Before the industrial revolution, work was 'task-oriented',[72] but the introduction of the factory system necessitated industrial discipline, involving the internalisation by workers of a concept of 'time-thrift'. This entailed regular attendance at the place of work and remaining there for the length of the shift, and the acceptance of 'regularity, routine and monotony quite unlike pre-industrial rhythms of work'.[73] However, the behaviour of some recruits in 1914 suggests that the influence of factory discipline has been exaggerated (see below).[74] Industrial workers would also have encountered supervisors (described as 'Non-Commissioned Officers' by historians),[75] and they would have been prepared for one of the more pleasant dimensions of military life if they had worked for a paternal company such as the Quaker-owned chocolate firms of Cadbury and Fry.[76]

The working-class soldier's previous experience as a civilian prepared him for the army: he was 'used to subordination and tedium, two of the principal features of military life'.[77] Moreover, the nuances of military society were not dissimilar from those of wider society, which was divided by economic and cultural factors into a number of mutually exclusive groups. Social mobility was possible in Edwardian Britain, but this tended to reinforce, rather than weaken the divisions between classes, as people who had recently risen to the middle classes took pains to distance themselves from the proletariat.[78] This did not necessarily lead to hostility between groups, but it did lead to ignorance. The majority of the urban upper and upper-middle classes only came across members of the working classes acting as servants of various descriptions, and such relationships were not generally conducive to intimacy. In rural areas landowners came into frequent contact with workers, but deference robbed these encounters of any real intimacy; the outward signs of deference included touching the cap to a social superior.[79] The parallels with the military salute are clear.

Deference, or respect for, and obedience to, 'leaders' of society by those in the lower reaches of that society, was one of the principal bonds of Edwardian Britain.[80] Deference was usually given to those of aristocratic or gentlemanly background, and, of course, employers. In rural areas, deference had both a 'social basis' and an 'economic basis in the dependence of farmers, servants and the labouring poor on the patronage or benevolence of individual landowners'. Similar considerations applied to the urban work force, who accounted for about three-quarters of the population. In 1911 there were one-and-a-half million domestic servants, who were by their very nature deferential.[81] Even the most paternalistic employer had considerable power over their workers' lives. There are many anecdotes of rural labourers being ordered to enlist by their employers in 1914.[82] However, an apparently deferential and paternalistic society such as prewar rural Norfolk disguised considerable tensions and conflicts, suggesting that attitudes were by no means uniform.[83] Deference was not, however, merely a pragmatic response to economic realities. It was also a way of life, engendered in part by the inculcation of deferential attitudes through education and religion.[84]

Deference was recognised as part of an interdependent, reciprocal relationship. The socially conservative working man of Edwardian Salford 'knew his place: he wanted that place recognised, however humble, and required others to keep theirs'.[85] This meant that social superiors should keep their side of the unspoken bargain by acting in a way that merited respect, and which allowed the 'respectable workingman', who was neither abjectly submissive nor revolutionary, to keep his self-esteem.[86]

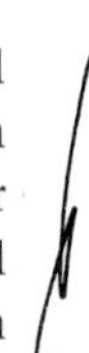

A perception of partnership between the classes, 'however bogus in reality', was the fundamental reason why the working man played his part in the deferential relationship. A further obligation of the social superior was paternalism (discussed above), which was the 'natural exchange' for deference.[87] In return, the individual worker gave loyalty, good service, and obeyed the not inconsiderable demands of factory discipline, while collectively accepting the social and political *status quo*. There is also some evidence that similar conditions applied to clerical workers. When the employer resorted to coercion, or where patronage was offensive to working-class sensibilities, or paternalism failed to live up to expectations, deference broke down. The social elite accepted their role in the deferential relationship partly because it made economic and social sense to substitute such a relationship, based to some degree on mutual trust, or at least on mutual interest, for one based on hostility and coercion.[88]

However, employers' philanthropy and sense of moral right and duty were also motives for paternalism. Much British industry was organised around the small workshop, the average workshop in 1898–99 employing only 29.26 male employees. Under such conditions 'a political affinity' and even camaraderie could be fostered between employer and employee. The personal relationship between master and man was all important,[89] although industrial conditions were beginning to change by 1914. Workers tended to be deeply suspicious of the representatives of the state; it was one thing to have a good working relationship with the 'boss', quite another to have one with the policeman, poor-law guardian or gamekeeper. Working-class people sometimes felt themselves to be helpless in the face of an apparently arbitrary and coercive authority, a situation that closely parallels the experience of many soldiers during the Great War.[90]

Imperialistic attitudes, which included a diluted form of the public school ethos, reached the working classes in a variety of ways.[91] The late Victorian and Edwardian period saw a military career growing in acceptability among the working classes. Plays and films depicted the army, and military life, in a positive way, school textbooks showed war in a beneficial light, and portraits of generals and admirals adorned classroom walls. A frequent feature of melodrama with an imperial theme was 'a cross-class brotherhood of heroism' in which white officers and white soldiers faced 'black barbarism' together. The message was the officer and the ranker were comrades, rather than class antagonists.[92]

The peacetime experience of the working-class male soldier thus prepared him for military life. The officer–man relationship of the army reflected to an exaggerated degree the reciprocal deferential/paternal relationship of civilian society. Thus working-class soldiers expected officers to have privileges. Moreover, by striving to protect their men from the worst excesses of military discipline, and by sharing some of their hardships and discomforts, junior officers fulfilled their part of the deferential bargain. Officers may not always fully have understood the nature of the 'unspoken bargain'. A temporary officer of 2/E. Lancs recorded that he regarded his daily inspection of his platoon's feet as 'an indignity for the men', but that his men did not see it in this light. His men's view is not surprising, since this attempt to prevent trench-foot was just one of the ways in which this officer demonstrated his paternalism.[93] It is no coincidence that middle-class rankers made some of the most eloquent attacks on military discipline and officers' privileges; working-class living conditions and thought processes were alien to these men. Nor is it surprising that Field Punishment No. 1 excited such disgust among rankers of all classes, since it offended against the idea of allowing the dependent class to retain their dignity.

The fact that the British lower classes were deferential and the higher classes were paternal meant that the two halves of the army had complementary attitudes. This helped to produce a loyal, hierarchically-minded and disciplined army with high morale. Some soldiers actually liked military discipline, since it gave a structure to their lives and removed the necessity to think for themselves.[94] The fears held by some prewar officers and politicians as to the reliability of the urban working classes in time of war proved to be unfounded.[95] Instead the wartime Adjutant-General was able to claim that 'the discipline and good behaviour of the forces that Great Britain put into the field were ... unequaled in the annals of war'.[96]

There was one major exception to the rule that the British army operated a rigid code of discipline during the Great War. For a number of reasons, some units raised at the beginning of the war were characterised by discipline that was akin to that of prewar Auxiliary units.

The British army's massive and unplanned expansion in the autumn and winter of 1914 caused considerable chaos. The commander of the newly raised 11th Division recalled that on arriving at Belton Park, he 'found a man surveying the ground' but 'no buildings of any sort'. Initially, only one battalion had uniforms.[97] The conditions endured by 11th Division were not untypical.[98] In October 1914, NCOs of 2/6 Lancs Fusiliers were reluctant to give the order 'pick up those feet' for fear of the clattering of clogs on the parade square.[99] With the best will in the world, it was not possible to impose strict discipline on men under such abnormal and slightly ludicrous conditions.

In spite of the impact of industrial capitalism, many recruits were profoundly ignorant of the basic demands of military discipline. There were many stories of men smoking on parade, falling out of the line of march to pick blackberries, or acting in a similarly unmilitary way.[100] Some recruits seem to have regarded the army as just another employer, without understanding the enormous difference between civilian and military methods. One officer commented:

> As civilians it had been no serious matter to take a day or a few hours off from their work.... So now there were cases of men being absent for awhile (*sic*) from duty or parade ... and they had not the remotest idea that this might be treated as a 'crime'.[101]

The RAMC private of 39th Division Field Ambulance who, tiring of army life, gave notice of his intention to resign, was acting perfectly correctly and honourably by his own code.[102]

Other civilian attitudes inimical to discipline took some time to subside, although, as will be demonstrated, many of them never vanished entirely. An officer recorded that many English recruits drafted to 10th (Irish) Division in 1914 were miners or other trade unionists who had 'acute suspicion of any non-elected authority'. Moreover, married men initially resented a deduction from their pay, which was sent to their wives, because they were not informed on enlistment that 'this stoppage was compulsory, and so they considered that they had been taken advantage of'. That newly enlisted trade unionists should think this way was less significant than the fact than an officer should go into print with a generally sympathetic account of their attitudes. He argued that a major factor in coming to terms with the army was the growth of respect for the paternal care of their officers.[103]

In many cases, the newly commissioned junior officers knew little more about the army than did the Other Ranks. One subaltern, on arriving at 1/6 DWR, found himself expected to command his platoon on his first day on parade. Fortunately, he had read up on drill on the train journey, and was thus able to keep one step ahead of his men.[104] Officers underwent much the same training as their men, and spent their valuable free time learning the additional duties of an officer.[105] Under these circumstances, officers were less inclined to insist on the niceties of military discipline, since they might not have personal experience of these niceties. In many units there very few Regular or ex-Regular officers or NCOs who might have offered guidance on such matters. Units of the first tranche of the New Armies, [K1], had first pick of the available Regular officers and NCOs. 10/DLI, for instance, had a Regular commanding officer, adjutant, and RSM. As the regimental historian commented, this alone counted for a great deal and gave them an advantage over units raised a little later.[106] Some dugout NCOs and officers caused more trouble than they were worth. An ex-Regular NCO of 8/R. Sussex frankly admitted that he did not understand some of the modern drills and dealt with dissatisfied recruits by threatening to fight them.[107]

The enthusiasm of many of the volunteers of 1914, their genuine willingness to learn and endure poor conditions should not be underestimated. It encouraged some officers – experienced Regulars, as well as temporary officers fresh from civilian life – to be tolerant and tactful in their handling of their troops. Indeed, contemporaries often cited the enthusiastic spirit of the troops and the tact of their officers as being among the primary reasons for the success of the New Armies in turning, under very difficult conditions, raw, untrained recruits into fighting units.[108] Officers attempted to balance the need to instill discipline with the necessity to recognise that recruits of 1914 were often of a very different stamp to those of prewar days. As the historian of 16/Manchesters (1st Manchester Pals) wrote, military discipline did not come easily to the civilian volunteer; its acceptance 'involve[d] a revolution in mental outlook, and the higher the intelligence and education of the men, the harder, perhaps the lesson'.[109] In many cases, officers recognised the need to create a more enlightened, 'auxiliary' style of discipline than that of the Regular army, a style that relied less on 'imposed' than 'self'-discipline.

A particularly vivid picture of the ways in which the officers of one freshly raised New Army battalion modified discipline emerges from two sources, an unpublished account written by a former temporary officer, P.G. Heath, and a short book published in 1915. The author of the latter was Capt. J.M. Mitchell[110] and the book described the process of training the men of 8/E. Surreys of 18th (Eastern) Division, disguised in the book as the '9th Blankshires'. Mitchell, a temporary officer, clearly brings out the officers' recognition that many of the recruits to the 8/E. Surreys were very different from the prewar class of recruit, and the discipline of the military unit needed to be moulded accordingly. They were not from the 'gutter and the pub'. Rather, they were men who 'counted for something in their civilian sphere'; Maj. A.P.B. Irwin, the battalion's Regular adjutant, described the men as 'most intelligent chaps'.[111] The battalion also included miners from South Wales who had 'a Celtic contempt for the red tape of discipline'.[112] The recruit was 'no automaton' like the prewar Regular recruit; rather, he had 'a trade at his finger-ends' and had experience of 'the free initiative which is part of prosperity in civilian life'.[113] However exaggerated and idealised this view might have been, it seems clear that the men of 8/E. Surreys were very different in character from those of prewar Regular units.

Inevitably, the process by which Regular and dugout NCOs and officers of the battalion came to terms with the changed circumstances was not entirely smooth. Some dugouts who had been retired since before the Boer War had to adjust to what Mitchell called 'the new discipline' which had evolved in the first decade of the century. The nineteenth-century army, Mitchell asserted, based discipline purely 'on the principle of authority' while new discipline 'incorporates with an equal rigidity of discipline the principle of individual responsibility and initiative'.[114] Although Mitchell did not say so, this new approach to discipline had emerged from the debates that followed the Boer War (see above). According to Heath, it even took Irwin, who played an important role in the development of the peculiar style of the

8/E. Surreys, some time 'before he realised that he was dealing with enthusiastic civilians and not regular soldiers'.[115]

Heath and Mitchell both paid tribute to the way in which dugouts responded to the need to adapt to the new type of soldier.[116] Mitchell argued that in the attempt to turn civilians without arms or uniforms into soldiers almost overnight, 'hard cases' were 'unavoidable'. A 'few men [were] punished unjustly' and some 'untactfully', but the punishments were minor, and officers regarded it as 'imperative whenever possible' to 'temper the wind to the shorn lamb'.[117]. Above all, Mitchell reveals the understanding and sympathy that the battalion's officers, temporary and Regular, had for the indiscipline of raw but enthusiastic troops.

Subalterns also began to forge links with their men based on admiration and paternal pride, a process aided by the fact that they were learning their trade together. Heath somewhat rashly told his men that he knew little more about military life than they did, and asked them to bear with him in the mistakes that he was bound to make. Heath, comparing army life to friendly rural gentry/villager relations, wrote that subalterns 'treated their men as friends and human beings ... in return the men seemed to like and respect their officers, and certainly gave them willing obedience'.

The 8/E. Surreys' officers were all too aware of the harsh disciplinary code of the army, in which 'The slightest trivial offence constituted a "crime".' Thus, Heath claimed, the officers sought 'to keep ... [the men] out of trouble provided that they were pulling their weight in the things that mattered'. Pte Bird, for instance, was five minutes late appearing on parade. Heath did not punish him, but ordered Bird's platoon sergeant to tell Bird's friends to ensure that he was more organised in future. Bird was a keen soldier, and by refusing to follow the official path of 'criming' him, Heath avoided alienating Bird.[118] This is just one example of a subaltern seeking to protect his men by bending the strict rules of the army, and it is also indicative of the way in which good relations were built up between officers and men long before units went on active service.

The officers of the 8/E. Surreys were not the only ones to adopt an enlightened style of discipline for a newly raised unit. Another 18th Division unit to do this was the 8/Norfolks, where in the early days discipline was aided by the virtual absence of crime, and the few malefactors were unofficially, but severely, punished by their peers, before officers could take action. An officer paid tribute to the battalion commander, Lt Col F.C. Briggs, 'whose tact had very largely contributed to the harmony and efficiency of the Battalion'. The efforts made during training seems to have borne fruit on active service, since in May 1916 Lt Col F. Maxwell, who recently joined the 8/Norfolks, noted in his diary that it was a battalion that was 'happy all through, with excellent officers and NCOs – a real good unit'.[119] Likewise, 22/R. Fusiliers, a Pals battalion raised by the Royal Borough of Kensington from Londoners and colonials, employed a liberal disciplinary regime, largely at the instigation of a prewar Regular officer, Major R. Barnett-Barker.[120]

Rankers as well as officers mentioned the use of 'auxiliary'-style discipline in many newly-raised units in 1914–15. A public school socialist, Sgt F.H. Keeling (6/DCLI), wrote that discipline in Kitchener units was different from that of Regular battalions 'because the conditions are different', but discipline also differed from battalion to battalion in the New Army, the variations being largely caused by the personal idiosyncrasies of officers commanding companies and battalions, and the distribution of Regular NCOs.[121] In 5/Wilts a soldier who went absent without leave by forging the signature of an officer on a pass (which would normally have been a serious offence) merely received ten days confined to barracks and ten days' stoppage of pay, apparently because it was realised that he had been treated unfairly in being denied leave in the first place.[122] C.F. Jones, a middle-class ranker of 2/15 Londons (TF) left an account of a warrant officer modifying discipline in training in England. When Jones was elected spokesmen for a group of men to complain about an NCO's 'tyranny', the CSM tore up the charge sheet against Jones (who had incurred the wrath of the military authorities) and appealed to 'a man of your age and experience' not to encourage 'these b— boys'.[123]

Undoubtedly, the tolerance of gross indiscipline in newly raised battalions did not last very long, and as training progressed, discipline improved. A ranker of 11/Suffolks (The Cambridge Battalion) recalled that discipline improved when they began weapons training.[124] On the whole, the volunteer soldiers of 1914–15 came to recognise the need for discipline, a process in many cases accompanied by the development of *esprit de corps*. The growth of trust of, and admiration for, their officers played a part in this process. Ivor Maxse, GOC 18th Division and a shrewd judge, commented in some confidential notes written in November 1915 on the 'excellent physical and moral' qualities of the subalterns of the division, who spent eight hours a day with their men and identified with their 'interests both on and off parade'.[125] A Regular dugout, Brigadier-General J.H. Poett, the commander of 55 Brigade (which included 8/E. Surreys), attributed the excellence of the brigade's discipline to the high quality of the recruits and to the battalion commanders 'who were sympathetic and human ... while insisting on strict discipline [they] handled their men with discretion and tact'. Poett stated that '[m]utual confidence between officers and men is the very essence of sound discipline and a happy battalion'. Under the conditions endured by recruits in the autumn of 1914, this growth of confidence between the ranks 'needed to be nurtured in every possible way'.[126]

Two more examples, both drawn from 18th Division, illustrate the development of *esprit de corps* in 'happy' units. R.A. Chell, an officer of 10/Essex, wrote of long training marches in England:

> one saw platoon pride and comradeship happily demonstrated. Platoon pride said 'no one must be allowed to fall out' and comradeship said 'two of us carry Bill's rifle and Jack's pack and they'll make it'.... We certainly were a happy team and a gentle, if really firm, discipline was the foundation of this.[127]

The fact that this battalion's all-rank Old Comrades Association continued in existence until the early 1960s offers some evidence that Other Ranks shared these feelings. An example drawn from 7/Buffs that indicates that the growth of comradeship between officers and men in 18th Division was not entirely wishful thinking on the part of the officers comes from the diary of Pte R. Cude, who was extremely critical of army life, particularly of the chaos that prevailed during the first weeks of the battalion. However, by the time he arrived in France, Cude had developed a strong admiration and affection for some (but not all) of his officers, and strong sense of *esprit de corps*: 'I am a Buff above all things.'[128]

The disciplinary regime of the British army was, in general, harsh, and as such it reflected the nature of British society. However, the twin concepts of deference and paternalism helped to reconcile working-class soldiers to military discipline. Moreover, at least some units raised in 1914–15 operated a liberal, 'auxiliary' disciplinary system during their training period in Britain. The extent to which liberal disciplinary regimes survived on active service is considered in Chapter 8.

6
Officer–Man Relations: the Officer's Perspective

At the beginning of the war the Regular British Other Rank was generally portrayed in the Press as a brave, dogged, tough, phlegmatic, soldierly working man in uniform who enjoyed an excellent relationship with his paternal officer. In the space of two days in September 1914 *The Times* printed an anecdote about a soldier who was less concerned about being wounded than the loss of his pipe, and quoted a sergeant's opinion that one of his officers had 'died one of the grandest deaths a British officer could wish for'.[1] A subheading of 12 September 1914 read 'Mutual Compliments of Officers and Men'. The following article quoted an artillery officer as saying that 'Our men and horses [!] are wonderful' and cited a sapper's opinion that 'the officers are grand. They do everything they can for our comfort ... I cannot speak highly enough of them'.[2] Conversely, the soldier–officer relationship in the German army was portrayed as being based on fear: 'machine-like' German soldiers were bullied in camp and driven into battle at gunpoint. Sometimes the contrast between the two armies was made explicit, on other occasions it was left unspoken.[3] Similar views on British and German officer–man relations can be found in British magazines and books of the time.[4] Broadly similar images of officer–man relationships appeared in print throughout the war years, even though the social composition of the army underwent significant change in this period.[5] Did officer–man relations in the British army of 1914–18 bear any resemblance to Fleet Street's version?

In November 1916 Rawlinson wrote that officers had to 'know and gain the confidence of their men'; this was 'the root of British discipline and leadership'.[6] The relationship began when the officer met his men for the first time. According to one authority, anyone can be appointed to a position of command, but an individual can only become a true leader when his subordinates ratify that appointment in their hearts and minds.[7] Bernard Montgomery, commissioned into the R. Warwicks in 1908, believed that

> The first thing a young officer must do when he joins the Army is to fight a battle, and that battle is for the hearts of his men. If he wins that battle and subsequent similar ones, his men will follow him anywhere; if he loses it, he will never do any real good.[8]

Many of Montgomery's brother officers agreed with these sentiments.[9] Most subalterns of the Great War were well aware of the importance of this first meeting. Alan Thomas, commissioned into RWK in 1915, was assailed by doubts before arriving at his battalion, fearing that he would not be 'up to it', and would be unable to win the respect of his men.[10]

Officers came to know whether or not they had been accepted by their men in a variety of ways. One officer of the 2/Wilts discovered that his men approved of him through censoring their letters on active service.[11] A Jewish officer of 4/N. Staffs became aware of his batman's approval when the latter took pains to cook bully beef instead of bacon, while G.H. Cole (1/20 Londons) 'realised my acceptance' when he heard a spectator call him by a (fairly respectful) nick-name while Cole was playing football.[12] Sometimes, of course, officers earned disparaging nicknames from their men, such as 'Ragtime', the contemporary adjective for absurdity or inefficiency.[13]

Officers who joined units while in training had a distinct advantage over those who joined a battalion on active service, particularly on the Western Front. A ranker officer who joined 2/Camerons in 1916 lamented that it took time get to know undemonstrative 'Jocks' but 'Almost before you have time to get to know their names some change is made, or you lose half of them in action.'[14] By comparison, Alan Thomas, who joined a reserve battalion in England, was able to spend weeks 'learning the strength and weakness' of each man in his platoon.[15]

Officers who knew and understood their men made the most effective leaders. Bill Slim, commissioned in 1914, 'had the inestimable merit of never having forgotten the "smell of a soldier's feet"'.[16] One officer wrote that subalterns should 'know personally' every man under his command: 'what he is good for, what he was in civil life, if he is married, etc. – and let the man know that you take an interest in him'.[17] Clearly many subalterns conscientiously followed such advice. Lt C. Meadowcroft (33/Londons), for instance, kept a record of the addresses of the next of kin of the soldiers in his platoon. Meadowcroft was a bank clerk, and thus unlikely to have absorbed the paternal ethos at a public school.[18] Another temporary officer, E.G. Venning of the R. Sussex, claimed to 'know the ways and peculiarities of every man of mine'. This seemed to be an effective leadership tactic. Two of his men refused promotion because this would have meant leaving Venning and moving to another company.[19]

The British army of 1914–18, dominated by prewar Regular officers, took no chances with the welfare of its lower ranks. Paternalistic leadership and man-management were institutionalised. Junior officers were given little opportunity to neglect their men. An officer who served in the 2/6 DWR in 1918 was 'always amused' to have to sign a certificate stating that his men's feet had been rubbed, their socks had been changed and they had eaten a hot meal, but he recognised that there was a serious purpose behind it all.[20] Even on the beaches of Gallipoli, 'endless returns', wrote a subaltern in October 1915, 'have to be made about one's men – health, clothes, equipment, arms, ammunition, etc....'[21] Official divisional trench orders issued in 1917 commanded that, among other things, at night the duty officer and NCO should frequently patrol the trench line, and arrange to provide a hot drink for the men between midnight and 4.00 a.m.[22] Other official documents ordered the officers to inspect men's respirators and apply 'Glasso' to the eye-pieces to prevent fogging,[23] to ensure the water they drank was pure,[24] and to drain trenches to prevent trench-foot.[25] At divisional level, staff conferences discussed 'the comfort and health of the men ...'.[26] In February 1918, 66th Division issued a memorandum in which the divisional commander drew attention to the wide discrepancies in sick rates of various battalions under his command, which were, he said, partly due 'to the men's moral (*sic*) and state of happiness, which are largely dependent on the thoughtfulness and care of their officers'.[27]

The much-derided obsession of the staff with apparent trivialities had its roots in concern that temporary officers, mere civilians in uniform, might neglect their paternalistic duties towards their men. Wyn Griffith, a temporary officer of 15/RWF, described the British army's bureaucracy of paternalism in these terms:

> every man above the rank of private is his brother's keeper.... This concern, this anxiety, and interest, minute and unceasing,... [is]

> characteristic of the British Army.... It can be harassing, and it often is, but it is omnipresent throughout the hierarchy of the command and the staff.[28]

Most temporary officers of the Great War confounded official fears by displaying an extremely high level of paternalism. The most practical expression of officers' paternalism was in welfare work. In contrast to the 1939–45 war, no centrally organised welfare service existed in 1914–18, although bodies such as the YMCA, Toc H and the Salvation Army did sterling work.[29] The primary responsibility of the officer was to ensure that his men were well fed and clothed and comfortable. There were many instances of officers taking this duty very seriously indeed. On one occasion during training in England, the men of 1/15 Londons received inadequate rations and a sergeant complained. Later, an officer's servant arrived with a packet of sandwiches for the NCO, much to the latter's embarrassment. What is more important, there was a general increase in rations.[30] Sometimes officers would forgo their own comforts to give their men luxuries. On one occasion, the all-important rum ration failed to arrive at a TF unit. The battalion commander promptly handed over six bottles of whisky, which were 'as precious as molten gold', so that the men should not go without their warming tot.[31] To a large extent, this was simply a continuation of the prewar convention that officers should place their men's comfort above their own.

In 1914 to 1916, it was extremely common for officers to use their own money to buy gifts for their men. Many officers' letters to their families included requests for cigarettes, clothing and chocolate for their men. Wealthy officers engaged in philanthropy on a spectacular scale. In December 1915 an officer of 2/Coldstream Guards sent home for 200 large mince pies for distribution to his men.[32] Other, less affluent officers kept their gifts to more modest proportions. Two gunner subalterns may stand as representative. The first wrote home in December 1914 for four pairs of gum-boots for his NCOs 'who would appreciate them highly'.[33] The second, concerned that the army was not issuing enough socks, asked his seven sisters to knit these essential items for his men. Each sock arrived with five Woodbines tucked inside.[34] The example of an officer of 9/Devons, who continued to send parcels to his men at the Front while he was recovering from wounds in England, and that of an officer who returned to his former unit to distribute cigarettes to men who were no longer under his command, indicates that such acts of generosity were sometimes, perhaps usually, carried out in a spirit of affection as well as pure pragmatism.[35] Not surprisingly, soldiers had affection for officers who took pains to attend to the needs of individual men, such as Maj. Heelas of 19th Division Artillery who obtained a pair of brown officers' boots for a soldier whose own footwear had worn out.[36]

Officers' regard for their men's welfare went beyond attending to their creature comforts. A revealing passage occurs in the memoirs of Bernard Martin, then an 18-year-old temporary subaltern of 1/N. Staffs. Martin wrote of his regret at not being able to keep a pipe he had taken from an enemy corpse: 'my men ... [would] have laughed at its swanky German appearance, tried smoking it turn and turn about ...'. This sentence, casually inserted into a passage concerned with other matters, demonstrates one of the unspoken assumptions that governed the conduct of many Great War subalterns: that keeping his men amused was an important part of the officer's duties.[37]

The soldier's soul and personal happiness were also the responsibility of the diligent officer. When there was no padre available, an officer of 8/ E. Surreys led a church service: 'Just the Confession, Gospel, Lord's Prayer, Creed, a tiny address on Easter & Lent, & a couple of hymns'.[38] In 1915 a Coldstream Guardsman broke down at his brother's grave. A fellow ranker noted that 'the officer has been very good to him and has got him sent to a convalescent camp for a time'.[39] By contrast, 2/Lt Bowker (9/W. Yorks) discovered while censoring a letter that one of his men had family problems and intended to confide in him. Bowker promptly went into 'a most horrible funk' and avoided the soldier until he was cornered. Bowker's youth and inexperience seem to have been at fault here, since he displayed otherwise impeccably paternal attitudes, buying food for his men to compensate for the inadequacies of the food onboard a troopship.[40]

An important part of the unwritten contract between officers and their soldiers was the defence of helpless men from higher military authorities. If many men felt themselves to be at the mercy of an impersonal and arbitrary coercive system, many officers tried their best to defend them from what they saw as unfair demands. In extreme circumstances, this could mean an officer jeopardising his career to defend his men. On the Somme in September 1916, the CO of 1/DCLI reported his nominally Regular battalion as 'unfit to go into action'. Two hundred men were sick, largely because of the poor conditions of their trenches and a logistic breakdown. To take such a stand was, as a fellow battalion commander of 5th Division commented, an act of 'great moral courage'.[41] A similar event had occurred in May of that year when the commanders of the 1/R. Berks and 22/R. Fusiliers had cancelled an attack on Vimy Ridge. Their corps commander contemplated trying them by court martial, although in retrospect their decision was clearly correct on military grounds.[42]

Protecting their men did not, of course, always entail commanding officers taking such outspoken and personally risky stands. Probably far more typical was the attempts of the commander of a home service unit, 6/Middlesex, to save a thrice-wounded private from returning to the front.[43] A rather similar incident occurred during the fighting in the spring of 1918 when the commander of a battalion in 51st (Highland) Division gave a man with a superficial wound permission to retire to a dressing station. A subaltern was surprised at this leniency, until he noticed that the soldier wore no fewer than six wound stripes on his sleeve.[44]

The language used by many junior officers indicates that their relationship with their men was based on something more than just a professional concern for their well-being. Indeed, the care and affection of the temporary officer for his men are constant themes running through memoirs, letters and diaries. An officer of 1/RWF spoke for many of his contemporaries when he wrote that 'the only way to run a company is by love'.[45] Another junior officer described the relationship in terms of a 'marriage' between the officer and his platoon.[46] A sapper subaltern used a slightly different metaphor, that of parenthood, to make much the same point.[47] All neatly capture the idea of a tender, caring relationship. It is possible that for some officers this relationship was tinged with homoeroticism, if not homosexuality, but for the vast majority of officers the relationship was entirely innocent.[48]

An officer need not be serving in a front-line unit to experience love for his men. In February 1916, G.S. Taylor, an ASC subaltern, put his feelings about his men into verse. In a poem he described his men (and horses!) as 'my friends', and showed much sympathy for their plight. He 'loved' his men, although he recognised that they were unaware of this, and aimed to win their love by his fairness as an officer and his 'trust' in them.

Shortly afterwards Taylor transferred to an infantry battalion, 2/Sherwood Foresters, where he began a love affair with his new charges. In April 1916 he wrote to his family that he had been appointed to permanent command of a platoon that he had previously led on a temporary basis: 'The men are ripping, and what more could one want'? A month later, Taylor referred to his men as 'absolutely splendid' and 'always ready to do anything that is wanted of them' despite their constant grumbling.[49]

Although his attitudes appear steeped in public school concepts of duty and paternalism, Taylor had actually attended a grammar school. He both idolised and idealised his men. The language he used to describe them, ('ripping', 'splendid', 'always cheerful') which might have come straight from the pages of a newspaper, indicates the degree to which Taylor identified with his men as their leader. Taylor would have agreed with Slim's view that the young officer's pride in his troops is such as would back 'his men' against 'the Brigade of Guards itself'.[50] Commanding men was, for Taylor, a fulfilling experience, and being of his generation, and of his social and educational background, he had no hesitation in using the word 'love' to describe his feelings for his men.[51]

Taylor's pride in commanding men helped cultivate a sense of responsibility for them and a desire to discover each man's needs. This is in accordance with modern leadership theory. Taylor also implied that he tried to protect his men from the unfair demands of the higher echelons of the army. Taylor tried to establish an informal relationship, mingling and conversing with them, lending a sympathetic ear to their troubles. Another officer believed when the men 'come to you with their private worries', this was a sign that the officer had won their confidence.[52]

Taylor sought to establish a relationship in which the parties were not equal, but which was characterised by mutual trust. He tried to gain the love of his men through his just treatment of them, while demonstrating that he trusted them not to take advantage of him. All of this, of course, looks at only one side of the relationship. Taylor's view of his men was probably highly romanticised and naive. Taylor's men might have had a very different view of their officer, although one sergeant took the trouble to write a letter to Taylor's family, nearly a year after his death on the Somme, in which he mentioned Taylor's concern for the men's welfare.[53] This brief case-study does, however, offer an interesting insight into the paternal attitudes of an unusually articulate subaltern. His views were not untypical, but while in most cases one can only find brief references to the attitudes of officers towards their men, Taylor eloquently expresses his feelings.

It would be rash to assume that officers' views of their men were always benign. A major source of annoyance was the apparent inability of the private soldier to help himself. This no doubt reinforced the officer's feeling of paternal responsibility, but it also added to his burden of work. A fairly typical comment occurs in a letter written by a subaltern of 1/King's, in late 1914: 'how like children the men are. They will do nothing without us ... You will see from this some reason for the percentage of casualties among officers.' Elsewhere, however, he expressed his admiration of his exasperating charges. This officer was writing of Regular soldiers, who at that stage of the war would still have been largely working-class in origin.[54] The men of the New Armies were little better in this respect. A journalist wrote in late 1916 that the

> dependency of the men upon their own officer, while it is an immense and unwitting tribute to that officer, is at times so absolute as to be embarrassing, and in these men, who have so many of them high intelligence and a retrospect (*sic*) of civilian responsibility, it is surprising.

The writer went on to argue that the problem arose from the surrendering of individual liberty when the men became soldiers. The men did not become 'automatons' but since the officers are appointed to think for them 'well, they are not going to deprive him of his job ... willingly, whole-heartedly they let him, and if he does it well they will follow him anywhere'.[55] This apparent helplessness was in part a natural consequence of civilian conditioning as well as the disciplinary system of the army. In sum, the institutionalising of paternalism created a sort of 'dependency culture' in which the soldier had little incentive to help himself.

Many junior temporary officers actively disliked the system that they thought treated 'their' men unfairly. John Nettleton, who served in the ranks of the Artists' Rifles and later as an officer in a Regular unit, 2/Rifle Brigade, commented on the lack of trust that Regular officers had for their men. Nettleton believed that Regulars' insistence on 'iron control' of Other Ranks at times placed the men's lives at risk, although he accepted and admired the paternalism of Regular officers.[56]

Another temporary officer who served in a Regular battalion, Robert Graves (2/RWF), went even further. Graves, although in many respects a far from typical subaltern, articulated the feelings of many of his peers when he wrote that he and Siegfried Sassoon believed that one of their most important duties was to 'make things easier' for their men. Graves believed that

> being commanded by someone whom they [the Other Ranks] could

> count as a friend – some one who protected them … from the grosser indignities of the military system … made all the difference in the world.[57]

An example of these 'grosser indignities' was encountered in February 1917 by E.C. Vaughan, a subaltern of 1/8 R. Warwicks, when his men had to stand in the snow for an unnecessarily long time as a result of bad staff work. Vaughan marched into the office of a staff officer and

> told him vehemently, 1) That my men had marched a long way and were tired, 2) That it was damned cold outside, 3) That he had kept them waiting nearly half an hour….[58]

For Vaughan, who lacked self-confidence and had made a bad impression on joining his unit, this was a morally courageous act in defence of his men's welfare. His action was not untypical of the way in which junior officers took unpopular stands against their superiors in defence of what they perceived as their men's interests, although the practical expression of officers' sympathy for their men usually took a less dramatic turn. Both junior officer and the private were trapped by the military system, and the officer could only modify it at the most local level.

Regimental officers were often reluctant to punish their men. The first level of (relatively minor) punishments were awarded by the company commander at 'orderly room' or 'Company Orders'. A temporary RMO wrote in 1917 that orderly room could be a tool of despotism. The avoidance of oppression was, in his opinion, not merely a matter of sympathy for the men, but it also made good military sense, because it could prevent unrest among the Other Ranks.[59] Similar sentiments were expressed by G.S. Taylor. In a letter of October 1916, written just after his battalion had come out of action, he confessed 'The thing I hate doing is holding Orderly Room & dealing out punishments such as 3 Days Confined to Camp. The men get a bad enough time anyhow.'[60]

The company commander presiding at Company Orders could, however, forward the case to the commanding officer, who could reduce men holding acting rank or take away privileges, impose stoppages of pay, or sentence the soldier to field punishment. More serious offenders could be sent for trial by Field General Courts Martial (FGCM). In the 1960s Charles Carrington wrote that many FGCMs tried military crimes that would have been 'no crimes at common law, and imposed sentences which seem shocking in retrospect'.[61]

Given the hazards of trial by FGCM, some officers turned a blind eye to what, in military terms, were serious crimes. According to Pte A.M. Burrage, a middle-class soldier who was often critical of the military system, officers were usually 'too decent to make a song about' exhausted sentries found asleep on guard in the trenches.[62] The trick was for the officer to wake the sentry up without acknowledging that the soldier was asleep. The officer could thus both adhere to the spirit of military law and save the soldier from the full rigour of the disciplinary code. An officer of 1/20 Londons on discovering 'one of my best men' asleep, fired a Very pistol to wake him up. As a ranker, Nettleton was discovered sleeping during Stand To, and was whacked across the behind with an officer's cane. 'Quite irregular' he commented, but he 'felt no resentment at the time'.[63]

W.R. Acklam, a ranker of 41st Division Artillery, recorded in 1917 that a drunken bombardier swore at an officer who was entering the canteen: 'Get out, get out, you bloody ugly bugger'. Fortunately for the soldier, the officer, Mr Mason, 'took no notice & went away'. Mason was acting in character, since he was an approachable man and, after an uncertain start, a steady growth of affection and respect for Mason is detectable in Acklam's diary.[64] This case neatly demonstrates the type of dilemma that paternal officers, perhaps with no great love for the niceties of military law, sometimes faced: did their duty lie in the strict enforcement of discipline, or in the protection of their soldiers? 2/Lt Bowker of 9/W. Yorks experienced a similar clash of loyalties. One of his men stole boots from a store, and Bowker knew perfectly well that the man was guilty. But Bowker thought that the thief genuinely believed that he was merely exchanging a pair of old boots for a new pair. Bowker decided that he had a duty to 'run him in' or the entire fatigue party 'would have absolutely refitted itself at the expense of ordnance', but he then protected the soldier by refusing to swear that the man was wearing stolen boots. This neat compromise – enforcing military discipline, in spirit if not to the letter, while protecting his errant charge from punishment – earned Bowker a stiff lecture from a senior officer.[65]

Another way that regimental officers could spare offenders from harsh punishments was to impose their own form of rough justice. Company commanders often gave offenders a choice: 'Will you accept my award, or will you go before the CO?' Commanding officers could similarly present an offender with the choice of accepting a relatively mild punishment or electing to go before a court martial, although some officers met their match in 'old soldiers' who knew their rights under military law.[66] In 1918 a company commander of 2/4 KOYLI dealt with a case of a self-inflicted wound by sentencing the offender – a youngster who had experienced a mental breakdown – to shoot the rabid dogs that roamed the area. If military law had taken its course, the boy would have been court-martialled and would have probably received a severe sentence.[67]

It would be tedious to list every other practical manifestation of the sympathy that many officers felt for their men. However, a private of 15/RWF wrote a fictional, but wholly convincing, vignette of a moment of leisure snatched in a busy day behind the lines that neatly captures one facet of the paternal subaltern:

> Platoon drill was tiresome, but Mr. Jenkins was kind. He used to sit by the little stream where they got their washing-water, and look into it for long at a time without moving, whilst they smoked under the turnip-stack, with someone watching to see if anyone – the Adjutant, or that shit Major Lillywhite, was anywhere about.[68]

One question was central to the officer–man relationship: how could officers demonstrate friendship for their men and yet retain their authority? Capt Hamond, in his unpublished treatise on officership, had some firm, common-sense views on this question. He stated that men would follow an officer who had a strong, attractive personality and who 'personally looks after their bellies and beds'. However, he continued, 'any form of familiarity that lowers your own position' should be instantly checked, 'but for God's sake don't always be thinking about your own dignity, it should be there without any possibility of mistake for everyone to see'. In sum, the officer needed to tread a narrow line between paternalistic friendship for his men, and undue familiarity.[69]

An interesting practical example of how easy it was to cross this line occurs in a book written shortly after the war. The author was Rowlands Coldicott, a company commander in a Territorial battalion, 2/21 Londons, in Palestine. He had no illusions about the limitations of Regular-style discipline when applied to his men: 'Nothing on earth ... could turn our clerks into members of a regular army ...'.[70] Yet Coldicott, a paternal officer with considerable sympathy for his men, felt it necessary to remonstrate with one of his subalterns, 'Trobus', over his relationship with his platoon. Trobus liked to surround himself with

> his admirers, to whom he would tell stories, reaping huge admiration thereby.... He was popular, and liked, but at the expense of something common to both parties that ought not to have been sacrificed.

Trobus's men, in Coldicott's opinion, were taking advantage of his good nature by marching sloppily, and falling out on the march:

> Assing along, telling tall stories to a sergeant, isn't looking after a platoon on the march.... Look at number ten [platoon]. Of course they detest Jackson, but he does manage to get them along when they're whacked. What the devil does it matter if the men like you or not now? They like the fellow in the end who pulls them through and drops on them occasionally.

Coldicott knew that the terrain that they were currently marching over was not as taxing as the hills of Judea that lay ahead of them. By failing to build up his platoon's stamina, Trobus was storing up trouble for the future. Coldicott required his officers to exercise 'lynx-like' supervision of his men and to be harsh towards laggards to protect them from themselves. Trobus' relationship with his men caused him to overlook malingering ('"Look at that great hulking chap Thunder, who pretended he couldn't carry his pack"'). Coldicott felt that, by opting for short-term popularity, Trobus was risking the lives of his men. In short, the thing which Trobus's relationship with his men had 'sacrificed' was the degree of detachment from those under his command that enabled the officer to make unpopular, but essential, decisions.[71]

This passage neatly demonstrates problems inherent in a close, informal relationship between an officer and his men. If all the Other Ranks could be relied upon to exercise self-discipline, and not to take advantage of an informal inter-rank relationship, the need for a social 'gap' between officers and men could largely be dispensed with; but in reality it could not. Coldicott was one of the few officers to articulate this dilemma. Another was 2/Lt R.W. McConnell (6/King's Own), who in late 1915 succinctly summarised what was probably the credo of most officers, temporary or otherwise: 'The men are all topping fellows. But one has to let them know who is master. First an officer has to be an officer, and then he may become a man.'[72]

The prewar ethos of paternalism and the influence of the public schools might help explain why officers regarded the welfare of their men as part of their duty. It does not entirely explain the enthusiasm that many officers showed for this task and the love and devotion that many felt for their men. There were also other factors, products of the peculiar circumstances of the Great War.

The age-profile of British regimental officers of 1914–18 tended to extremes. Regimental officers of the prewar army were fairly young[73] and the war exaggerated this trend. The expansion of the army normally ensured fairly swift promotion for those Regular officers who survived. The casualty rate among junior officers was extraordinarily heavy. To give but one example, in September 1915 normal wastage rates for officers on Gallipoli were 20 per cent per month.[74] By September 1917 the casualty rate among officers resulted in most battalions being commanded by officers of 'not more than' 28 years old and most companies by men no older than 20.[75] Conversely, the raising of the New Armies in 1914–15 also brought a number of older men into the army as overage subalterns. In 22/R. Fusiliers in 1914 approximately one-half of all officers, and one-third of all subalterns, were aged 31 or over, although some of these men did not go overseas or did not remain long with the battalion on active service.[76]

Many subalterns were thus of an impressionable age. Young officers enjoyed the excitement and comradeship of war, revelled in the new-found freedom after school or university and the novel responsibility of commanding men. 'Leadership, the most heady and intoxicating draught for a young man' wrote one of them, 'became a duty and a delight.'[77] F.A. Shuffrey, an officer of 11/DWR, spent his 21st birthday on the Western Front. He later wrote that the experience of war was often frightening, but 'outweighing the fear', for 'very young' men such as himself, was the fact that 'the war was still an adventure which thrilled us, stimulated as we were by comradeship much more vital than is ever found in peace'.[78] This comradeship was sometimes able to transcend the barriers of rank.

Many rankers were also young. It was not uncommon for units on the Western Front to receive drafts of 'men' of 18 years old.[79] It is perhaps unsurprising that boys of similar ages made friends across the rank and class divide. Some of the older officers had an almost literally paternal attitude towards soldiers who were of an age to be their sons. Ernest Parker, who served in the ranks of 10/DLI and was later commissioned, remembered with affection 'our dear old Bombing officer', Capt. Pumphrey, who said 'just the right things to encourage a youngster like myself'.[80] Some older officers sympathised with men of their own age who were enduring conditions in the ranks (see below).

Many officers came to admire the fortitude and stoicism displayed by their men in enduring the conditions of trench warfare while enjoying few of the comforts available to officers. Officers' writings often contain expressions of admiration: 'A private's life out here is a very rotten one', an officer of a New Army unit, 9/DWR, confided to this diary in 1916, '– the more one thinks about it, the more one admires the men – they're absolutely wonderful to stick what they do stick....'[81] Capt. Shuffrey thought that the spirit of the ordinary soldier in enduring the hardships of military life was a major factor in winning the war.[82] In retrospect, one officer admired the men for enduring life at the bottom of the military hierarchy, while retaining 'their individuality and their courage'.[83]

The rationale behind the disparity in the army's treatment of officers and men was that, having greater responsibilities, officers were entitled to more comfort. However, very occasionally, one detects in officers' writings a twinge of guilt that they had privileges denied to their men.[84] Officers who had previously served in the ranks in particular knew only too well what ordinary soldiers had to endure. 'When I think of the men who had none of my 1000 comforts', a subaltern wrote in February 1916, 'I am glad that I didn't stay in the [ranks of the] 16th Middlesex.'[85] In the retreat in March 1918, a self-confessed 'stony-hearted' RMO (Regimental Medical Officer) discovered for the first time what it was like to be utterly exhausted, and as a consequence his attitude towards men reporting sick 'changed entirely'. Significantly, the adoption of a more sympathetic approach by this RMO did not alter his basic attitude towards his men, for he had always taken 'an intense parental pride' in his battalion, 8/Queen's.[86] Another RMO felt that the disparity between leave for officers (once every six or seven months) and rankers (once in every 15 to 18 months) was grossly unfair.[87]

If anything, the disparity grew more pronounced as the war went on, for in 1918 the army introduced a scheme that gave junior officers the opportunity of serving for six months in Britain.[88] The only equivalent for Other Ranks was the granting of leave to prewar Regular and Territorials whose period of service would, under normal circumstances, have expired.[89] Many officers had particular sympathy for educated, middle-class rankers who under other circumstances might have found themselves in the officers' mess (see below).

Although officer–man relations were undoubtedly good in many newly raised units in 1914–15, it was often said that such relations were, in general, much closer on active service than at home. An ASC private, a former miner, believed that officers he encountered at a base on the Western Front and their counterparts in England were very different creatures. The latter had been mostly Regular officers, who regarded wartime soldiers as mere cannon-fodder to be trained 'as quickly as possible'. In France, by contrast, temporary officers had experience of active service, were sympathetic to their men, and largely replaced traditional discipline with '"esprit de corps"'.[90] Later in the war, convalescent temporary officers also served in camps in England, but an infantry officer who first arrived in France at the end of 1917 expressed very similar sentiments. Commenting shortly after his arrival on the high level of *esprit de corps* in his battalion (2/4 DWR), he noted the very different atmosphere in the BEF, where there was a spirit of comradeship between officers and men, from home service units, where there was not.[91]

In contrast to the situation that existed in Britain, where the important role played by the NCO in training minimised contact between officers and Other Ranks, the conditions on the Western Front were exceptionally favourable to the establishment of close relationships between leaders and led. Regimental officers and Other Ranks alike lived in rat and vermin-infested holes in the ground, although the officer's hole was usually better appointed, and all shared much the same discomforts caused by weather. Shells and bullets were no respecters of rank, and indeed officers were more likely to become casualties than Other Ranks. Yet it is important that the gulf between the officer and the ranker should not be underestimated. This is graphically illustrated by a 1/N. Staffs officer's comment that his working-class soldiers were 'bilingual'. Among themselves, the rankers spoke a language virtually incomprehensible to a middle-class officer. When addressing an officer they used 'pukka talk'.[92]

While the officer's lot was better than that of the Other Rank,[93] the regimental officer's experience of warfare had far more in common with that of the private than it did with of the general. Constant patrolling of the trenches and supervision of various activities ensured that there was constant contact between the officer and men in the trenches.[94] One subaltern noted that even 'the most taciturn' sentry would talk when visited at night by an officer.[95] Capt. B.G. Buxton (1/6 DWR) recalled that

> I would go round the lines, perhaps between two and three a.m. on a soaking day, and a sentry would turn on the fire step and make a remark. I would get up by him, and he would tell me of some

> problem at his home – a child ill; anxiety about the loyalty of his wife; whatever it might be: and we would talk it over together. It was a wonderful human relationship, not least seeing that I was often ten or twenty years younger than he.[96]

Under some circumstances, such as when holding 'elastic' defensive positions in the spring of 1918, small parties of men commanded by an officer might have to man isolated posts for days at a time. In 1915–16 trench warfare 'was carried on by small detached units, companies split into platoons and parties, who seldom saw their own battalion headquarters ...'.[97] Inevitably under such conditions officers and men were forced into a degree of intimacy, and opportunities arose for fraternisation between what one temporary infantry officer described as the men who took 'nine-tenths of the risk and ... [did] practically all the hard work in the Army ... the private and the subaltern ...'.[98] An example of informal fraternisation occurred in August 1917 when a gunner of the RGA was trapped in a dugout by intense shelling along with some officers, who

> began to talk of guns and all sorts of things. Then tea was made and an officer went out and fetched it from the cooks (*sic*) dugout. Had good feed.[99]

One officer declared that the 'characters of any officer or man when under fire were laid to the bare'.[100] After spending eight days in the line in July 1916, enduring heavy shelling, Lt R.E. Wilson (1/4 York and Lancs) wrote that 'Officers and men have suffered and worked together and have come to regard each other as *real men*.'[101] Active service, a Regular officer of the Devons noted, highlighted the importance of 'looking after' his men.[102] Some officers came to believe that they formed what might be termed a 'community of the trenches' with their men (see below). It is also relevant to note that there were many occasions on which temporary subalterns were out of the reach of their superiors. Under such conditions, junior officers were under less pressure to conform to Regular norms of behaviour towards their men than would have been the case behind the lines. It is significant that discipline in small units such as machine-gun companies and trench-mortar batteries, commanded by captains, was in many cases more relaxed than in larger units such as infantry battalions (see Appendix 3).

In infantry battalions, commanding officers usually held the rank of lieutenant-colonel while artillery and trench-mortar battery commanders were majors, these ranks often being held on a temporary basis. The ideal CO was a benevolent paternal figure. One padre in 1918 went as far as to describe the Almighty as 'a gallant and fatherly Colonel who went over the top with his men'.[103] Some men did indeed like and admire their COs. A private of 1/4 R. Sussex declared '... I would do anything for our colonel as he is a real gentleman & leader of men & is liked by all ranks'.[104] An NCO of 5/Welsh noted in his diary the general sense of loss at the death of Lt Col Pridham, whom he described as a 'good man'.[105] However, comments such as these must be balanced by other, less complimentary remarks by soldiers of other units for whom their CO was a remote figure who had little apparent impact on their lives, for it took a particularly dynamic personality to impose itself on a unit as a large as an infantry battalion.[106] One such man was Lt Col J.H. Dimmer, VC, an ex-ranker, whose battalion, 2/4 R. Berks, became known as 'Colonel Dimmer's battalion'.[107]

Moving from inspecting one training unit to another in May 1916, Maj. Gen. G.G.A. Egerton found himself 'in a different world altogether', the principal difference between the two units being the characters of the respective COs.[108] On active service the influence, for good or ill, of the unit commander was even more important. In the autumn of 1917 the poor leadership of the CO of 11/Argylls demoralised the men, who made obscene remarks about him within earshot of their officers, who shared their men's opinions. His successor possessed all the qualities he lacked; 'efficiency', 'sense of duty' 'common sense' 'good humour', and this new broom revitalised morale in the unit.[109]

Popular, paternal commanding officers were not necessarily efficient soldiers. A Liberal politician, Sir George McCrae, raised 16/R. Scots in Edinburgh and commanded it on the Western Front. Following the battalion's failure in an action in August 1916, his divisional commander reported that although McCrae was personally gallant and always cheerful, and his men were devoted to him, his deficiencies as a commander (which possibly included a reluctance to accept casualties) rendered him unfit to command a battalion on active service. McCrae was appointed to command a reserve unit.[110]

The sheer power wielded by the CO over the lives of his men was impressive; he could modify the military system, or conversely, enforce the disciplinary code with utmost rigour. In some cases the attitude of the CO to officer–man relations and discipline could set the tone for all the officers of the unit. Since commanding officers were very often Regular soldiers, at least for the first half of the war, they provided a valuable thread of continuity with the prewar army, passing on the Regular tradition of paternalism to temporary officers.

Neither chaplains (or padres), nor Regimental Medical Officers were, formally, combatant officers, and thus their relations with rankers were inevitably slightly unusual. RMOs could have a great influence on the fate of individuals. A soldier coming to an RMO with an ailment might be treated sympathetically, which could mean a man being excused trench duty, or he might receive 'medicine and duty'. Thus the RMO constantly had to judge whether a soldier was shamming sickness. Rankers often judged individual RMOs by their compassion (or, to put it another way, their leniency). The men of 8/Queens regarded their RMO, Capt. Lodge Patch, as callous and nicknamed him 'Iodine Dick', a judgement that receives some confirmation from Lodge Patch himself (see above).[111] By contrast, a private recorded in his diary his admiration of the RMO of 10/Essex, who had worked unstintingly with the wounded after an action.[112] It seems that rankers tended to judge RMOs, like line officers, by the criteria of paternalism, courage and fairness.[113]

Padres had to overcome a number of difficulties to establish close relationships with rankers. There was inherent tension between the padre's spiritual role and his position in the military hierarchy. In the words of one divisional chaplain, the padre's status as an officer was 'a hindrance to be overcome; it removes him to a distance'.[114] Padres could, however, use their anomalous position to good effect, moving 'between the ranks with diplomacy and understanding'.[115] Edward Thompson, who served as a padre in Mesopotamia, wrote approvingly of the chaplain who could talk to 'an unknown tommy ... without being suspected of patronage (*sic*) or deliberate will to be Christian'.[116] In some cases, officialdom prevented padres from visiting the trenches, which automatically set up a barrier between soldiers who had experienced the stress of trench warfare, and padres who had not. Those padres who did visit the trenches could form rewarding relationships with rankers.[117]

As one chaplain commented, there was a temptation for the padre to neglect the spiritual side of their work for the pastoral: 'looking after their material comforts, writing their letters and procuring cigarettes for them'.[118] Some soldiers were impressed by the spiritual side of the padre's work, the more the padre shared the soldiers' conditions, the greater the attention paid to his message. Often, rankers judged chaplains by criteria unrelated to their effectiveness as priests. The men of 32nd TMB admired the courage and paternalism of their chaplain; one ranker, a Yorkshire miner, paid him a considerable compliment: '"he's a man is Our Padre"'.[119] The most effective padres were those who allied courage and paternalism with the ability to overcome the barriers of rank and class. E.P. St John, the first padre of 22/R. Fusiliers, was a jovial, soldierly, paternal figure, popular with all ranks. The men and officers largely ignored, however, his successor, C.E. Raven, a noted theologian. Raven's letters reveal that he was a lonely and isolated figure, who failed to adjust to the demands of ministering to an infantry battalion on the Western Front.[120]

Some soldiers of all ranks believed that generals and staff officers of the BEF were remote from the front-line troops, callous, and incompetent.[121] This was far from universally true. Plumer, commander of Second Army, was renowned as a 'caring' general who did not waste his men's lives.[122] Further down the chain of command, two brigadier-generals, F.W. Lumsden VC and R.O. Kellett, commanders of 14th and 99th Infantry Brigades respectively, were well-thought-of by Other Ranks. Rankers admired these men for their bravery, paternalism or both. A private of 5/6 R. Scots described Lumsden as:

> one of the very best. Wearing his cheesecutter with red band ... just as if he had been on the parade ground, he went into the front line trenches amid all the mud, barbed wire etc....[123]

Rather than generalising about soldiers' attitudes to 'the generals', each case should be examined on its individual merits.

Inevitably, staff officers' duties made it difficult for them to demonstrate courage or paternalism, although the term 'staff officer' is a broad one and the position of, say, a captain on the staff of a brigade could at times be highly dangerous. Inevitably, some saw the staff as gilded popinjays, living in the lap of luxury, in stark contrast to front-line soldiers.[124] In reality, staff work was unglamorous and exhausting. Animosity between fighting soldiers and the staff had a long tradition that certainly predated the Great War.[125] However, when the infantry were victims of bad staff work, their bitterness came to the fore.[126] Pte J. Woolin described a different sort of rear-echelon officer, a Railway Transport Officer, as 'omnipotent. The serving soldier was to him merely a unit to be entered on the forms he filled.'[127] Ironically, Woolin himself, as a member of the ASC, would have been both envied and despised by many infantrymen as a soldier with a 'cushy' job.

A Regular officer once stated that, from the perspective of the front-line soldier, good generals and staff officers exhibited 'man mastership' (i.e. man-management) skills. Those who visited units in the trenches during dangerous periods or after a unit had returned from battle were 'amply rewarded by increased confidence on [the part] of the regimental officers and men'.[128] Unpopular staff officers and generals perhaps helped to define relations between regimental officers and men. The colonel of a battalion was usually the highest ranking officer who was regarded as 'one of "us"', as part of a community.[129] Defending their men against the unfair demands of higher command gave some regimental officers a focus for their paternalism, while the knowledge that their officers were behaving in such a fashion enhanced soldiers' morale. By appearing to both regimental officers and Other Ranks as an enemy against which they could unite, unpopular generals and staff officers helped to turn military units into cohesive communities.

The broadening of the social base of the officer class did not meet with universal approval. In the early years of the war there was some fairly predictable snobbish criticism of 'temporary gentlemen'.[130] Some temporary officers adopted this pejorative name out of a perverse form of pride: Dennis Wheatley, a man of some social standing, used it in the title of his war memoirs.[131] A temporary officer could be a man with a public school education who had initially enlisted in the ranks, or he could be a man of lower-middle-class or even working-class origins. There were certainly occasions when temporary officers lived up to their unflattering image by committing *faux pas* such as arriving at the exclusive Sudan club riding a camel, or using red or green ink to apply for their commissions.[132] However, some Regular officers genuinely believed that men lacking a traditional social and educational background would also lack paternalism, and would therefore make poor officers.

In 1917 a newly commissioned old Etonian subaltern, J.E.H. Neville, of the socially prestigious 52nd Light Infantry (2/Oxfordshire LI), described an officer of 23/R. Fusiliers as 'a most temporary gentleman'. Claiming that 'A man like that gives himself away badly', Neville wrote

> he shows at once that he has not got the spirit of his regiment, or he would not run down another better than his own. Heaven help the Army if chaps of his kidney are going to be its officers of the future.[133]

A more measured but essentially similar opinion was that of R.T. Rees, a public schoolmaster serving as a temporary infantry officer. He was to write that although non-public school officers 'often made good', it took time for them 'to acquire the sense of responsibility and faculty of leadership' fostered by the public schools: 'We had some anxious moments with them at first because of the lack of these qualities'.[134] Much the same fear was implicit in the views of the Adjutant-General at GHQ, who in May 1915 suggested that Regular officers serving in New Army and Territorial units should not command companies or 'expose themselves unnecessarily, as it is impossible to find regular officers to replace them'.[135]

Although these assumptions died extremely hard – controversies occurred along similar lines during the Second World War and in the 1960s[136] – these views were not held by all Regular and public school-educated officers of the Great War. In the 1930s Maj. Gen. Sir Ernest Swinton roundly denounced views that the officers of 1918 were 'poor in quality' as 'largely rubbish'. Swinton conceded that in the circumstances it was not surprising that officers were only half-trained but 'they did their best, and what more could a man do?', implying that the performance of such men was all the more remarkable for their lack of training.[137] Hubert Essame, who served as a Regular subaltern in the 2/Northants in 1918, emphasised the importance of the junior officer in the battles of 1918, but stressed that the officer corps formed a rough meritocracy. In 1918 regimental officers were drawn from a number of social classes, forming a society 'based on mutual loyalty and trust from which distinctions of class had long since vanished'. However, temporary officers adhered to the Regular code of officer–man relations, with all that implied in terms of paternalism.[138] The British official history contains some criticisms of the tactical skill of the, mostly temporary, regimental officers in 1918,[139] but recent historians have taken issue with these findings, one suggesting writers have underrated 'the standards of junior leadership in the BEF'.[140]

Most temporary officers were received into their units with the same degree of pragmatism that lay behind their commissioning in the first place. There were some exceptions, particularly at the beginning of the war, in which some lower-class ranker officers were patronised and even insulted by other officers.[141] In mitigation, the treatment of newly arrived temporary officers in 1914–15 in Regular units was in many ways a continuation of the treatment of their prewar predecessors who received the unflattering sobriquet of 'warts'. In addition, the rapid expansion of the army at the beginning of the war undoubtedly resulted in commissions being given to men who were unfitted to hold them.[142] Crozier estimated that one in three of the officers that first went to France with 9/R. Irish Rifles were 'duds', and that the proportion was probably higher in other battalions.[143] However, Regular officers, whatever their private feelings, seem to have avoided out-and-out rudeness.[144]

In the latter years of the war, when officers were selected on merit and there were fewer Regulars serving as regimental officers, prejudice against temporary officers seems to have subsided. A ranker officer who joined the 2/RWF in June 1917 commented that there was not 'another Regular battalion where the Temporary Officer had as much fair play if he had anything in him', which presumably means that the officer gained acceptance if was he was congenial company and attempted to behave in a officerlike fashion.[145] The latter factor was all important. 'One did not want to spend one's life in a barrack room atmosphere' wrote a former Regular officer: 'Others who may criticise this statement did not have to do so.'[146]

Potentially, an even more delicate situation arose when an ex-private or NCO returned to his former unit as an officer, and in fact most ranker-officers were posted to new units on receiving their commission. However, John Lucy, who joining 2/R. Irish Rifles (a Regular battalion) as a second lieutenant in 1917, was one of many ranker officers who received a warm welcome on joining their former unit.[147] Temporary officers were accepted readily enough as platoon, company and even battalion commanders, but it was commonly believed that Regular prejudice and jealousy never vanished entirely. The GSO 1 of Second Army complained that 'We had the best brains of the Empire at our disposal, and we failed to make full use of them.'[148]

By one obvious criterion, the officers of 1917–18 may appear to have been less paternal than their predecessors. It is fairly rare to find references to them buying gifts for their men, with the odd exception of atypical officers such as chaplains and prewar Regulars.[149] This does not necessarily indicate that temporary officers lacked paternalism. The British civilian population was short of food in 1917–18, which led to some alarm in official circles at the possible impact on the morale of the BEF.[150] Soldiers received small quantities of food as late as the autumn of 1918[151] but it is likely that bulk orders of 500 large mince pies would have presented greater difficulties.

Moreover, many officers commissioned in 1916–18 simply could not afford to buy large numbers of gifts for their men. In 1916, a second lieutenant's pay was 7s 6d or 8s 6d per day plus various allowances.[152] Expenses could eat up much of this pay. An officers' field kit from Humphreys and Crook of Haymarket cost £7 1s 9d. In July 1915, a subaltern's mess bill while in England amounted to £3 16s 4d, after a messing allowance had been deducted from the original bill.[153] On joining a Regular battalion in France in 1916, an officer discovered, to his horror, that his first week's mess bill exceeded his pay.[154] While some officers may well have been reasonably well-off on active service,[155] others, particularly poorer men who sent money home to their families, would not have had sufficient spare cash to buy gifts in bulk for their men on a regular basis. That was the prerogative of the moderately affluent officer, of whom there were decreasing numbers in the British army of 1917–18.

By 1917–18 the British army had largely abandoned its prewar criteria for officer selection and was commissioning men because they had demonstrated leadership or leadership potential on the battlefield.[156] A wealth of evidence, examined below, suggests that there is a connection between leadership and morale, and clearly the BEF's morale remained sound throughout the war. This suggests that the 'traditional' view of the attributes of an officer – the possession of 'a certain style of dress, behaviour and speech which had to be quite different from that of the rank and file'[157] – was, by 1918, outmoded. The average temporary regimental officer of 1918, unlike his public school-educated predecessor of 1914, was a 'professional' in the sense that he had earned his commission on the battlefield, rather than attaining it through the possession of social and educational advantages. He was no less imbued with paternalism, however. This last point is illustrated by reference to a revealing passage in an autobiographical novel which displays the thought processes of a 'temporary gentleman'.

'Phillip (*sic*) Maddison', newly commissioned into the 'Gaultshires' from the ranks of a Territorial unit, is invited to drink tea with some of his men:

> [He] had made a point of speaking to each man by name.... He must imitate the Duke's way, in the Gaultshires, according to 'Spectre' West, [a senior officer] of asking questions about their homes, encourage them to speak.... He wanted to leave while the good impression of him remained. Should he say Goodnight, men, as was correct, or Goodnight, you fellows? Which? 'Goodnight, boys!'
>
> 'Goodnight, sir!' in instant chorus.... [He was jubilant] that at

> last, he felt that he belonged to the men of his platoon, as they to him....[158]

The career of the author of this passage, Henry Williamson, closely paralleled that of his fictional alter ego. The passage illustrates the effort made by a ranker-officer, who had an inferiority complex about his lower-middle-class social origins, to conform to expected standards of paternalism by consciously imitating the behaviour of Regular officers, and his elation as he felt that his efforts were rewarded.[159]

Thus the newspaper stereotype of British officer–man relations was accurate in that the vast majority of officers, Regular, Territorial or temporary, shared a common belief in the need for paternal care of their men, and in many cases paternalism developed into deep affection for their men. The next chapter examines officer–man relations from the perspective of the Other Ranks.

7
Officer–Man Relations: the Other Ranks' Perspective

> Officers fell into two categories. If they passed dirty rifles, handled a spade, or carried a bag of cement, they were 'aw reet'. If not, they were 'no bloody bon.'[1]
>
> Pte W.V. Tilsley, a 'Derby' infantryman of 55th Division

Other Ranks did not respect their officers merely because they held the King's commission.[2] Rather, the soldier's respect had to be earned by the officer, who had to demonstrate a number of leadership qualities. Working-class rankers tended to judge officers by a simple set of criteria. The views of working-class soldiers in 2/5 Glosters support Tilsley's comments:

> A bad officer, that is, a bully, is a —! A good officer, that is, a (*sic*) considerate, is 'a toff'. 'I'd follow him anywhere'. 'The men's friend;' or simply, but in significant tones, a 'gentleman'![3]

Other Ranks tended to judge officers almost entirely in terms of the deferential dialectic. Expressed more simply, the ranker's view of the officer was largely determined by the way the officer behaved towards him.[4] Officers had to juggle two aspects of their duties. They had to be both militarily efficient and also protective of their men, and these two roles could sometimes conflict. Inevitably, a ranker's view of his officer could vary according to the circumstances. A ranker recalled that on one occasion hungry, cold men on a long march took a dim view of a normally popular officer, but that attitude changed to one of genuine gratitude when a surprise Christmas dinner was provided for the men.[5]

Other factors were far less important in determining a soldier's perception of an officer. Strict disciplinarians were not necessarily unpopular, as they could also possess other qualities, such as leadership, of which the men approved.[6] An officer's youth was not necessarily a barrier to winning his men's approval. In later life, Lt W.R. Bion (Tank Corps) wondered if anybody, 'outside of a public-school culture, believe[s] in the fitness of a boy of nineteen to officer troops in battle?' The answer was that the non-public school classes of 1914–18 accepted 19-year-old boys as military leaders provided the latter behaved in an officer-like manner. An incident in Bion's career suggests that a form of reverse paternalism could exist, in which rankers made concessions to the youth of officers. When his tank broke down in action in 1917, Bion was calmed by the 38-year-old 'grandfather' of the crew who showed him photographs of his family.[7] Pte Clarkson of 5/6 Royal Scots recalled that green young officers were inclined to try to teach old sweats their business, but nevertheless he admired their courage.[8]

Pte A. Jobson (39th Division Field Ambulance) placed officers into three categories: 'Good, Bad and perfectly B ... y'.[9] While this over-simplifies the ambiguities inherent in the officer–man relationship, Jobson's view may mean that good officers fulfilled their paternal role, bad officers did not, while 'perfectly B ... y' officers were those who were deliberately unpleasant or oppressive towards the men. Broadly speaking, there were three major reasons why officers were disliked by Other Ranks: failures in paternalism; failures of leadership; and deliberate unpleasantness.

Possibly the most important factor in determining a soldier's attitude to his officer was the extent to which he cared for the well-being of his men. The diary of a ranker of 27th Division Ammunition Column shows a direct correlation between his low morale and poor conditions and food, for which he blamed his officers: 'Rotten lot of officers – they fare alright but they don't mind about us.... Fed up'.[10]

Rankers also expected their officers to show leadership qualities in battle. Pte S.B. Abbot (86th MG Coy) condemned one of his officers (nicknamed 'The Orphan') as a 'thruster', prepared to endanger his men's lives by unnecessary displays of excess zeal in 'strafing' the enemy positions, while simultaneously appearing to be over-concerned for his own safety. Abbot implicitly compared The Orphan with another officer, referred to respectfully as Mr Street, who was 'a splendid man', a paternalist who was mourned as 'our brave and kind officer' when he was killed in April 1917.[11] The essence of leadership is diverting the cohesion of the group to the ends desired by the military hierarchy; but this example demonstrates that if officers are perceived to be too eager to take risks, and thus jeopardise their troops, at the very least they forfeit the respect of their men. This seems to have happened, temporarily at least, in 2/R. Sussex after the battle of Aubers Ridge in May 1915. According to one sergeant, the men blamed the officers for adopting tactics which resulted in heavy casualties.[12] Conversely, in the eyes of his men, an officer's courage could compensate for other failings. A group of rankers, discussing their officer, were heard to say 'No that little one don't know much, but he's always about when it comes on to shell.'[13]

In general, a middle-class Territorial ranker wrote, officers' 'outward and visible standard of courage' was higher than that of the Other Ranks.[14] 'Windy' officers were usually regarded with some disgust. Both senior and junior non-commissioned ranks felt contempt for an officer of 1/13 Londons 'for showing his fear in front of the men he was supposed to be leading', by ducking on hearing shells explode, the RSM going so far as to shout at him to 'keep his head up'.[15] An officer of 22/R. Fusiliers was once found cowering at the bottom of a trench at the beginning of an attack; his platoon sergeant swore at him, and physically bundled him over the parapet.[16] George Coppard (37th MG Coy) mingled his disgust for an officer who refused to emerge from a dugout with pity for his physical and mental condition.[17] Although one ex-ranker wrote of men covering up the 'deficiencies' of 'dud' officers, this attitude does not seem to have been typical.[18] Many soldiers appeared to have shared Lord Moran's view that courage was very much a matter of character and willpower, that everyone felt fear, but only cowards gave way to it.[19] Officers were expected by their men to set an example of courage. Cowards had, in the eyes of the Other Ranks, forfeited all right to commissioned status, and the privileges that went with it.

Rankers also expected their officers to behave in a fitting, gentlemanly manner when out of action. Genteel disgust at the loutish behaviour of some 'temporary gentleman' was shared by some rankers. An interesting insight into this is given by Pte Eric Linklater (4/5 Black Watch). One evening, Linklater was sitting in an estaminet with some sergeants when the peace was disturbed by a drunken, argumentative and visibly sexually aroused temporary officer chasing the hostess. The sergeants, working-class slum-dwellers in civilian life, were 'incensed by such behaviour in an officer of their regiment'.[20] Officers did not have to make an exhibition of themselves to be condemned as ungentlemanly by their men. Passages in the diary of the officers' mess sergeant of a TF unit, 1/5 Buffs, indicate that he respected the original officers of the battalion, who were gentlemanly and paternal, but he disliked their replacements who lacked these qualities. The sergeant was greatly aggrieved when his pay was reduced because the six surviving officers judged that he had less work to do: 'A gentleman's thanks', he commented sarcastically, 'for what you have done for them.'[21] This sergeant was reacting to his hierarchical superiors' failure to keep their side of the deferential bargain.

While failures of paternalism and leadership might be ascribed, by charitably-minded soldiers like Coppard, to the frailties of human nature, deliberate unpleasantness on the part of officers was deeply resented. Pte A.J. Abraham came across two officers who were regarded as petty tyrants. One, at a training unit, was nicknamed the 'Black Bastard'. He was 'a mean type and we hated his guts'. The other, Abraham's platoon officer in 8/Queen's, made a decision which long rankled with Abraham, when he refused to allow the men to wear greatcoats or groundsheets in heavy rain. This failure to improve the conditions of the men was just one of many reasons why Abraham had a low opinion of this officer. However, Abraham had a very different attitude towards others: 'Some of our officers were born leaders, men we instinctively trusted and respected.'[22]

It is rare indeed to find a blanket condemnation of officers in soldiers' memoirs, diaries or letters. A furious denunciation of one officer is likely to be followed by a complimentary reference to another. Pte Frank Dunham of 1/7 Londons was scathing about one officer, nicknamed 'Nellie', but wrote about Capt. K.O. Peppiatt in glowing terms. Peppiatt was 'a sport', a 'fine soldier…, [who] was not afraid to take his share in any of the risky jobs'.[23] In fact, it is uncommon to discover an officer who was actively hated by his men, as opposed to one who was criticised for neglecting his men or for thoughtlessness. One such was a Northamptonshire Yeomanry officer, known as 'the Bloody Bastard', described by one ranker as 'the most detested and hated officer I ever met in two world wars'.[24] The interesting point is not that this officer was despised, but that he suffered by comparison with the officer whom he had replaced, who had been popular with the men. Because most officers were paternal and lived up their side of the unspoken deferential bargain, officers who did not conform to the general pattern of officer–man relations were regarded with especial distaste by rankers.

Favourable references to officers can often be found in the writings of Other Ranks, although not as frequently as complementary references to men occur in officers' letters and diaries. In part this was a reflection of the differing perceptions of the relationship. It was also a product of the generally healthy state of officer–man relations. Only if an officer was exceptionally good, or exceptionally bad, or if a particular officer suddenly came to mind, if he was killed or wounded for instance, was he likely to be mentioned in the letters or diaries of an Other Rank. To take one instance, the first fatal casualty mentioned by name in the diary of L/Cpl Joe Griffiths (1/KRRC) was 2/Lt Bentall, 'who was only 18 a real good sort & was liked & respected by his men'.[25] His sense of loss prompted Griffiths to record his appreciation of this officer which otherwise would have been unknown.

Officers' privileges were resented by some, mostly middle-class, rankers. One was a private of the London Scottish who objected to the greater opportunities for leave available to officers.[26] His complaints were echoed three years later by a conscript Pay Corps private.[27] The artist Stanley Spencer, who served as a ranker with 7/Royal Berks, slipped an oblique comment into his painting *The Resurrection of the Soldiers*. In among scenes of dead soldiers rising from their graves and shaking hands with their mates is a glum-looking officer – identified by his brown boots – cleaning his own kit.[28]

These criticisms were fairly exceptional. Pte Coppard had no doubt about the reason why most soldiers accepted the disparity in privileges without complaint: 'the Tommy accepted it as the natural order of things', although they might joke about the differences, for example by referring to 'Old Orkney' whisky as 'Officers Only'.[29] Provided that an officer behaved in a certain way, his privileges were not resented by the ordinary working-class soldier. If an officer behaved in an 'unofficerlike' way, by acting unfairly, neglecting his men or acting in a cowardly manner, in his men's eyes he forfeited his rights to his lifestyle.

This point is illustrated by an incident that occured on a troop ship *en route* to the Dardanelles in August 1915. On two days officers were allowed ashore while the men were kept on board ship. Several revealing remarks about this appear in Pte G. Brown's diary. First, he commented that the officers 'didn't play the game with us'. Secondly, while admitting that to send a large number of men on shore leave presented difficulties, he argued 'the OCs should have been sports and tried some arrangement'. The use of public school sporting imagery reinforced the sense of unfairness experienced by these rankers. Whether in the trenches or on board a troopship, ordinary soldiers accepted that the officer might retire to a well-appointed dugout or cabin, but only after he had ensured that his men were fed and made as comfortable as possible. In this case the officers had neglected their paternal duty and officer–man relations suffered as a result: '[There was] Bad feeling about the business and officers were booed leaving.'[30]

In 1916, an upper-class gentleman ranker wrote of a temporary officer who had joined a New Army battalion at the beginning of the war, knowing as little about military life as the men he commanded. Gradually he trained as a soldier alongside his men. Little by little he learned the character of each individual soldier of his platoon. By his kindly and tactful handling of the men, he won their confidence, affection and love. The troops grew to feel that they belonged to him, and he belonged to them. His smile 'was something worth living for, and worth working for', while 'his look of displeasure and disappointment was a thing that we would do anything to avoid'. In the trenches, the men worried for his safety, and they mourned him when he was killed. In the final paragraph, the 'Beloved Captain' appears alongside Christ in heaven.[31]

The author, Donald Hankey, despite his upper-class origins, served in the ranks of 7/Rifle Brigade for a year in 1914–15. Later, as an officer in 1/Royal Warwicks, Hankey does seem to have been brave and paternal.[32] His idealised portrait of 'The Beloved Captain', which first appeared in the *Spectator*, reflects, in exaggerated form, the feelings of many rankers towards good officers. It would be ludicrous to claim that all rankers regarded all officers in this way, but some soldiers, working-class and middle-class alike, certainly had a very high opinion of some of their officers. Some younger soldiers hero-worshipped their officers, just as other youths idolised sportsmen or popular masters at school.[33] More mature men respected officers for their courage and their demeanour. Ernest Shephard, a prewar Regular NCO of 1/Dorsets, described Capt. Algeo as 'a real example of the Regular 'Officer and Gentleman'.... Absolutely fearless and [whose] first and last thought [is] for the men'.[34] A private of 1/15 Londons wrote that his company commander

> held the devotion and respect of all who served him.... His officers and men were his family. He knew their foibles and most of their hopes and fears. They executed his orders explicitly and confidently.[35]

Pte Giles Eyre (2/KRRC) also wrote of men defending the honour of their officer against a rival platoon:'There ain't no one in the Batt. like

Mr. Walker, and you can swank as much as yer likes. We know's 'im and wouldn't swap 'im for nuffink.'[36]

Just as the Beloved Captain's platoon throve on his smile, it does seem that small acts of kindness and friendship on the part of officers had a disproportionate effect on rankers' morale. In a letter of July 1915 a lance-corporal of 7/Norfolks, who, interestingly, was of middle-rather than working-class background, and an artist in civilian life, mentioned that he had attended an early-morning Communion service. His former platoon commander, a fellow scoutmaster, 'came up and spoke to me afterwards, which was very decent of him'.[37]

Rather more practically, in mid-1915 an officer of 2/Rifle Bde told his men who had been selected for a working party that it was unfair for them to be called upon 'to do fatigues while we were at rest, and told the men not to work too hard'.[38] There are two points of particular interest about this incident. First, it appears in the unpublished memoirs of J.W. Riddell, who was not a sensitive middle-class artist but a hardbitten prewar Regular NCO. Second, the officer's advice was well-intentioned, but if the troops had taken it, they would have been condemned to a longer spell in the trenches. The fact that Riddell bothered to record the incident in his postwar memoirs, which were extremely critical of military authority, indicates that he appreciated the officer's kindness and concern for his men, and his desire to protect them against the unfair demands of the military system. It also illustrates the gulf in perceptions between the commissioned and non-commissioned ranks.

How common a figure was the 'Beloved Captain'? A partial answer occurs in an interesting analysis of the officer–man relationship which appeared in 1938. Its author was an anonymous former ranker. This article drew attention to the ambiguities in the officer–man relationship. When he tried to recall his officers, he wrote, a trick of memory produced a composite figure:

> boyish and middle-aged, cool and reckless, grave and humourous, aloof and intimate; a martinet lapsing into an indulgent father; a thwarter becoming an aider and abetter; an enemy melting into a friend.

This ex-ranker's analysis of the attributes of the good officer, interestingly enough, had many points in common with the 'official' view of military leadership discussed in an earlier chapter. He regarded the officers' battlefield role as important: '[we] despised some for their deficiencies on parade, while admiring their impeturbability under fire'. However, other attributes of the 'good' officer were perhaps less likely to be approved by the powers-that-be: 'no officer was good who had not learned when to be deaf, dumb, and blind – and when not to be'. Most officers, the writer asserted, acquired these skills on active service. They also learned to question both Rudyard Kipling's opinions of the private's 'psychology and character', which were, after all, some forty years out of date by the 1914–18 war, and also textbook views on 'the behaviour of men in the mass'. In the field, officers learned man-management, and their effectiveness in this sphere greatly influenced their men's opinion of them. The ideal officer, in the writer's view, would have been a man of all-round talent. However, a paternal officer who genuinely cared about the welfare of the troops under his command would be forgiven many sins of omission and commission by the ordinary soldier. One of the writer's officers was renowned for his ineptitude on the drill square 'yet this officer was the best in the battalion for the care of his men in the trenches'.

'Looking back', this writer argued,

> with a better appreciation of their difficulties than we then had, at the officers under whom we served, we can have nothing but admiration for almost all of them – admiration with a tinge of affection.

Officers who fell short of the ideal in some way, 'we can afford to forgive':

> We do not need to be reminded that if in civil affairs we could get as square a deal and as much consideration from our superiors as we got from officers when we were in the Army, the world would be a pleasanter place to live in than some of us are finding it.

Thus the writer was suggesting that most regimental officers were effective man-managers who possessed, in some measure, the attributes and attitudes of the 'Beloved Captain'. This view lacks the sentimentality of Hankey's idealised portrait, depicting instead officers as fallible human beings. However, like Hankey's article, it expresses the rankers' admiration of brave and paternal officers, and recognises the officer's role making life bearable for the soldier. The impact of the officer on the morale of the private perhaps only became apparent in retrospect. Back in civilian life, former soldiers who were now unemployed, or who worked in dangerous or unrewarding jobs, had no paternal subalterns to look after their interests.[39]

Traditionally, Regular officers believed that working-class soldiers preferred to be commanded by gentlemen rather than by officers of humble origin who had been promoted through the ranks.[40] How, then, did ordinary soldiers regard the large numbers of officers of lower-class origin commissioned during the Great War?

The traditional view of ranker officers was slow to disappear. Pte John Tucker (1/13 Londons) recalled the lower ranks of this class corps disdaining a subaltern because he was a former bank clerk and spoke 'with a slight cockney accent'. Interestingly, Tucker, who recognised in retrospect that this snobbish prejuduce was ludicrous, was himself a city clerk before the war.[41] A.M. Burrage, a middle-class journalist turned embittered private soldier, wrote scathingly of some officers he encountered who

> judging by the[ir] manners and accents ... were nearly all 'Smiffs', late of Little Buggington Grammar school, who had been 'clurks' in civil life ...[42]

In 1917 Pte R. Cude (7/Buffs) commented that some newly-arrived officers were only commissioned because of the manpower shortage: 'Pon my word, if this is the best that England can do, it is time she packed [up].' However Cude, who seems to have been an artisan in civilian life, also described his platoon officer as 'a thorough Gentleman'.[43] He made this comment in September 1915, before his unit had taken heavy casualties and replacements for the original public school subalterns arrived.

Some commentators attempted to rationalise the dislike of Other Ranks for lower-class officers. G.W. Grossmith's evidence supports the traditional view of ranker-officers. He believed that rankers preferred officers to be recognisable as such by their speech and behaviour, and once heard a ranker comment that his new platoon commander was 'only one of us'.[44] Grossmith served in the ranks of 7/Bedfords and was later commissioned into a Regular battalion, 2/Leicesters. Such views may have been typical of Regular units, for a Regular RSM of 1/HLI believed that humbly-born temporary offices, not being 'born and bred' to leadership, did not command the same loyalty given by the men to public school-educated officers.[45]

Others offered more specific reasons for the common dislike of lower-class officers. A temporary officer of 1/6 RWK believed that ranker-officers were unpopular with the men because 'they knew their job' and were aware of the various tricks and dodges employed by the ranks; in other words, they were poachers turned gamekeepers, and as gamekeepers they were rather too effective for the men's liking.[46] A working-class private of 23/R. Fusiliers thought that former NCOs found it necessary to assert themselves with officious behaviour.[47] Burrage held a similar view:

> Quite the worst type of officer was the promoted sergeant-major.... Whatever rank they achieved they were still warrant-officers in spirit. They could never be anything else.[48]

An ex-Regular NCO who served as an officer in 2/Camerons seemed to fit this pattern. According to a fellow officer, writing in 1916, 'like most rankers, but not all, [he is] not too well liked by the men. He is apt to be fussy and bullying in matters of detail'.[49] This opinion is of interest not least because the writer was himself a ranker-officer, although having served in the ranks of the London Rifle Brigade, a Territorial class corps, he clearly regarded himself as being in a very different category from a former Regular NCO.

It is not surprising that attitudes such as these should be so widespread, given the degree of class consciousness in British society and the assumptions underlying the deferential/paternal relationship. A study of the Leeds Rifles (1/7 and 2/7 W. Yorks) concluded that the men of these Territorial units insisted on gentlemanly officers, and would not accept officers who were not gentlemen, although this may not have been an attitude which was typical of the Territorial Force in its entirety.[50]

Other wartime soldiers thought differently. J. Gibbons, who served in the ranks of a London TF unit, believed that working-class replacements for public school officers were just as effective as their socially elite predecesssors.[51] M.L. Walkington, a grammar-school boy who served as a ranker in a TF class corps (16/Londons) before being commissioned, believed that competent but poorly educated NCOs who received commissions generally made valuable officers. The prospect of officer status gave 'great encouragement to young NCOs who developed ambition'.[52] The usual practice was for newly commissioned officers to be posted to units other than the one in which they had served as rankers, but some cohesive 'family' Territorial and New Army units preferred to take back their 'old boys'.[53] This practice can also be found in some Regular units throughout the war. CSM Sayers of 4/Middlesex was commissioned in the field in October 1914 and served with the battalion until his death in 1915, while Sgt Fenner (3/Rifle Bde) was commissioned in his battalion in 1917.[54]

Commanders of units such as these presumably considered that the discipline and cohesion of their battalion or battery was strong enough to overcome any problems that might have resulted from allowing ranker officers back into their original unit, although often such men were posted to different companies. One such officer, G.H. Cole, commented that he had no problems adjusting to officer status because he 'grew up' as a ranker in his battalion, 1/20 Londons. Cole also saw the matter from the ranker's perspective. As a private, his company commander was a man who had been in his form at school. 'In public, of course,' Cole wrote, 'No-one would have known that we had ever met.'[55] Although there was some prejudice against ranker-officers among Other Ranks, it is rare indeed to find criticism of a specific officer whom a ranker had known in his previous incarnation as an ordinary soldier.

Even outside 'family' units, soldiers meeting friends who were now commissioned officers seem to have observed the spirit, if not the letter, of discipline. Other Ranks sometimes talked informally with officer friends but rarely took advantage of this relationship.[56] The British army could have followed the Australian practice and allowed more ranker-officers to return to their old units. Generally speaking, the self-discipline of Other Ranks was strong enough to ensure that military efficiency did not suffer from the commissioning of officers within a unit. It may even have enhanced it, rather as Walkington suggested, by encouraging rankers to strive for excellence, in the knowledge that they would not have to be posted away from their battalion on becoming an officer.

By the end of the war the officers of the British army were drawn from a wider social spectrum than ever before. It is possibly significant that Tucker's comments quoted above refer to 1915, a time when lower-class officers were somewhat rarer than was to be the case later in the war, for if mistrust of working-class and lower-middle-class officers had been as widespread as some have claimed, officer–man relations should have been poor throughout the army by 1918. Indeed, following this argument through to its logical conclusion, the British army should have disintegrated in 1917–18 because Other Ranks would have refused to follow the lower-class officers commissioned to in place of the 'gentlemanly' officers who had been killed.

Of course, this did not happen: officer–man relations remained generally cordial throughout the war.

Ultimately, an officer's relations with his men were determined not by his social class, or by his previous service in the ranks, but by his competence, leadership skills, paternalism and courage. It is true that some former Regular NCOs did not find the transition to commissioned rank easy, and that some lower-class officers had some difficulty in establishing their credibility with their soldiers. However, it should not be forgotten that officer training was remarkably effective in educating ranker-officers in the ethos and methods of the Regular officer class, and that from early 1916 onwards most commissions had to be earned on the battlefield. A newly-commissioned officer had to give practical demonstrations of his paternalism and leadership qualities in the trenches and on the battlefield, and this compensated for any lack of social standing, whatever misgivings private soldiers might originally have had about the social origins of an officer. Confirmation of this theory comes from a surprising source. That scourge of the temporary gentleman, Pte A.M. Burrage, concluded that officers

> who came from shops and offices, with little education and less tradition, did their job somehow and did it well. I hated being jiggered about (we used a slightly different phrase) by people that I considered my inferiors ... but I who was a private, and a bad one at that, freely own that it was the British subaltern who won the war.[57]

There is a very useful phrase of Great War vintage: 'On parade, on parade; off parade, off parade', meaning 'what was permissible on certain occasions might be a military crime on others'.[58] This phrase aptly describes the relations of many officers with their men; 'regimental' on some occasions, informal on others. In the trenches, relations between officers and men were generally characterised by a greater degree of informality than was the case behind the lines. Officers and men quietly dispensed with much of the pomp and ceremony. In one extreme example, an 18th Division private reported (in a scandalised manner) that officers of a 32nd Division unit 'were known to the men by their Christian name'.[59] More commonly, some officers used soldiers' nicknames.[60] Such informality was not always appreciated by higher military authorities, the lack of 'regimental' soldiering in XI Corps in 1916 leading, in the view of Corps staff, to a dangerous slackening of discipline.[61]

In the trenches, it would often be difficult for the casual observer to tell officers and men apart. A newly-commissioned ranker-officer was helped to play the part of a gentleman by his uniform, which was 'the khaki equivalent of hunting dress', very different from the 'shabby garb of the artisan' worn by the private.[62] However, in some units, officers carried rifles and packs and wore privates' uniforms, the rank badges on the sleeve replaced by unobtrusive pips on the shoulder.[63] While this adoption of rankers' dress as a protection against snipers was not universally popular, some officers arguing that it was wrong that men could not easily recognise their officers,[64] it aptly symbolised the decrease in formality in inter-rank relations that generally occurred in the line.

Coppard somewhat cynically referred to the decreased gap between officers and men in the trenches as 'a temporary attempt at chumminess'.[65] In some units it might be the case that only in the trenches were junior officers, out of sight of their superiors, able to establish informal relations with their men. However, in other units officer–man relations achieved a degree of informality out of the trenches. A sergeant of 2/6 Lancashire Fusiliers recalled that in June 1917 D company was 'one great happy family. After parades discipline was relaxed and we were at liberty to spend most of our time in our own way.' There was a 'close bond' between officers and men, a 'very dear thing in the throes of war'.[66]

Coppard also was not unsympathetic to officers. He commented on the weight of responsibilty that they bore for their men's lives. One mistake could kill the men of their platoon: 'The nervousness, strain and irritability of his officers could be responsible for a lot of what Tommy had to put up with'.[67] Similarly, the stress of waiting to go into battle caused one artillery officer to verbally abuse the officers' servants.[68] Coppard also made an important point about the way in which one of the artificial barriers of rank was reduced on active service. He believed that he became less scared of officers as time went on, not because officers became 'any more friendly, but because we youngsters were growing up'.[69] In action, officers could not hide behind their status and rank. They had to prove themselves as leaders, and inevitably some made mistakes and demonstrated that they were far from omnipotent. A private of 32nd Field Ambulance saw this process in operation on 7 August 1915, at Suvla Bay:

> You could see the spreading dismay as the ordinary Tommies recognised their own fear and hesitation in the eyes of these one-pip

> striplings [second lieutenants]. Men under fire ... watch each other with nerves on edge. 'Blimey! even the bloody officers are lost!...'[70]

Such comments suggest that Capt. T.M. Sibley was to some extent correct when he wrote in June 1916 that the gulf between officers and men was 'a very important part of the British Army system' and soldiers would lose their respect for some officers if they came to know them. This remark gives a salutary reminder of the difficulties of generalising about inter-rank relations in an organisation as big as the British army.[71]

In the words of a subaltern of 2/KOYLI, 'the horizon of the Infantryman in the Great War was small, but his philosphy was straightfoward'; the war had to be fought, and if mail, food and cigarettes were available, the war was going well.[72] One private was not untypical in regarding himself as belonging first to his platoon, then to his company, and then to his battalion.[73] For the most part, higher formations meant little to the private, although some divisions such as 18th (Eastern), 51st (Highland) and 56th (London), did acquire a measure of divisional *esprit de corps*. Junior officers and rankers alike shared this narrowness of vision.[74] In this tiny, insular world, it is not surprising that men turned in on each other for affection, or that minor acts of benevolence were greatly appreciated. Many officers regarded it as part of their duties to write letters of condolence to the families of soldiers who had been killed or wounded while serving under their command. While this could be interpreted as just another aspect of military paternalism, there are also many examples of NCOs and privates writing to the families of their officers. It was not uncommon for soldiers on leave to visit the families of their officers, or officers the families of soldiers.[75] Indeed, one historian, citing the correspondence of a ranker with the widow of his officer, has suggested that 'mourning for the same man created a strong bond' between disparate individuals.[76]

Apart from demonstrating the affection and comradeship felt by men for their officers, and vice versa, such letters also helped to relieve one of the principal factors that undermined the morale of fighting men: worry about their families. The soldier could face death knowing that their loved ones would receive some comfort, however small. Letters of sympathy from ordinary rankers were perhaps especially comforting to officers' families, because they gave evidence of the effectiveness of their military leadership, of a duty performed unto death, of a sacrifice nobly given. In many cases, it took a real effort for ill-educated privates to write a formal letter of this sort. This obviously did not apply to Pte S. Brashier of 22/R. Fusiliers, who wrote to the family of the late Capt. G.D.A. Black:

> To us he was life itself, and the confidence we placed in him was great. Really we used to say – 'He knew no fear' and so though we greatly miss him we realise what a sorrow and grief it (*sic*) has come to you, and so our thoughts go out to you in your great sorrow.

This letter was copied and circulated among Black's family. It obviously did provide some comfort, since it has been treasured in the family down to the present day.[77]

Some point of mutual interest, such as common regional loyalties or language, helped to break down barriers between the ranks. Edmund Blunden actually found it easier to get on with his soldiers, fellow Sussexmen, than with some of his brother officers.[78] Welsh-speaking officers and men of 15/RWF talked freely together; English was regarded as 'the language of the Army, Welsh the language of friendship and companionship' and the use of Welsh formed 'a bond of unity, that sense of being an enclave within a community'.[79] Likewise, in Scots units, enthusiasm for bagpipes, 'which were played by Scottish gentlemen', 'reinforced the bond' between rankers and officers'.[80]

Some close relationships develop between officers and men when a soldier emerged from the khaki mass. NCOs would sometimes find themselves alone with officers, and mutual respect could blossom into greater intimacy. This happened in, of all units, the South Persia Rifles, where a middle-class officer (formerly a ranker in the 7/R. Dublin Fusiliers) was thrown into the company of a British sergeant.[81] Similarly, Anthony Eden (21/KRRC) wrote movingly of nights spent on watch in Plugstreet Wood, when he would hold long discussions with a platoon sergeant, Norman Carmichael, whom Eden counted as a friend.[82] (See below for a further discussion of officer–NCO relations.) The authors Stephen Graham and Wilfred Ewart served together in the Scots Guards, and struck up a friendship, even though the former was a private and the latter a captain. Ewart's fellow officers apparently disapproved of the relationship.[83]

Soldier-servants and officers could become friendly within the bounds imposed by rank and class. A public school officer of 1/N. Staffords summed up his relationship with Tidmarsh, his working-class 'old-soldier' batman, in these words:

> We were not exactly friends because of the differences of social class, but, accepting these differences, we were not separated by them. Each regarded the other as a personality to be respected.[84]

Soldier-servants had a unique opportunity to get to know their officers, 'warts and all'. One prewar soldier-servant's duties commenced each morning at 6 a.m., when he had to take a glass of whisky to his officer's bedroom, followed by two boiled eggs and more whisky.[85] Soldiers had good reason to to be friendly to their officers. As a private of Worcestershire Yeomanry pointed out, being an officer's servant 'is much better than being in the troops' since he received many luxuries and was excused night guards.[86] It is certainly true that many servants had a privileged position. These privileges might take the form of physical comfort – a company commander of 2/21 Londons shared a tent and pooled rations with his servant while on campaign in Palestine[87] – but there were also more subtle benefits to being an officers' servant. This is indicated by the obvious delight of a mess cook who summoned other servants to watch the spectacle of a newly-arrived subaltern making a fool of himself.[88] This incident, in which the officer had to be disentangled from coils of wire by the grinning cook, also indicates that soldier-servants were allowed a certain amount of licence, an aspect of the relationship which is beautifully captured by some of the comic scenes in Sherriff's play *Journey's End*.[89]

An unfriendly or surly servant, let alone an incompetent one, ran the risk of being returned to normal duty and forfeiting his privileged existence, so it was in his own interest to be pleasant. However, some genuine friendships developed between officers and servants. An officer of 11/Cheshires 'witnessed a most touching farewell' between the battalion commander and his old servant: 'they embraced and both shed tears'.[90] Pte Harry Adams (6/Queens) developed a 'real attachment' to his officer, Mr Jefferies, and experienced 'great grief' when he heard of his death in 1918.[91] Capt. V.F. Eberle (48th Division RE) commented that 'the relationship between a good batman and his officer is often no mean criterion of the latter'.[92]

Other rankers who emerged from the anonymity of the ranks also enjoyed more than usually intimate relations with officers. Pte Clarkson, a runner for a company commander in 5/6 R. Scots, wrote that mutual respect was high and that he learned to trust his officer. In a common act of friendship, the officer would often give Clarkson extra rum on cold nights.[93] Another soldier with a semi-independent existence was Sgt Jones, 'of Jones's water dump', on Gallipoli, whom an officer compared to a friendly 'inn-keeper'. Officers and men alike would congregate in Jones's dugout to hear the latest rumours.[94] All of these examples indicate the type of informal, friendly relations which could develop between officers and men when circumstances allowed individuals to get to know each other as men.

However, it is clear that for the most part, circumstances did not allow rankers and officers to develop this sort of relationship. The restraining hand of the NCO was one of the factors why inter-rank relations did not often grow from friendliness into real intimacy. This is well illustrated by a scene in an autobiographical novel by a ranker-officer, where a newly-arrived subaltern briefed his men and then asked them if they had any questions. This was clearly regarded as unusual, and to 'continue the feeling of part-intimacy with the officer' a private took advantage of the invitation and actually asked a question. On receiving a polite and informative answer the private was emboldened to ask others. However, the private was well aware of the disapproval of his sergeant, who suspected insolence, although none was intended. The moral of this episode was even if the private and the subaltern were prepared to establish an informal relationship, the NCO, who in many ways had the greater influence over the life of the private, was capable of being less broad-minded.[95]

The relationship between the non-commissioned officer and the officer deserves special consideration. The NCO played a crucial role in the maintenance of discipline, and the administration and management of military units. During the Great War, as before and since, NCOs were the 'backbone' of the British army. They formed the crucial link between the officer and the ranker, passing orders down the chain of command and performing, as a contemporary commentator noted, the 'grave and all-important task of enforcing that prompt obedience to orders that is the life's blood of an army'.[96] As noted above, the NCO, rather than the officer, was often the figure of authority who had the greatest impact on the life of the ordinary soldier,[97] although it is fair to say officers and men came into contact more frequently on active service than in peacetime. NCOs varied greatly in status. They included the lance-corporal, 'one who has position, but no magnitude',[98] an appointment which was only one step up from a private and was often held in an acting and unpaid capacity. For our purposes it also included the senior non-commissioned rank in a unit, the regimental sergeant major (RSM). The RSM, technically a warrant officer (WO), was, in contrast to the unfortunate 'lance jack', a powerful and often respected figure. The British NCO of

1914–18 deserves a major study, but here we can reflect only on those aspects of their role that directly affected officer-man relations.

There were two basic species of NCO. First, there were the Regular NCOs encountered by all soldiers at training establishments throughout the war. Some of whom were the 'old soldier' types immortalised by C.E. Montague: men who preferred drinking to training, who were open to bribes, and who stole army property.[99] Many were older men, reservists who were medically unfit for active service. Second, there were NCOs appointed from the ranks of wartime volunteers and conscripts. In the earlier part of the war, some difficulty was experienced with such NCOs, as attempts to enforce the separation from the rank and file deemed necessary by the army were not always successful in New Army and Territorial units.[100]

NCOs had reached positions of responsibility because officers believed they were more intelligent than privates and had the ability to adminster, and indeed accept, discipline[101] although the degree of trust reposed in NCOs varied from unit to unit, depending on the personality of the commanding officer.[102] A former farm labourer, serving as a junior NCO in a trench mortar battery, summed up the relationship between NCOs and men in these words:

> it dose (*sic*) not do for us [the NCOs] to sleep with them [the men] for we are like Masters on a farm and the men under us you see how the thing works.[103]

Like civilian foremen, NCOs ensured smooth running of a unit, keeping a finger on the pulse of a complex organisation. Thus the word of an NCO was usually taken at face value, even if it conflicted with that of a private. Reinforcement of the NCO's authority was seen as being of greater importance than the strict administration of justice. NCOs were known to impose punishments which were illegal but nonetheless tacitly condoned by officers.[104] Thus it was vital that NCOs could be trusted by their officers. Capt. Hamond, in a typically forthright sentence, wrote that an NCO who was a liar or who manufactured evidence 'must be destroyed at once'.[105]

The NCOs' duties were not simply concerned with discipline. They had a vital role in training, both on active service and at home. One gunner commented that he did not come into contact with a single officer during his training in England, for NCOs carried out all the work.[106] C.S. Lewis, an officer of 3/SLI, a Special Reserve unit based in England, wrote in October 1917 that all the training was carried out by NCOs; 'All you do is to lead your party onto parade, hand them over to their instructor, and then walk about doing nothing at all.'[107] On active service, a whole host of other duties came the way of the NCO, including ensuring a fair division of food when in the line, and also responsibility for kit, arms and equipment.[108] Less formal duties included protecting soldiers against higher authority, and inculcating regimental traditions.[109] On the battlefield, NCOs had to lead men, to command platoons if the officers were killed or wounded, and promote and sustain morale.[110] It is not surprising that one ex-ranker wrote 'Platoon sergeants – what would the War have been without them? Why, they ran the thing! At least, that was the impression we received.'[111]

The NCO's role therefore overlapped with that of the officer. Although the military hierarchy imposed 'distance' between the private and the NCO, it was not as great as that between privates and officers. Junior NCOs, for instance, often shared many of the living conditions of Other Ranks, and some NCOs operated the principle of 'on parade, on parade; off parade, off parade' with their men.[112] NCOs could also do things that an officer officially could not, such as physically lay hands upon the men. Most importantly of all, since NCOs were usually appointed from within the unit, they were in a position to gain more detailed knowledge of the men than even the most paternal and informal officer could ever hope to obtain. In 1917, RQMS Young of 2/17 Londons reflected on his methods of command: 'By a word, I can hold them in check, when they get unruly, because I know them and their East End spirit.'[113]

Given the wide range of types and functions of NCOs, it is difficult to generalise about the state of relations between the NCOs and privates. Driver R.L. Venables, for instance, served under two very different battery sergeant-majors. The BSM in his battery of 31st Division artillery was a foul-mouthed 'nasty piece of work' from 'the metropolis gutter', while the BSM in his previous battery, in 32nd Division, was 'first-class', never using foul language on parade. Morover, Venables believed that the discipline in the 32nd Division unit was superior.[114] Broadly speaking, the relationship between privates and NCOs within the unit was more often than not characterised by respect.

It can be seen that many of the NCO's duties, responsibilities and experiences were paralleled by those of officers:

> ... I am learning how to mix discipline and persuasion.... I have

> got to know the roughs in our platoon pretty well.... You never get to the stage of really trusting them, but you can establish working relationships with them by expedients which seem almost childish, silly jokes and a kind of assumed (for me) music-hall, pub-loafing heartiness. It's acting, of course, but I come to feel more and more that all leadership is in a way acting, conscious or unconscious.[115]

This passage could easily have been written by a subaltern but it was in fact penned by a Wykehamist NCO of 6/DCLI. The experiences of Sgt. C.F. Jones of 2/15 Londons also have many points of comparison with those of officers. In his time he defended a new draft of boys, 'as a lioness its whelps' against what he perceived to be the unfair demands of a higher authority, in this case, the orderly sergeant. The good NCO, Jones believed, could play a vital role in 'getting the best out of his men' by seeing that rations were fairly distributed.[116] Clearly, the good NCO, like the good officer, was a paternalist. According to an officer of 1/RWF, the acid test of 'good' and 'useless' NCOs was their behaviour during a 'working party in the rain'. The useless NCO would take shelter. The good NCO would help the men with their tasks.[117]

In practice, NCOs could become the junior partners of regimental officers in running a platoon, company or battalion. Frederic Manning's fictional RSM concisely expressed the importance of the relationship between the officer and the NCO:

> [W]hen you're an officer you won't know your men. You'll be lucky if you know your NCOs, and you'll have to leave a lot of it to them. You'll have to keep them up to the mark; but you'll have to trust them, and let them know it.[118]

The fact that in the latter part of the war many officers had served as NCOs undoubtedly aided the building of good working relationships. Sgt R.H. Tawney (22/Manchesters), writing of the moments just before going into action on the Somme in 1916, noted that his platoon officer 'had enough sense not to come fussing round'; sense gained, it is implied, as a result of his previous experience as an NCO.[119]

Wyn Griffith, a company commander in 15/RWF, left a pen-portrait of his relationship with his company sergeant-major. Just as a company commander would often hold an informal 'board meeting' with his subalterns,[120] Griffith and his CSM would relax together over a glass of whisky and a pipe in the company officers' mess and gossip about the men of the company. Griffith made two revealing remarks about this relationship. First, 'Our life thrust us close together; his [the CSM's] position was in its way as solitary as my own.' Both had responsibility for their men. Both needed to strike a delicate balance between being part of the company 'team' and being slightly aloof from it. Second, the gossip allowed Griffith to find out incidents in the life of the company 'unknown to the least unapproachable of company commanders, unguessed at in spite of the close contact of life in the trenches'. For example, 'Had I heard what Delivett said when a pip-squeak blew some mud in his mess tin...?' In short, the CSM provided an important link between the private and the company commander. In this case, and many others, the NCO and officer worked together as a harmonious team.[121] Similar relationships could exist between other grades of NCO and officer, but in all cases, they had to be founded upon mutual goodwill and carefully nurtured.

It is instructive to compare Griffith's relationship with his CSM with the comments of Sgt S.F. Hatton (Middlesex Yeomanry) concerning an officer who tried to

> court popularity by being over-friendly with the sergeants, and coming into the sergeants' mess to stand drinks.... In fact, you have to be just the right type of officer to ever receive an invitation into the sergeants' mess, to be able to drink with them, and preserve their loyalty and your own dignity.... [A] sergeant no more wants a young and inexperienced officer in the mess than a man really wants a woman in a public-house.

This passage neatly encapsulates the problem that faced officers who wished to demonstrate respect and friendship for their NCOs. Hatton's subaltern breached some of the important ground rules, recognised by officers and NCOs alike, as essential for the maintenance of discipline. To buy drinks for NCOs could be interpreted as an attempt to buy loyalty. In addition, the good officer understood that the NCOs were entitled to privacy in their mess, their home. No matter how friendly an officer might be, it was impossible for a subordinate to be completely relaxed in a superior's company. While an experienced officer would know enough not to abuse the privilege of admission to the sergeants' mess, to talk to the sergeants in an appropriate way and to make a tactful withdrawal, this inexperienced officer clearly out-stayed his welcome on a number of occasions. The fact that the Middlesex Yeomanry contained a large number of middle-class men, and enjoyed very informal relations with their officers, makes Hatton's insistence on the rights of the NCO all the more striking.[122]

More generally, it may be suggested that for the most part privates and NCOs did not want their officers to be too friendly, but rather preferred them to maintain a certain social distance, to avoid role-ambiguity. Even before a man left his unit to go for officer training, a subtle change came over his relations with his comrades, impending promotion 'already dividing him from them'.[123] It is in fact very rare to come across an officer misguided enough to endanger his authority by becoming over-familiar with his soldiers. One suspects that service in the ranks and training at an OCB gave most subalterns a firm grasp of the correct way to treat their men.

Many officers relied heavily on their NCOs. This was especially true of young subalterns, fresh out from England, with no previous war experience. The steady, experienced NCO supporting the 'green' subaltern with a whispered word of advice is almost a commonplace. In late 1915 one gunner officer wrote that subalterns fresh from Woolwich 'know very little about the interior economy of their batteries. They step into the machine and glide along with a first class B.S.M. and Q.M.S. behind them.'[124] For the sake of discipline, it was important that the position of the NCO should not be undermined. In 1915, Sgt T. Boyce (1/10 Londons) was rudely treated by his CO in front of his men; this incident still rankled with Boyce fifty years later.[125] In fact most senior officers were well aware of the importance of the NCO in the smooth running of a unit. If a subaltern was to undermine the position of an NCO, for instance by swearing at him or rebuking him in front of his men, a senior officer was likely to take the part of the NCO.[126] Experienced NCOs were invaluable, while subalterns were all too easy to replace. The chastening experience of one young Territorial gunner officer underlines the relative importance of the newly arrived subaltern and the battle-hardened NCO. On one occasion in 1917, a working party was unloading wagons under shellfire. Lt P.J. Campbell called to Sgt Denmark to come over to him. Denmark flatly disobeyed, demanding, with a 'face of thunder', 'Who's taking charge here, are you Sir, or am I?' Campbell was left feeling humiliated and crushed. Denmark's appreciation of the situation was correct, as Campbell apparently recognised in retrospect; the NCO was carrying out a dangerous task that needed to be completed as swiftly as possible without interruption. Campbell did not even contemplate making a disciplinary issue of Denmark's insubordination, fearing that even to confide in a fellow officer would only result in Campbell looking even more foolish. Instead, Campbell worked to try to win his sergeant's respect.[127]

A case could be made that the NCO corps was damaged by the wholesale commissioning of corporals and sergeants who showed leadership ability. One temporary officer believed that the commissioning of warrant officers was a mistake, because an RSM enjoyed much greater prestige than a mere subaltern.[128] A number of Regular NCOs had poor opinions of their New Army and Territorial counterparts.[129] One officer's belief that most NCOs were ineffective under shellfire and the exceptions 'ought to be officers', while no doubt a broad generalisation, is indicative of the general belief that the place for those soldiers with leadership qualities was as officers, not as sergeants.[130] Many NCOs were held in high regard by their officers, as men, as leaders, and as partners in the administration and management of military units. After a battle in 1917, Lt R.L. Mackay of 11/Argylls described Sgt McQuarrie as:

> one of the bravest and best gentlemen I have ever met. He has been utterly invaluable to me on this job.... I have more respect for this man than for any other dozen I have ever met.[131]

The language used by Mackay is revealing. McQuarrie was a courageous 'gentleman', perhaps not by birth, but certainly by behaviour, who had earned Mackay's respect. In short, he fulfilled most of the criteria demanded by Other Ranks of their officers. McQuarrie had, one might say, 'leadership qualities'. Similarly, the picture that emerges from the diaries of CSM Ernest Shephard, a prewar Regular soldier of 1/Dorsets, is of a man who 'nursed' inexperienced officers. who acted as a rock of stability and continuity after the battalion had taken heavy casualties, and who admired, and had good relations with, various officers.[132] One of the major factors in maintaining the cohesion of the British army through the long years of attrition was the presence of Regular, Territorial and New Army NCOs like Shephard, Denmark and McQuarrie.

During the war years, there was much talk among civilians about the positive effects of war service on social cohesion.[133] In 1916 the Bishop of London spoke of a 'brotherhood' being 'forged of blood and iron' in the trenches, which should be maintained into peacetime, thus ending the class war between 'Hoxton' and 'Belgravia'.[134] Subsequently historians have pointed to the growth of solidarity among front-line soldiers of all nations as a reaction to the politicians, capitalists and shirking or striking workers on the home front, and as argued above, generals and staff officers.[135]

Is it, then possible to talk about the existence of *Graben-kameradeschaft*, a comradeship of the trenches, which united British front-line soldiers, regardless of rank, into a common fraternity? Many officers believed that it was. 'Through all their ordeals and sufferings' wrote one, 'they knew they had become a brotherhood of all ranks ...'.[136] The padre of 12/HLI argued, from personal experience, that men who had fought in battle had 'proved our manhood to ourselves and to one another', the 'bond' of a shared experience of battle being

> finer and more intimate than could be forged by any other association ... we shall for ever have in common a host of dearly-bought memories, sacred and incommunicable.[137]

I have argued elsewhere that war experience did make an impact on 'officer-class' perceptions of the working classes, a phenomeneon which had considerable repercussions for postwar British society and politics.[138] But how far, if at all, did Other Ranks regard themselves as sharing a common war experience with their officers, an experience which transcended rank?

At one level, men of whatever rank who had undergone the experience of battle shared an experience denied to everyone else.[139] The working-class private who wrote to Edmund Blunden after the war to say that *Undertones of War* had put his war experience into words was also testifying that, even if the officer and the private had nothing else in common, they shared the experience of battle.[140] Combat had the ability to dissolve the formal bounds of rank, at least temporarily. Capt. E.G.D. Living (2/19 Londons) wrote of returning from an action in Palestine. A ranker marched beside him

> and, officer and man, we opened our hearts to one another as everyone else in the stumbling fours in front of us was was doing, and as only those can who have been through terrible experiences together.[141]

Studies on other twentieth-century armies drawn from western industrialised societies suggest that the small cohesive group, offering mutual support and affection, is of vital importance in sustaining morale in war.[142] A private's view that the 'set of mucking-in pals' was 'the true social unit of the army' of the 1914–18 war would tend to reinforce this view.[143] Some very deep relationships were forged between soldiers during the First World War, especially on active service. The commonly held view that, in war, life and human relationships were especially vivid was held by a very ordinary private of 2/4 Londons, Jack Mudd, who wrote to his wife of the importance of comradeship in the trenches:

> Out here dear we're all pals what one hasn't got the other has we try to share each others troubles get each other out of danger you wouldnt believe the Humanity between men out here.... Its a lovely thing is friendship out here.[144]

There is much evidence from the writings of Great War soldiers that comradeship was indeed of vital importance in maintaining morale.[145] Conversely, men who were excluded from primary groups usually had a miserable time, and this was an important factor in the disillusionment of specific individuals.[146]

Primary groups could transcend social class, for although some middle-class rankers could be rather uncomfortable serving alongside working-class soldiers,[147] others happily 'mucked-in' with their proletarian comrades. An artist serving in the ranks of 8/Rifle Bde noted that:

> I have gained a knowledge of the 'workers' point of view, opinions & workings of his mind, that would be invaluable if I were going to do anything in the political or sociological line![148]

In his diary, a middle-class conscript infantryman referred to 'The splendid qualities of the men with whom one is associated'. Later he wrote:

> It is very educative to mix among these men, whose ideas and characters are as diverse – sometimes as grotesque – as the burrs or drawls of their speech.... They are all very nice to me....[149]

These quotations sit neatly alongside similar ones from socially privileged officers such as Alec Waugh (MGC), who wrote in the 1960s that

> for many young soldiers, certainly for me, there came a newly awakened social consciousness.... The young officer began to feel differently about the men he led in action.[150]

J.R.R Tolkien (11/Lancashire Fusiliers) described 'Sam Gamgee', a character in his novel *The Lord of the Rings*, as a portrait 'of the English soldier, of the private and batmen I knew in the 1914 war, and recognised as so far superior to myself'.[151]

Could this process be taken a stage further? Could officers, as well as middle-class Other Ranks, form a comradeship group with working-class soldiers? Rank and discipline placed considerable barriers in the way of uninhibited friendship between leaders and led, but some individuals came close to breaking these down. R.C. Foot, a temporary gunner officer of 62nd Division claimed that:

> Officers shared the same food and slept in the same ditches as their soldiers; about the only thing they [the Other Ranks] could not share was their responsibility, and the soldiers recognised this.

It is possible that Foot exaggerated the closeness of inter-rank relationships in his unit, but he certainly seems to have formed a bond of mutual friendship and trust with an NCO. Long after the war Foot was visited by the daughter of his old sergeant. This lady had a personal problem, and she had been told that she could refer to Foot in time of trouble, but as he wrote,

> that incident, some twenty five years later than her father's service and friendship with me, rather took my breath away at the time.

Foot went on to argue that 'such friendships, based on mutual individual respect' and the comradeship engendered by a male society made it possible to endure the horrors of war.[152]

A similar incident occurred in 1969, when W.M. Jenner, a former ranker, wrote to the family of Capt. Peter Blagrove, after seeing his old officer's obituary in a newspaper. Jenner wrote that 'To me he was a friend as well as a superior officer', and said that all of the men of his trench mortar battery were proud of Blagrove, who was regarded as 'a real gentleman and a very brave man'. One of the things which endeared him to Jenner was that, when short of labour, Blagrove helped the men with hard physical work. In the eyes of his subordinates, Blagrove displayed the traits of a 'Beloved Captain', being gentlemanly, courageous and paternal. Blagrove and Jenner last met in December 1918. For a ranker to treasure the memory of an officer for over fifty years is evidence that *Grabenkameradeschaft* existed in this particular case.[153]

Perhaps Maurice Bowra, who served as a temporary gunne tern, captured the essence of many relationships between officers and soldiers when he wrote that his dealings with his men 'were more formal but in the end hardly less intimate' than his relations with his brother officers. The men looked after one another, and Bowra, with 'protective care' and 'In moments of danger or excitement or even of frustrating tedium they would relax their restraints and tell me about their families and their jobs in time of peace.[154]

The fact that junior officers and rankers shared much the same dangers in battle was important. Charles Crutchley, who served in the ranks of 135th MG Coy in Mesopotamia, captured the way in which shared danger could forge men into a community, if only temporarily, regardless of rank:

> Thousands of rounds of empty ammunition cases were strewn around a deserted machine-gun emplacement. 'Nasty bit of goods' said our officer.... 'I wonder how many they got with that little lot'.
>
> The look on his face made me wonder if he were (*sic*) also thinking of our own 'nasty bit of goods' ... We squatted around: a mere handful of us, on a lonely ridge in the desert.... Dreamy said it was his twenty-first birthday, and my officer fished out a flask [of whisky] from his haversack.
>
> 'Pass it around, sergeant,' he said.... That drink, taken from the same flask, cemented our comradeship.[155]

Clearly at least *some* officers were regarded by *some* men as comrades; even if rank and the disciplinary structure prevented the uninhibited friendship possible between two privates. Without a host of case studies of individual units and individual soldiers, it is impossible assess how widespread was this sense of inter-rank comradeship. Nevertheless the resolutions passed by the 'Soldiers and Workers Council, Home Counties and Training Reserve Branch' held at Tunbridge Wells on 24 June 1917 offers an important clue. These resolutions were for the most part closely akin to trades union demands, calling for an increase in separation allowances, relaxation of the Defence of the Realm Act, and so on. However, two of the resolutions read:

> 5. That the general treatment of soldiers be brought into line with the spirit of the Officers and men in daily contact. As things stand,

> the Army Council continually issues orders which have the effect of reducing the organisation to a cross between a reformatory and a lunatic asylum. Only the goodwill and tolerance of the Officers and men make life endurable. We be neither dogs, criminals, or children.
> 6. We ask for a more generous treatment of younger Officers who, out of a daily casualty list of over 4,000, suffer the heaviest proportionate burden.[156]

This document gives a clear indication of the general state of officer–man relations, although it is fair to note that soldiers involved with this council were obviously atypical. Although far from revolutionary in its aims,[157] the very existence of this body represented a direct challenge to the formal hierarchical and disciplinary structure of the army. Yet Resolution 5 demonstrates that the council members drew a sharp distinction between senior officers, who were seen as inflicting a humiliating disciplinary system on the men, and regimental officers who were 'in daily contact' with the men and who did their best to modify the system. Resolution 6 not only offers evidence of the sympathy that existed for junior officers among some other ranks, but can also be interpreted as recognition that a community of interest existed between between soldiers and regimental officers, many of whom had risen from the ranks. It was in the interests of those striving for better conditions for the ranks to do the same for junior officers, because some of the rankers would eventually receive a commission. By 1917, it was no longer valid, if indeed it had ever been, to think of officers and men as belonging to two distinct, watertight groups, possessing no knowledge of each other's conditions.

Only a minority of wartime soldiers joined ex-service organisations after the war,[158] but the existence and longevity of an Old Comrades Association (OCA) can offer a broad hint as to the *esprit de corps* and state of officer–man relations of a unit. While not all cohesive units formed an OCA, and some units and formations such as 66th Division had associations for officers only,[159] many OCAs seem to have been organisations in which former officers and Other Ranks could meet on approximately equal terms. In the interwar years, the OCA of 32nd Division Trench Mortar Battery met once a year for dinner. This OCA, wrote its Honourary Secretary, a former ranker,

> spells brotherhood first and last, and class distinction is taboo'd (*sic*). The old Tock-Emma [i.e. trench mortar soldier] is welcome for

what he did 'out yonder', and not necessarily for what he is today.[160]

Many OCA members probably had little in common apart from a wish to share and rekindle memories of wartime service in a particular unit, to fulfil a deep psychological need. Former soldiers usually dwelt on the humour and comradeship, rather than the horrors, of war.[161] Some OCAs, and other veterans' organisations, such as the British Legion and the Old Contemptibles Association, had a charitable function.[162] In these bodies the paternal pattern of the war-years was extended, with ex-officers and ex-soldiers working together to provide financial and other help for poorer members and their families. This was especially important since 'in the immediate post-war period "unemployed man" and "unemployed ex-serviceman" were close to synonymous'.[163] Some OCAs continued in operation for many years. The 22/R. Fusiliers' OCA existed from 1919 to 1976, while the Machine Gun Corps OCA had a similar lifespan.[164] Nostalgia for comradeship and paternalism, which contrasted starkly with many ex-soldiers' (and indeed ex-officers') experiences of the harshness of life in a land which was far from 'fit for heroes', was undoubtedly a factor in the popularity of OCAs.[165]

Even in the absence of a formal unit OCA, former members of a unit could continue to demonstrate comradeship and respect in time of peace. When, in the 1920s, a former officer of the Accrington Pals died, his chief mourners included five members of the battalion, four of them rankers, the officer having 'no family'.[166] Former officers and soldiers of many units, particularly locally raised battalions, also met to commemorate the dead, whether at memorial services in Britain or on 'pilgrimages' to the battlefields.[167] Unit histories, especially those of disbanded service units, were another means of commemorating the dead and recapturing wartime *esprit de corps*. Those produced between 1918 and 1923 in particular, 'although not overtly consensual in tone' often portrayed officer–man relations in terms of a community of interest.[168] Although contributions from Other Ranks were often included, most unit histories were written by officers.[169]

All members of the Old Contemptibles Association were, somewhat artificially, referred to as 'chum', regardless of rank.[170] Regimental journals, particularly those produced by OCAs of service units, were full of obituaries, articles and reminiscences written by former soldiers of all ranks, which stressed, consciously or not, that a spirit of comradeship which encompassed all ranks had existed and continued st.[171] In sum, the postwar activities of veterans of all ranks offers ...er evidence that rankers could, and did, regard officers as comrades.

Taking all this evidence into account, one is led to the conclusion that it is indeed valid to talk of a British 'war generation' who shared a common experience. In Janet Roebuck's words,

> Under battle conditions class lines came to be overshadowed by the shared experiences of combat and the mutuality of death.... The conditions of war made contact between upper-class officers and lower-class soldiers inevitable and gave them a set of common experiences which neither group shared with civilians of their own class.[172]

There is much to be said for Marc Ferro's idea that a 'special 'ex-serviceman's' outlook grew up from bitterness and nostalgia'[173] leading to postwar idealisation of the war years although, in the case of Britain, he underestimates the degree of continuity with wartime relationships. Clearly, it would be wrong to assume that all Other Ranks regarded all officers as comrades. It is likely that some of the more sweeping claims made by officers about the existence of a community of the trenches which united soldiers of all ranks contained a large element of wishful thinking; we return to the fact that Other Ranks tended to judge their officers on an individual basis, rather than giving their loyalty to officers as a group.

Some politicians attempted to capitalise on their war service in an attempt to win veterans' votes. One such was Sir George McCrae, a Liberal MP who had raised and commanded 16/R. Scots. In a 1923 election address he claimed to be 'an Ex-service man' who, having 'shared their dangers and hardships' would support the fight of former soldiers for fair treatment.[174] It is instructive that men like McCrae and two future prime ministers, Clement Attlee (described as 'Major Attlee' between the wars, partly in an attempt to stress Labour's respectability) and Anthony Eden (who used a photograph of himself in uniform on the cover of his 1922 election address) were members of the three major established parties.[175] No 'military party' emerged as a force in British politics. Mosley's British Union of Fascists, which promoted militaristic values and attempted to appeal to ex-servicemen, was electorally unsuccessful. Ex-servicemen's organisations had a minimal political impact. All this suggests that the British war generation was a rather different beast from its German counterpart.[176]

But a British war generation did exist, in the form of individual relationships between officers and men, forged in the face of hardships and dangers shared, to a greater or lesser extent, by all ranks. Many of these relationships continued after the war through the medium of ex-servicemen's organisations. Memories of wartime relationships between officers and men were treasured long after the war, even if, like Capt. Blagrove and Gunner Jenner, they lost contact in 1918. Writing nearly fifty years after the event, ex-L/Cpl S.A. Boyd of 10/R. Fusiliers stated that 'My lasting impression of the Somme battle is the fine young officers who led us so well. They were extremely brave but so young, many under the age of 20.'[177]

Just as the character, ethos and experience of no two military units was the same, war veterans reacted to peace in different ways. Cohesive 'family' units were probably more likely to establish and maintain OCAs than other units. Nevertheless, as the evidence of soldiers referring in affectionate terms to officers with whom they had lost contact long ago suggests, the British war generation should not be located solely in the reunion dinners and magazines of OCAs of disbanded Pals battalions. The British war generation was characterised by general, if unquantifiable friendly feeling between ranks and classes. Although unquantifiable, it was nonetheless real.

The failure of British veterans to create cohesive political organisations did not mean that their war generation was politically insigificant. In their study of French, German and British literature on the Great War, Bessel and Englander concluded that the war generation 'existed only for so long as it remained under fire', and that on demobilisation 'it appears to have disintegrated into its constituent parts'.[178] This interpretation ignores the many ties of affection and comradeship that continued to bind former soldiers of all ranks in peacetime Britain. G.H. Roberts, a trade unionist MP and Minister for Labour, noted after a tour of the Western Front in September 1918 that not only were officer–man relations 'excellent', but that officers wanted 'conditions at home' to improve for their men after the war. Men had come to 'respect their officers' while officers had come to

> know and appreciate the lives of their men at home. They have been taught to give every consideration to their comfort in the field, and many of them evidently regard it as their duty to do the same for them at home when the war is over.[179]

This wartime comradeship and concern was not simply abandoned or forgotten at the Armistice. As noted above, some veterans' organisations were an extention of wartime paternalism by other means, and more importantly, in John Keegan's words, many officers conceived

> an affection and concern for the disadvantaged which would eventually fuel that transformation of middle-class attitudes to the poor which has been the most important social trend in twentieth century Britain.[180]

Recently Gerard J. DeGroot has argued that a 'myth' about the Great War has arisen which sees 'class antagonism' as the product of 'ignorance, which the trench experience eradicated'. While he correctly argues that 'paternalism should not be confused with ... equality', he seriously underestimates the extent to which inter-rank barriers came down on active service. Class antagonism was not eradicated by the trench experience, but it was modified. The wartime officer–man relationship and its impact on postwar class relations was rather more subtle than DeGroot's analysis indicates.[181]

To try to assess something as nebulous as the social attitudes of such a large and diverse group as British veterans of the Great War over twenty or more years is a difficult task, to put it mildly; but by the 1940s the idea that a total war entitled the participating population to a 'decent existence' was firmly established. This stood alongside the belief that after 1918 'the rank and file of the nation had been denied their entitlement'.[182] This of course is a prime example of Andreski's 'military participation ratio' in action, and Keegan's 'process of discovery' undoubtedly played a part in this phenomenon.[183]

The argument that the concept of the war generation as such should be left behind in favour of analysis of 'those constituent parts which temporarily comprised it' has much to commend it, but to abandon the idea of a war generation altogether is to risk throwing out the baby with the revisionist bathwater.[184] Although this subject is in need of further examination, it is safe to state that many British officers and rankers believed themselves to be part of a war generation, united by comradeship and the shared experience of combat.[185]

8
Officer–Man Relations: Morale and Discipline

In 1968, R.C. Sherriff, who served as an officer in 9/East Surreys from 1916 to 1918, wrote that his highly successful war play *Journey's End* had been criticised because 'there was too much of the English public schools about it'. Sherriff retorted that 'Almost every young officer was a public school boy' and if he had omitted them from *Journey's End*, 'there wouldn't have been a play at all'. Furthermore,

> Without raising the public school boy officers onto a pedestal it can be said with certainty that it was they who played the vital part in keeping the men good-humored (*sic*) and obedient in the face of their interminable ill treatment and well-nigh insufferable ordeals.[1]

In a recent book based largely on trench magazines[2] and other published sources, J.G. Fuller has criticised these views. He argues that public school officers, educated to believe that they were natural leaders of society, could not have been objective in their assessment of their role in maintaining the morale of their men. In addition, he suggests that officers generally tended to overrate the esteem in which they were held by their men; the institutionalised gulf between the ranks was too great to be bridged; and the rapid turnover in personnel precluded the establishment of close relations.[3] All of these arguments have some validity; the question is, do they have enough to nullify the general thrust of Sherriff's thesis?

The arguments advanced in this present work, supported by much evidence drawn from unpublished sources, offers a broad measure of support for Sherriff's claims. However, Sherriff, who was commissioned from the Artists Rifles, was indulging in hyperbole by neglecting to mention the large number of non-public school officers

officers that served in the army. In fact, in an account of his early days with his battalion Sherriff mentioned a number of ranker officers, including one who had great influence on the subalterns, and others with whom he shared a railway carriage on a long journey. Is it impossible to imagine a 'public school' subaltern discussing life in the ranks, and the soldier's view of his officers, with such men?[4] However, in a more general sense, Sherriff was absolutely correct to insist on the importance of public school values in the maintenance of morale, since these were inculcated during officer-training. Indeed, his views are given added weight by the fact that Sherriff was painfully aware of his own inadequacies as an officer.[5]

By the end of 1915 officers' messes were being filled with men who had seen active service in the ranks, and who had first-hand experience of the importance of paternalistic leadership to the ordinary soldier. One such man was a middle-class lance-corporal of 1/4 Glosters who wrote of the insights he had gained in the ranks. These served him in good stead when he was commissioned into 16/Manchesters, his divisional commander paying tribute to his understanding of, and sympathy with, his men.[6] In August 1914 a private noted in his diary his gratitude to his paternal CO who provided soft drinks at the end of a long route march. Small wonder that in September 1918 we find the same soldier, now a company commander, distributing to his men cigarettes which had been delivered up by the padre.[7]

There were several other ways in which officers could come to know the 'character and thoughts' of their soldiers.[8] Simply working alongside the men could be beneficial.[9] Junior officers had to read and censor their men's letters. This was an activity disliked by both officers and men, even if they reluctantly recognised its necessity.[10] A number of officers commented that reading these letters gave them an insight into the mind of the writer. Charles Douie (1/Dorsets) entered 'a new world, with interests and standards of which I had previously no experience'.[11] Harold Macmillan (4/Grenadier Guards) gained 'an insight into the lives of his men' which contributed to his lifelong interest in, and sympathy for, working-class people. Indeed, as an MP in the interwar years, Macmillan believed his relationship with his constituents in Stockton to be similar to that of a company officer and his troops.[12]

It may be argued that censoring his letters was not the best way to understand a man. The letters were often 'similar and childish',[13] and one historian has described them as 'a demotic literary genre' which disguised the true feelings of the writer.[14] Some were pure fiction, sent as a joke, or even to win money from gullible newspapers.[15] Against these arguments must be set the fact that many of the rankers' letters surviving in archives are very frank.[16] Some rankers went to considerable lengths to avoid censorship by sending letters home by unofficial routes, such as giving them to men going on leave to post in England.[17] Some officers were less than whole-hearted in their censoring duties: at one stage, 1/5 Black Watch even issued envelopes pre-stamped with the censor's mark.[18]

Officers were sometimes struck by the openness with which men displayed their emotions in their letters home. After 2/4 DWR had been engaged at Cambrai in November 1917, Capt. B.D. Parkin censored his men's letters and gained 'a peep into the soul of those who had acted so bravely for three days of battle'. These letters contained few exaggerations, and were frank and 'full of relief' at having survived the ordeal. They contained both good and bad news; and interestingly from the point of view of officer–man relations, many made favourable references to the pre-battle address delivered by their popular brigade commander, Brig. Gen. R.B. Bradford.[19] Whatever their normal diffidence about committing their thoughts to paper, these men treated their letters home as a form of catharsis. Although many other ranks' letters were indeed uninformative and stereotyped, it was possible for officers to gain real insights into the minds of their men by reading them. Furthermore, the weight of evidence presented in this book strongly suggests that, officers, public school or otherwise, were in a good position to assess the importance of leadership in maintaining the morale of the other ranks, not least through consultation with NCOs.

Dr Fuller's analysis of inter-rank relations forms only a very small part of his book, which breaks new ground in emphasising the importance of working-class culture in maintaining the morale of British troops on the Western Front. His work on this topic is not incompatible with the notion that officer–man relations played an important role in the maintenance of morale. It is not suggested that officer–man relations were solely responsible for keeping morale high. Food, drink and tobacco; recreation; comradeship; humour; reminders of home, such as letters from family and friends; trust in leaders and belief in their cause; success in battle; and sheer stoicism were all of vital importance.[20] Correlli Barnett has argued that the hardships and poverty endured by the working classes in their everyday life, and the danger ever present in many of their peacetime occupations, prepared them for the conditions endured by soldiers on the Western Front. He asserts that many soldiers were 'better off in the trenches than at home'. In France, unlike at home, the working man 'had the support – moral and material – of an immense organisation' which looked after his welfare. It might be added some middle-class soldiers felt that military life was an improvement on their civilian existence, the 'prison-like life of a city office', as one described it.[21] The thrust of Barnett's thesis has been echoed by other modern writers such as Michael Howard and J.M. Bourne.[22]

Barnett's work is important, not least because he places the experience of the working-class soldier at centre-stage, rather than generalising from the writings of a tiny number of officer poets, as too many literary specialists and cultural historians have done.[23] There is little direct evidence from working-class soldiers to support his central argument, although it is hardly surprising that poorly educated men were not inclined to analyse their state of morale in sociological terms. However, occasionally one comes across comments such as that of the soldier who recalled that one 'got used to' being 'wet cold and hungry'; this stoicism can perhaps be related to his poverty-stricken background.[24]

Many contemporary observers testified to the stoicism and powers of endurance of working-class soldiers. Writing of the men of 2/DLI, many of whom were miners, a ranker in a London class corps noted that the life of the Regular private 'is a pretty hard one':

> He gets little enough consideration from his officers, he is paid very little ... in fact, he has every encouragement to down tools and strike, instead of which he plods on steadily, grousing and grumbling, always kept hard at work, getting drunk as often as possible, but ready and willing at the final call – faithful unto death.[25]

Similarly, after watching his men play two games of football a week after his battalion had taken heavy casualties, an officer of 11/Argylls confided to his diary that 'the recuperative powers of the British Tommy are wonderful'.[26]

The major weakness in Barnett's argument lies in the fact that not even the most dangerous of civilian jobs could prepare a man for the reality of battle, and many working-class occupations, such as that of servants, for instance, were not especially dangerous. It should also be noted that stoicism was not confined to working-class soldiers. Army life, particularly on active service, blunted the sensibilities of many middle-class men. An infantryman of the Second World War believed that the secret was to 'conquer the abnormal by the simple device of treating it as if it were normal'.[27] There is something of this in one middle-class NCO's tongue-in-cheek theory that men who were killed 'had failed to keep up their spirits'; the spirit, he felt, could divert shells from their course.[28] One veteran believed that soldiers 'chose to make fun of their situation'; had they acknowledged the reality, 'the men would have "cracked" and collapsed'.[29] Some soldiers became fatalists, accepting the prospect of death and becoming reconciled to it. One such was an MGC private, a prewar clerk, who wrote a moving letter to be posted to his family in the event of his death in which he spoke calmly of dying, and which ended with a prayer.[30]

Nevertheless, the efforts of individuals to keep up morale were greatly enhanced by the 'immense organisation' to which Barnett refers, in which regimental officers played a crucial, although underrated, part. The role of popular culture was, as Fuller convincingly demonstrates, of considerable importance. The replication of patterns of civilian leisure – sport, concert parties, social drinking – helped to keep the reality of service on the Western Front at bay.[31] While rankers were perfectly capable of making their own entertainment, the importance of officers in providing organised leisure at unit level should not be underestimated.

To take several battalions at random, officers of 9/R. Irish Rifles encouraged their men to 'indulge in their favourite pastimes',[32] while 2/6 King's boasted an orchestra, raised from former professional musicians.[33] 1/16 Londons provided a regimental canteen, directed by an officer, where very cheap food, beer and tobacco were sold to the men, profits being used to supplement official rations. While training in 1917 most afternoons were devoted to sport, with trips being arranged to a swimming pool. There were also horse shows and sports days, complete with the Divisional band, the 'Bow Bells' concert party, coconut shy, tea and beer. The unit historian noted that 'It may appear somewhat incongruous, in a war history, to devote space to a sports meeting; but such events have a direct bearing on the development of that *esprit de corps* and the will to win' which brought about victory. Such 'trivial' pleasures should be seen in the context of recent grim experiences. The sports engendered 'A spirit of enthusiasm and lightheartedness, which is difficult for those who have not experienced the ups and downs of war fully to realise …'.[34] As a Guards officer commented, 'the simplest change gave the men pleasure'.[35] The importance of the provision of amusements was well understood by officers.[36]

Officers performing in unit concerts provided a valuable safety mechanism. Officers stepped down from the pedestal of rank and hierarchy, albeit briefly, allowing rankers to cheer or jeer them with impunity. Surprisingly few of these officer 'turns' seem to have undermined discipline. Indeed, they may have enhanced it. An undersized officer of 2/6 Lancashire Fusiliers provided some impromptu cabaret at a concert at Christmas 1915 by colliding with the bass drum. Thereafter, this officer, a favourite with his men, was often greeted by calls of 'Who fell thro' t'big drum? Little Tich!'[37] Not only does this incident demonstrate the way in which officer–man relations could be enhanced by sharing in entertainments, but the fact that it was recorded in the battalion history, written by two former officers, speaks volumes about the relaxed nature of officer–man relations in this TF unit.

Morale-enhancing infrastructure notwithstanding, few men could undergo the strain of front-line service indefinitely. While psychiatric casualties were not unknown in earlier wars,[38] they occurred in 1914–18 on an unprecedented scale. Lord Moran, who served as RMO of 1/R. Fusiliers, believed that a man has only a limited 'bank' of courage or 'will-power' 'and when in war it is used up, he is finished'. There was no such thing as getting used to battle. Men, like clothes, simply wore out. The time taken to expend the 'capital' varied with the intensity of combat, but if a soldier fought for long enough, his courage will be exhausted.[39] Moran's basic theory, although modified by more recent scholars, is still influential.[40]

It is difficult to establish precise figures for rates of 'shellshock' (which was in 1914–18 a generic term for psychiatric disorders brought about by combat).[41] In 1939, some 120 000 men were receiving pensions or had received a final pay out for 'primary psychiatric disability' as a result of service in the First World War, which represented 'about 15% of all pensioned disabilities'.[42] Shellshock has aptly been described as the '*mal de siècle*' of the interwar period, influencing everyday life and culture.[43] Even fictional aristocratic detectives and hobbits could suffer from it.[44] Statistics do not tell the whole story. A distinction can be usefully drawn between 'battle stress', which is experienced by all soldiers in action, and 'battleshock', stress so severe that it renders the soldier a casualty. Clearly, many men suffered from battle stress which was serious but not severe enough to incapacitate. In August 1916, 86th Machine Gun Company reported that about eight per cent of the unit had been admitted to hospital 'owing chiefly to the strain of the previous 5 weeks'[45] while a doctor believed that 'practically every man' evacuated from Gallipoli to Lemnos in late 1915 'was neurasthenic, whether he was supposed to be fit or not'.[46] One infantryman frankly admitted that a week of fighting on the Somme in 1916 had deprived him of courage. To use Moran's terminology, he had made a sudden withdrawal of capital, which all but closed the account. However, a period spent in a rest camp or in a quiet sector could help to restore an individual's morale.[47] At the very least, it can be said that psychiatric casualties were something which most officers and men were likely to encounter on the Western Front.

The widely held belief that shellshock was the product of weakness led to a reluctance to admit to the existence of psychiatric casualties.[48] However, some front-line soldiers disputed the view that victims of shellshock displayed 'weakness', and traditional views also began to be challenged within the medical profession.[49] By the end of the war, psychiatric casualties were beginning to be treated according to the principles recognised by modern armies.[50]

Many Regular officers believed that there was a close link between the incidence of psychiatric casualties and the state of a unit's morale, discipline, and leadership. These views were enshrined in the 'Summary of Findings' of the 1922 Shellshock Committee, which baldly stated that 'A battalion whose morale is of a high standard will have little "shell shock".'[51] A modified version of this view is held by most modern scholars, who suggest that membership of a cohesive unit with high morale and good leadership can retard – although not prevent – the onset of mental collapse.[52]

The role of the officer in minimising psychiatric casualties was twofold. First, officers could spot men approaching breakdown, and remove them from front-line duties, to allow them to replenish their stock of courage. This technique, practised by contemporary armies, does not seem to have been much used in 1914–18, although it is possible that it was carried on unofficially. The RMO of 1/10 King's, Capt. Noel Chavasse, is one of the few officers who is recorded as adopting this approach.[53] In the Guards Division, men with 'considerable war service' were put into 'works' units which served away from the front line to give them 'a chance of resting their nerves'.[54]

Secondly, the officer could minimise the physical and mental factors that contributed to stress by exercising a paternal concern for the welfare of their men, such as ensuring that the men were well-fed. Although the rations of the British soldier tended to be monotonous and, as an American sergeant ungratefully noted, unappetising,[55] regular supplies of food, drink and tobacco were all important in maintaining the morale of soldiers of all social classes. One ranker believed that 'Nothing changed one's spirits from buoyancy to utter despondency more quickly than a shortage or surfeit of rations.'[56] In a quiet moment, the usual reaction of the British soldier was to light a cigarette and brew tea. Letters home are full of requests for cigarettes, for as one ranker of 1/21 Londons put it, 'it is rotten to be without a smoke'.[57] At one stage in the Palestine campaign, noted a sergeant of 5/Welsh, the lack of tobacco reduced men to smoking tea leaves.[58] John Brophy, a ranker, claimed that tea, 'even when made in onion-haunted dixies and stewed over a smoking fire ... conserved and yielded a delicate fragrance, an exquisite suggestion of civilisation'.[59]

The officer also had an important role in the daily distribution of rum, which was supposed to be drunk in the presence of an officer to prevent it being hoarded.[60] The rum issue was undoubtedly popular, and gave some men 'dutch courage' before battle. Pte B.F. Eccles (7/Rifle Brigade) admitted that he went through battle 'far better than I ever imagined' as a result of drinking rum.[61]

Brophy's comment, quoted above, indicates one of the principal reasons for the popularity of caffeine and nicotine: they represented a slender thread of continuity with normal life, and reminders of home are important in reducing stress and maintaining morale. This fact underlines the importance of the practice of distribution of gifts of cigarettes and other luxuries by officers, and the considerable efforts made by regimental officers to ensure that their men had hot meals and tea. The regimental officer had a vital role in ensuring that soldiers benefited from the bureacracy of paternalism, which was dedicated to the upkeep of the soldier's morale. This role helped to cement comradeship between the ranks. In that it was also an example of officers adhering to the unspoken assumptions inherent in the deferential relationship, it provided further continuity with the civilian experience of working-class soldiers.

Conditions at the front were an important source of stress. A Guards private recorded that some men appeared to rather more afraid of the ever-present rats than they were of shells and bullets.[62] Bad weather, mud, lice and boredom also ranked high among the stresses of war.[63] At the front, most soldiers were permanently tired. In the trenches, sleep was a rare and precious commodity. Men who managed to ignore the cold, rain, noise, bustle and rats would, before long, be roughly shaken awake for their turn on sentry-go or to stand-to. Even the periods out of the line, which were quaintly termed 'rest' by the army, were punctuated by exhausting bouts of physical labour. Men were not infrequently reduced to a drugged, dream-like state, unable to react even to threats to life and limb.[64]

On marches, and sometimes on the battlefield, the soldier was heavily weighed down with a pack and equipment which weighed as much as 60 lb. One private complained that the military authorities never seemed to realise that 'our spirits rose with the lightening of our bodies'.[65] In fact, many regimental officers sympathised with the rankers' struggle to endure route marches bowed down with what L/Cpl. R. Mountford (10/R. Fusiliers) described as 'a cruel, unnatural weight that no man should be called upon to carry'.[66] There are many examples of officers (and NCOs) putting paternalism into practice by carrying men's rifles and packs on long marches. G.H. Cole (1/20 Londons) would have his groom bring up his horse on leaving the trenches but he rarely rode it, using it instead to carry the packs of weary soldiers.[67] Pte R.H. Sims (1/4 R. Sussex) wrote of a four-hour route march under the heat of an Egyptian sun. The march was a sore trial but, for Sims, the one redeeming feature was the conduct of his officers:

> our company captain is a true gentleman & a brick & so are all of the officers as those who had horses carried chaps (*sic*) equipment & those who were marching carried the mens' rifles, our captain alone carrying two ...[68]

A rarely considered aspect is the mental and physical strain endured by officers as they strove to set an example on long marches by suppressing signs of weariness.[69]

The role of the officer in providing mental comfort for the ranker, by countering fears that the individual soldier was impotent in the face of the military moloch, has been discussed above. Here, the fate of one group of men who were especially vulnerable to stress will be considered. During the war the army received many men who were unsuited to the rigours of life in the ranks through age, infirmity, or physical unfitness.

Life in the army could be grim for such men. One, Pte James Williams (29th Labour Coy, Queen's) may stand as representative. He was conscripted at the age of 38 in March 1917, having spent 20 years working as a clerk. Although he enjoyed some aspects of military life, he did not adjust to life at the bottom of the military hierarchy. Williams resented discipline, describing being ordered to be inoculated as 'Being treated like So Many Dogs', and he disliked his working-class fellow rankers. Having worked for so long in a sedentary occupation, he found it difficult to cope with the physical demands of military life. Much of his work consisted of heavy loading.[70] Williams once wrote that he 'Turned in thoroughly worn out & worked up to a pitch of intense hatred to (*sic*) the life out here. What with the filthy mouthed men one has to work with.'[71] Williams did not have the compensation of service in a cohesive unit with high *esprit de corps*. Neither did he have a particularly sympathetic NCO or officer, although at least one officer seems to have been paternal in a general way, buying beer for the men, which Williams characteristically 'declined with thanks'.[72]

Williams's experience can profitably be compared with that of Pte Frank Grey (8/R. Berks) a middle-class journalist who was conscripted at the age of 37. Grey benefited both from service in a cohesive unit and the leadership of paternal officers. Grey was given an easy job on account of his advanced years and showed his gratitude by dedicating his war memoirs to his commanding officer, Lt Col R.E. Dewing, whom he described as 'a part (or the victim) of a system, but who possessed a fine knowledge of men, was humane, kind, and courageous, and so remained to the end'.[73] The difference in cohesion between an infantry battalion and a labour unit is worth noting; the former had the experience of combat to bind the ranks, the latter did not. Many other similar examples could be cited. Pte Birdsall of 2/4 DWR was appointed as an officer's servant because he was older than most other privates, and was physically weak, having been wounded.[74] In 7/KSLI, according to 2/Lt N. Hughes-Hallett, older men were often given jobs as sanitary men. The work was unpleasant but responsible and such men were excused most parades.[75]

Younger middle-class soldiers were also treated with sympathy by some of their officers. A former London clerk, Pte R.D. Fisher (1/24 Londons) shared a keen interest in serious music with his officer, Lt Poll, and sometimes they would talk briefly about the subject. Occasionally, Poll would slip a copy of a musical magazine into Fisher's pocket. Such relationships were not without their difficulties. The need to avoid overstepping the bounds of 'familiarity' were obvious: Fisher referred to Lt Poll as 'a friend in spirit rather than fact, owing no doubt to the difference in rank'.[76] Traditionally minded NCOs could be horrified by such relationships. In Aldington's novel *Death of a Hero*, 'Winterborne', an educated ranker, is singled out by an officer who asks him to take a commission, and then they shake hands, 'to the impressed horror of the NCOs'.[77]

As so often, Manning's great war novel *The Middle Parts of Fortune* provides a valuable insight:

> although the conventions which separated officers from men were relaxed to some extent on active service, between men of roughly the same class they tended to become more rigid. Even when momentarily alone together, they recognised, tacitly, something a little ambiguous in the relation in which they stood to each other....[78]

Nevertheless, in the opening scene, in which Pte Bourne encounters a subaltern in the aftermath of a battle, Manning captures a moment in which the two men meet in a dugout, share a drink from a whisky bottle, and talk almost as equals. The formal barrier of rank has temporarily dissolved and is replaced by a relationship characterised primarily by mutual respect. Although both men dispense with formalities, neither man tries to take advantage of the situation, and the fundamentals of discipline are not challenged. They then leave the dugout where they resume normal relations. In this instance, as in so many others, the army adage 'on parade, on parade; off parade, off parade' can be usefully applied.[79]

Manning's portrait of relations between privates and NCOs and privates and officers has been criticised as unrealistic.[80] The suggestion that Bourne is on 'unusually informal terms with his officers and NCOs'[81] is more accurate, but informal and friendly relations between officers and men were by no means uncommon in the British army of 1914–18, even in units which did not employ a liberal disciplinary regime. Manning's portrait, based on his experience in the ranks of 7/KSLI, although at variance with the stereotype of inter-rank relations, should be seen as entirely realistic.

The experiences of middle-class rankers such as Williams and C.E. Jacomb (23/R. Fusiliers), who complained that he had been 'spoken to and treated like a dog by practically everyone of higher rank than myself'[82] cannot be ignored, but at the very least it is possible to say that some older and middle-class rankers were singled out for sympathetic treatment by their officers. The effect of good, paternal leadership in alleviating stress and thus enhancing the morale of rankers of all social backgrounds should not be underrated. Rankers knew that they were not simply at the mercy of the military system. As a conscript private of 14/Argylls noted, while recalling a friendly and paternal officer, 'It makes all the difference when one is treated with kindness and consideration by one in authority.'[83]

The battlefield role of the regimental officer can be divided into command and leadership functions. The former, which falls outside the scope of this book, encompassed tactical decision-making, allocation of resources and the like. The leadership function can be summarised as the need to ensure that the goals of the primary group are congruent with the goals of the army. One of the most important methods used by the British regimental officer in carrying out this function was to act as a leader in the most literal sense. The disproportionately heavy casualty rates for junior officers suggests that Crozier's belief that officers had to have a 'three seconds lead' over their men in the advance, so they could say 'come on' rather than 'go on' to their men, was widely held.[84]

Social relationships between soldiers lay at the core of British combat motivation in the Great War: soldiers did not want to appear cowardly in front of their comrades.[85] However, given the closeness of inter-rank relationships, fear of letting down an officer could be a source of motivation.[86] It should also be noted that a regimental officer had to strike a balance between leading from the front and being regarded by his men as a 'thruster', all too ready to sacrifice his men's lives.[87] In the attack, the officer had to overcome his men's fear and lead them into battle; on the defensive, he had to keep them from running away. Training and discipline helped the soldier to choose the 'fight' rather than 'flight' option, but an officer's example was also important.

First, the officer had to suppress his own fears, although the experienced soldier came to learn how to judge the degree of danger in everyday trench life.[88] Waiting to go into battle was particularly stressful. A conscript private of 6/Queens believed that 'only those who have faced the ordeal can form an adequate conception of the anxiety of those long minutes which sometimes seemed like hours'.[89] More specific fears included the fear of death and pain, fear for the soldier's family at home and fear of weapons such as gas.[90] Even the noise of friendly artillery could be immensely stressful.[91] In battle, the soldier's emotions were 'numbed', while their senses were heightened; this is the biological reaction of the human body to severe stress. When the immediate danger had passed, the senses 'gradually returned to normal and we looked around us like men awakened from a nightmare'.[92] Manning wrote of the sensitive way officers treated men who were beginning to reassert normal patterns of life and come to terms with the death of their comrades, although long exposure to danger could reduce the impact of such deaths.[93] Nonetheless, as an officer wrote in 1916, 'The majority of men ... are more or less scared stiff all the time.'[94]

The constant battle of the officer to suppress fear, and set an example to his men, was described by Capt. Hanbury-Sparrow of 2/R. Berks, who was trying to nerve himself to look over a parapet:

> For very shame's sake pull yourself together, man.... Set them an example. With a dozen pairs of eyes watching you, you unstrap your field-glasses and, kneeling, look over the parapet.[95]

Many officers referred to the strain of command: 'you owed it [to the men] to stick it too', as one phrased it. As this officer had previously served in the ranks, he would have been well aware of the importance to the private of the officer's example.[96] By contrast, other officers were exhilarated by command. One claimed that responsibility for his men's lives actually reduced his fear in battle, because he had no time to concentrate on himself.[97]

There is plenty of evidence that Other Ranks of all social classes were influenced by the example of their officers on the battlefield. A young private of 7/Rifle Bde commented that 'Our officers were splendid and showed great coolness' during the battle of Arras.[98] Pte John Gibbons, in a judicious assessment of the role of the officer, implied that the officer did indeed set an example: their 'outward and visible standard of courage' being superior to that of the other ranks.[99] Pte Frank Dunham (1/7 Londons) claimed that 'One thing only helped to keep our spirits up' while serving at Ypres in September 1917: 'a daily visit' from Capt. K.O. Peppiatt: 'He set us an excellent example in cheerfulness and good humour ...'.[100] Neither Gibbons nor Dunham were uncritical admirers of the British officer corps. Other officers were remembered for their leadership in specific situations. Pte Jimmy Walton (11/Suffolks) admired an officer for his courage in rallying men in Lochnagar Crater on the Somme on 1 July 1916:

> 'Gentlemen, we're going to be faced with a counterattack. We stand and we fight'. He might have been on a barrack square – so calm, so collected.[101]

The presence on the battlefield of a few men with the ability to impose their will on their fellow soldiers played a vital role in preventing men from running away.[102] Such individuals were not necessarily officers or even NCOs: a private of 1/8 West Yorks recorded how the battalion commander used veteran soldiers to keep up the morale of youngsters during their first attack,[103] but such formal leaders were trained to set an example. One was Lt J. Proctor (10/Lincolns), who was decorated for his leadership during the battle of Arras on 9 April 1917, when his company was held up by uncut wire in front of an enemy position. Such an event could be fatal for the impetus of an attack, therefore Proctor's actions were particularly significant:

> He ordered his men to lie down, as snipers were active, while he looked for a gap. He walked up and down making jokes and imitating Charlie Chaplin, keeping the men laughing. The officer was wounded at this point but continued to display the utmost *sangfroid*.[104]

Proctor's actions, although eccentric, epitomised the leadership traits that characterised the British junior officer. He led from the front, searching for the gap in person, rather than sending an NCO. He demonstrated his courage by walking about despite the threat of snipers and by maintaining a calm demeanour after he had been wounded. Proctor showed concern for his men's lives by making them lie down. Finally, he attempted to keep up the men's morale by his words and actions in a situation in which the men must have been gripped with fear.

In 1918 junior leadership by example continued to be vitally important, particularly in the March Retreat, which was very much a 'soldier's battle'. The major difference between leadership in 1918 and in previous years was that it took place in a context of increasing tactical sophistication. By this stage most units had discarded the crude 'waves' of 1916 in favour of the 'blob', or dispersed section column. Yet in an action of 1 September 1918, Captain Griffiths (2/4 Londons) 'strode steadily ahead of the advancing line ... waving his walking stick above his head and blowing his whistle'. Griffiths showed a 'light-hearted contempt for death'. 2/4 Londons had adopted the 'blob', illustrating that tactics had changed by September 1918 but concepts of junior leadership had not.[105] Conversely, on some occasions in the British advance of 1918, leadership from the front by officers who were themselves inexperienced was necessary because of the poor training of the infantry.[106]

A document issued by Reserve Army in October 1916 highlights two other practical applications of paternalism to the battlefield. First, company commanders were to allocate platoons to the first (and hence most dangerous) wave of attack 'alternatively, so as to avoid any sense of injustice'. Second, the document stressed the importance of providing successful troops with food, water and sleep, on capturing a position, so they were fit to repel counterattacks. The officer had to strike a fine balance between consolidating trenches and resting men.[107]

Leadership by example on the battlefield must be set against a darker side of the officer's role. It has been claimed by Dave Lamb that officers acted as 'battle police', forcing reluctant soldiers into battle at gunpoint.[108] I have dealt with the subject elsewhere, and here it will suffice to say that while battle police were deployed on occasions (although not all soldiers believed that they existed),[109] they were usually drawn from a unit's regimental police, not officers. Battle police had various official tasks, including traffic control, rounding up stragglers and arresting deserters, but summary execution was not one of them.[110]

That is not to say that summary executions did not occur. At moments of extreme crisis, officers and indeed NCOs and ordinary soldiers did threaten, or even carry out, summary executions. On 6 June 1915, on Gallipoli, 2/Lt G.R.D. Moor (2/ Hampshires) shot perhaps four British infantrymen, and thus prevented a retirement from becoming a rout.[111] However, while a tiny handful of officers did make a practice of summary executions,[112] this practice was never sanctioned by GHQ. Indeed, when on one occasion of extreme crisis on the Aisne in May 1918, the commander of 19th Division attempted to speed up the process of executions by requesting from GHQ – permission to 'confirm and have carried out' death sentences on stragglers, permission was refused.[113]

It is significant that the fire-eating Brigadier-General F.P. Crozier, who is used by Lamb to illustrate the use of officers as battle police, was described by a fellow officer of 36th Division as having 'unbalanced' views. Furthermore, he claimed that the divisional commander hesitated to promote Crozier because of his 'roughness and the ruthless way he handled his men'.[114] Crozier was not, in this respect, a typical officer. If Other Ranks had lived in constant fear of being shot by their own officers, it would have destroyed the element of trust which lay at the heart of the officer–man relationship. Further research might uncover more examples of policies of summary executions being instituted at a local level, but the notion that fear of the officer's revolver was a major factor in officer–man relations can be safely dismissed.

In 1918 an artillery sergeant, informally chatting to his officer, told him that there were two reasons why the men remained cheerful in appalling conditions: comradeship and fair treatment by their officers and NCOs.[115] With the latter half of this statement in mind, it is useful to examine some negative evidence about the importance of officer–man relations; that is, occasions on which poor officer–man relations and leadership resulted in low morale. British rankers were by no means blindly obedient cannon-fodder. They were perfectly capable of conveying their displeasure in a number of different ways if they thought themselves unfairly treated.

The most drastic, and risky, way was to assault or murder an officer. In separate incidents two soldiers were executed on the Western Front for striking a superior officer, even although both assaults were minor in nature.[116] Capt. Hamond, in his treatise on officership, wisely urged officers to avoid placing themselves in a situation in which even an accidental blow could be laid on an officer.[117] It would have been relatively easy to murder an unpopular officer or NCO under the cover of battle, although naturally it is impossible to ascertain how prevalent this practice was. The only new piece of evidence on this subject which has emerged during the research for this book is the testimony of a former private of 2/Bedfords who claimed that an officer of the battalion was murdered early in 1915 by a grenade thrown into his dugout.[118] Between 4 August 1914 and 11 November 1918 only one soldier serving on the Western Front was executed for the murder of an officer (this figure excludes Chinese and Cape Coloured labourers). This murder was apparently motiveless, and was committed during a rifle inspection behind the lines – hardly the classic scenario for a 'fragging'.[119] Six men were executed for the murder of NCOs. The paucity of rumours concerning the murder of officers is another factor to be considered when assessing the nature of the officer–man relationship.

To argue with or insult an officer was also a serious offence, but insults were sometimes shouted from the anonymity of the ranks. While blanket condemnations of officers did occur – shouts of 'death to our officers' were reported from the camp cinema at Catterick in January 1918[120] – most protests of this sort seem to have been aimed at individuals who failed to live up the standards of behaviour that soldiers expected of their officers. A draft of Lancashire Fusiliers leaving for foreign service in 1915 called for 'three cheers' for one officer; but then followed it with 'three boos' for another, who was called 'a slave-driver, a coward, a pig, and several other names not fit to mention'.[121] A variation on this theme occurred when 2/17 Londons, drawn up on parade to witness the presentation of medals, refused to cheer an unpopular officer who had been awarded the MC.[122]

Such methods were used not merely to relieve feelings but also to convey to officers that the men considered some behaviour to be unacceptable. On occasions, when senior officers agreed with their men, steps were taken to rectify the situation. A newly arrived officer of a TF unit, 2/15 Londons, was regarded as officious by his men, an opinion which was shared by his company commander. With the latter's tacit sympathy, the men communicated their views to 2/Lt 'Counterjump', as he was nicknamed, for instance by loudly commenting on the poor type of man currently being commissioned. Counterjump's authority was undermined by a conspiracy between the CSM, men and company commander. According to an NCO, Counterjump's' career was 'an object lesson on the powerlessness of the officer when his men are against him'.[123]

Such incidents were not confined to TF units with relaxed standards of discipline. During the winter of 1915–16, a mounted officer of a crack Regular unit, 2/Rifle Bde, 'very foolishly' shouted 'Left–Right–Left' to men emerging from a trench. As Sgt. Riddell noted, 'This is all right at the right time' but to shout orders to exhausted men who had spent six days in muddy trenches, who were marching in thigh boots and were slipping about 'was asking for trouble'. The men shouted abuse from the safety of the darkness. This incident damned the officer in the eyes of his men. Riddell commented that 'if men are beat the same as we were, then was the time to encourage them, even if he had to walk'. Interestingly, Riddell hints that the company commander disapproved of the officer's actions, but for discipline's sake 'had to side with the officer'. One can easily imagine the latter receiving an uncomfortable lecture from his superior.[124]

Mutiny was defined as 'collective insubordination, or the combination of two or more persons to resist lawful military authority'.[125] Most mutinies of the Great War were not ideologically motivated.[126] Even after being socialised into military life, working-class temporary soldiers did not abandon civilian patterns of behaviour or thought, and were liable to react to what they perceived as unfair treatment by their military superiors by going on strike.[127] A number of minor mutinies occurred during the training of Kitchener and TF units in 1914–15 when conditions were poor and men were relatively unaccustomed to discipline, although such incidents were sporadic rather than widespread.[128] Tactful handling by sympathetic officers could often defuse such strikes.[129] Only on one occasion did a major mutiny of British (as opposed to Dominion) take place on the Western Front, at Etaples base camp in September 1917.[130]

At this date Etaples base camp contained many of the classic ingredients for mutiny.[131] Food and conditions were poor, and there were few leisure facilities in the camp. It is noteworthy that the mutiny began on a Sunday afternoon, when crowds of men had gathered in the camp with nothing to do. Moreover, many of the men who were subjected to a fairly brutal form of recruit training in the 'Bull Ring' had front-line experience, and were returning for a further spell after recovering from their wounds. Men who had previously undergone this rite of passage, and were accustomed to being treated with a measure of respect by NCOs and officers with whom they shared the dangers of active service, were especially resentful of the military police and the 'canaries' (instructor NCOs who wore yellow armbands) who were perceived as 'column dodgers'. Pte A.F. Sheppard (11/R. Sussex), who trained at Etaples in early 1917 after having fought on the Somme, spoke for many when he dismissed the canaries as all 'out of the same box of soldiers competing with each other who could grow the longest moustache and shout the loudest'.[132] In short, there was general resentment that combat veterans were being bullied by men who, unlike regimental officers and NCOs, had not experienced the danger of the trenches.

At Etaples, the army's disciplinary regime was expanded to nightmarish proportions. Writing of a similar camp at Rouen in 1915, a ranker supposed that at a base where 'thousands of men' drawn from many regiments 'are herded together' there was 'bound to be a lack of esprit-de-corps (*sic*) and sense of comradeship'. He went on to write that the result was that everyone hated drill, implying that drill carried out in the parent unit was at least tolerable, and that *esprit de corps* could offset bullshit.[133] Likewise, at Etaples two of the pillars of unit cohesion, *esprit de corps* and paternal junior officers who could intervene to modify the disciplinary system and protect their men, were absent. Officers were kept well away from the men, and the usual unit-structure did not exist. The creative tension that existed between the army's disciplinary system and the paternalism of the junior officers did not exist at Etaples, where unalloyed discipline ruled supreme.

Capt. J.H. Dible, a medical officer at Etaples, blamed the disturbances on those officers who treated citizen-soldiers in an inflexible manner more appropriate to long-service Regulars. Although frequently officers realised the necessity of a different approach, Dible believed that one bad officer could negate the work of twenty good ones. This problem was aggravated by the fact that there was disparity in the conditions of officers and men, which was the fault of the staff rather than the regimental officer.[134]

Although one mutineer was executed for inciting a picquet to attack its officer, there is little other evidence of mutineers expressing resentment of regimental officers, providing they did not attempt to defend the military police or 'canaries'.[135] The war diary of the commandant of Etaples base states that on the first day of the mutiny 'Feeling in the crowd was only against the police and Officers were treated respectfully. The officers gradually got the men back to camp, and by 9.45–10pm all was quiet.' On 10 September it was recorded that 'The demeanour of all crowds towards Officers was perfectly good.'[136] This evidence is corroborated by a middle-class woman ambulance driver who recalled the mutiny as 'an orderly affair, with officers mingling with the men in sympathy and keeping order'.[137] Likewise, a ranker officer of the 2/R. Fusiliers stated in his memoirs that the mutineers 'had no quarrel with fighting officers ... we went over to Paris Plage [a nearby resort] every day unmolested by the mutineers ...'.[138] The 'canaries' and military police, not regimental officers, were the targets of the mutineers' hatred.

Two historians have seen the Etaples mutiny as marking a major change in officer–man relations, with generals no longer prepared to trust the Other Ranks. They also dismiss junior officers' opinions on the essential reliability of their troops as wishful thinking.[139] The evidence presented in this book casts doubt on both of these statements. If the Etaples mutiny is seen as the military version of a strike, there is no need to question the reliability of the army as a whole. At unit level, a basic reserve of goodwill between the ranks, together with a measure of tact and man-management skill on the part of the officers and the essential willingness of the men to obey, providing, of course, that the unspoken code of the deferential relationship was observed, served to defuse most problems. In the very different conditions prevailing at Etaples, rankers began what Julian Putkowski has termed 'collective bargaining in khaki'.[140] This was not dissimilar to workers' behaviour in prewar industrial disputes. The British working class was heavily trade-unionised, with a strong tradition of both collective bargaining and the protection of skilled workers' prerogatives.[141] There are obvious parallels with 'collective bargaining by riot' in the eighteenth century.[142] The men took direct action by breaking out of camp and enjoying the forbidden fruits of Etaples town, but the mutineers' aims were limited, not revolutionary.

The Etaples strike resulted in the removal of an unpopular Assistant Provost Marshal and Base Commandant, the town of Etaples being thrown open to the men and, ultimately, the training function being removed from Etaples and made a Corps responsibility, where more sympathetic treatment for the men might be expected. Capt. Dible recorded that officers were divided into two schools of thought over the appropriate response to the events at Etaples. Older, Regular officers were in favour of repression, while others were in favour of discovering the men's grievances and rectifying them. To adopt the former response, Dible argued, would be to risk a further outbreak, which might escalate into much greater disorder.[143] In fact, the military authorities, by making relatively minor concessions, succeeded in halting the single most serious disturbance of the war, and removed the basic ingredients which might have caused a second major mutiny. In sum, the only major British mutiny on the Western Front highlights the importance of good officer–man relations in maintaining military morale and cohesion, and the generally sound state of officer–man relations in the BEF.

A number of minor mutinies also occurred on active service. In 1917 Alan Thomas, an officer of 1/6 R. West Kents, was faced with a strike of men on a carrying party. Thomas, who had some sympathy with the men's grievances, threatened them with his revolver but in retrospect believed he should have accepted the protest as a *fait accompli*, rather than risking a test of his authority. One should imagine many such situations occurring but going (for understandable reasons) unrecorded, being defused either by immediate disciplinary action, or the threat of it, or through tactful handling.[144] The latter course was adopted in 1915, when a battalion was ordered to carry out stamping exercises to help prevent trench-foot. The men reacted angrily, but their officers exercised tact and turned the order into a joke, thus securing 'the fulfilment of the orders without much ill humour'.[145]

Some officers took this style of tactful, consultative leadership a stage further. On taking command of 12/Middlesex in 1916, Lt Col Frank Maxwell anticipated the leadership style of Montgomery in 1942. Maxwell had all the men of the battalion gather around informally, and gave them a 'pep' talk, spicing his remarks with a number of jokes. His words were matched by his deeds, for he arbitrarily abolished Field Punishment No. 1 in his battalion. Maxwell firmly believed that a man should be treated as 'a human being with intelligence'; too many officers, he believed, did not have the knack of engaging the men's interest when giving orders.[146] Not all officers went as far as Maxwell or Barnett-Barker of 22/R. Fusiliers, but many used tact and a few basic man-management techniques.

The extent of such leadership by negotiation should not be exaggerated. Pte J. Cuthbert (9/Cheshires), executed in 1916 for disobeying an order to go into No Man's Land with a wiring party, seems to have been the victim of an attempt at negotiation which went disastrously wrong.[147] The scale of such activities, and the circumstances under which they took place, must await further detailed research. However, the very existence of leadership by negotiation is an important corrective to the idea that the Other Ranks were blindly obedient cannon-fodder. British officers learned that in dealing with citizen-soldiers, the application of 'Regular' methods did not always produce the most effective results.

The continued existence of 'auxiliary' styles of officer–man relations and discipline on active service depended on the survival of personnel to pass on the traditions and spirit of the original unit to replacements, or on receiving replacement officers sympathetic to the original ethos. Under the Left Out of Battle (LOOB) system, which was introduced in 1915–16 (some units may have adopted it earlier) 10 per cent or so of personnel were held back from action to form a cadre on which a unit could be rebuilt.[148] An officer of R. Dublin Fusiliers argued that the LOOB system was important 'in preserving the tradition and experience so essential after heavy losses ...'. The LOOB system helped to offset both the effects of the mass casualties of the Somme offensive and the cross-posting of drafts which also seems to have begun in July 1916. This unpopular system resulted in, for example, London Rifle Brigade men arriving at the London Scottish and vice versa.[149] Occasionally, it seems that transfers of men between battalions were arranged, but in general they were prohibited.[150] The integrity of many units was dealt a further blow with the wave of amalgamations and disbandments in February 1918. The angry reaction of all ranks of 11/Argylls when it was announced that they were to be absorbed by 1/8 Argylls was typical of many men facing the disappearance of their unit: 'It will completely destroy the esprit de battalion of a good mob.'[151]

The nature of a unit's disciplinary regime could be judged by whether soldiers were forced to attempt parade-ground levels of smartness on active service. Attitudes to 'spit and polish' could vary greatly between units of similar types. Sgt L. Davidson, a sapper, on moving from one Regular division to another, found that much more emphasis was placed on smartness in the 8th Division than in the 2nd Division.[152] Cleaning and polishing was usually left to rest periods, although they were not unknown in the trenches. Such activities were resented by many soldiers, because they ate into the little time available for relaxation, and were seen as essentially pointless. In one war-novel written by a former NCO, it is implied that a unit's combat performance was actually enhanced by disregarding bull.[153] New Army and TF units tended to be more likely to forgo bull in the trenches than Regular units, but this rule was not infallible.[154]

In most cases, direct recruiting into the TF came to an end after 11 December 1915. Thereafter, men enlisted for general service. TF battalions thus began to receive drafts of men without any experience of the Territorial ethos, although the occasional draft of men might 'fortuitously strengthen the territorial element'.[155] Thus in 1918 1/5 DLI received a draft of 'splendid material' from the yeomanry.[156] More typical was the experience of 1/7 Middlesex. Continuity with the prewar unit was sufficiently strong for the companies to be referred to by territorial titles: A Company was known as 'Hampstead and Highgate' and B as 'Enfield', until heavy casualties in September 1916 brought about 'practically the end of the original 7th Middlesex'.[157]

The impact of heavy casualties on a cohesive unit was discussed by an officer of 1/6 W. Yorks. This unit, whose prewar disciplinary code was discussed in Chapter 2, was 'from April to December, 1915 ... a self-conscious Territorial unit'. Prewar experience had produced a cohesive unit with excellent *esprit de corps* and officer–man relations, but it was a 'closed corporation'. Reinforcements were regarded as 'interlopers' and 'ragged unmercifully, or, worse still, left severely alone'. However, the fighting in 1916 'smashed up a good deal of this "Territorial" influence'. Large numbers of original men and officers became casualties, and their replacements came from all over the country: 'what was narrow and "local" in the Battalion died out in the "blood bath" on the Somme'.[158] Heavy casualties could bring about the destruction of a community which predated the war, a point graphically illustrated by the grief of an NCO of 1/5 Buffs after their first major action had cost the battalion 250 casualties.[159] Similarly, an officer of 1/4 R. Berks remarked that during the Somme campaign the nature of the battalion underwent a profound change. However, those original members who were wounded but later returned to the unit were able to restore some of the Territorial spirit of the early part of the war.[160] In the case of 1/4 Loyals, the TF spirit survived combat experience which caused heavy casualties but led to the battalion regarding themselves as 'Storm Troops', an attitude which enhanced morale.[161]

The replacement of Territorial officers by more experienced Regulars was bound to cause problems, even if the 'basic wisdom' of such postings were accepted by TF units.[162] When such Regular officers made no attempt to understand the ethos of the Territorial unit, considerable bitterness could result.[163] By contrast, Lt Col R.N. O'Connor, a Regular appointed to command 2/HAC in May 1917, was surprised by its 'auxiliary' style of discipline, but in reshaping the battalion (which had recently taken heavy casualties) managed to tighten discipline without transforming the ethos of the unit. Initially, a junior NCO of the battalion admitted, the introduction of some Regular ways was resented, but in the long term they were 'a very good thing for morale and discipline'. Significantly, some Territorial ways remained. The CSM of C company, 'Gee' Grose, 'did his best to make life endurable for [the men]' and

> never shouted an order except when on parade. Off parade he ASKED you to do something and if it was a lousy job invariably did so with his arm across one's shoulder.[164]

It seems that the peculiar disciplinary code of TF units was steadily eroded on active service. An officer of 1/20 Londons recalled that early in the war, discipline was based on a 'spirit of comradeship and pride', founded on a common 'volunteer' spirit and local feeling, most soldiers and officers being drawn from the same area of south-east London. Inevitably, in time the local character became diluted, and the battalion atmosphere 'changed and depreciated'.[165] Similarly, in 1/7 R. Warwicks, as the original personnel were replaced, discipline became more 'Regular'.[166]

However, the prewar ethos survived for a considerable period in a number of units. W.H.A. Groom, a ranker of 1/5 Londons (LRB), recalled that in October 1916 the officers and Other Ranks were still largely drawn from the same class, and discipline was based on 'good understanding and *esprit de corps*'. A 'definite change' in the disciplinary system did not occur until April 1918, when 'orthodox army discipline' was introduced, because of the influx of men who lacked experience in the ways of the battalion.[167] However, Aubrey Smith, another ranker, placed the modification of discipline much earlier, in autumn 1916, although he gave similar reasons for the change.[168] Even so, Pte S. Amatt, who arrived at 1/5 Londons in the second half of 1916, noticed a considerable difference between the disciplinary regime of his old unit, 2/7 Essex, and his new one, where less emphasis was placed on 'bull', and there was a 'matey, chummy' relationship between officers and men.[169] Discrepancies between the different pieces of evidence is probably explained by the fact that Smith had come out with the battalion in 1914 and thus would have known the unalloyed auxiliary discipline of a class corps, while Groom and Amatt sampled a modified, but still 'un-Regular', version.

Other TF units also retained the Territorial ethos during the war, largely because of the chance survival of key personnel. 1/7 Manchesters (discussed in its prewar form in Chapter 2) which was fortunate to be commanded by one of its prewar officers, falls into this category, as does 4/5 Black Watch.[170] 47th (London) Division's historian claimed that it remained 'to the end what it was from the beginning: a division of London Territorials' retaining 'homogeneity' and 'civic patriotism'. This is almost certainly an exaggeration, but the many Territorial officers who commanded battalions and brigades, and the Division's Regular officers who had experience of the prewar TF, undoubtedly assisted it to maintain its ethos.[171]

Yeomanry regiments clung to their traditions and character particularly tenaciously. This was symbolised by the adoption of a broken spur as the sign of 74th Division, which was formed from dismounted yeomanry units. As late as June 1916 an officer joined the R. Gloucestershire Hussars in Egypt to find that 'The regiment was essentially a landlord and tenant affair ... a Farmer could lose his son and his landlord in the same battle',[172] while on joining 25/RWF, which was formed from yeomanry units, in June 1918 an officer found 'an entirely different atmosphere' from any other unit he had served in.[173] In 1916 a party of New Zealanders arrived at the 1/1 Lincs Yeomanry, fearing they would be subjected to the 'repressive discipline' of the British army. An officer took them aside and 'told them something about the Yeomanry and told them that in the case of difficulties, they must come straight to me'. This regiment was converted to a battalion of the Machine Gun Corps in April 1918, but the peculiar yeomanry ethos survived, thanks to a sympathetic commanding officer: 'he understood and was happy with Yeomen'.[174]

Some Regulars had initial doubts about the paternalism of TF officers. A Regular officer of 51st (Highland) Division argued that, during the early period of training, paternalism had not yet become second nature:

> They had not been taught to carry the rifles of the weary, to encourage the footsore.... They did not yet know that the welfare of their command must always be their first consideration.[175]

On active service, most Territorial officers seem to have matched the paternalism of the Regulars, even in the early months of the war: the CO of 1/6 Londons, for example, attempted to get extra sugar for his sweet-toothed soldiery in May 1915.[176] There are, however, some examples of poor man-management in TF units. Pte Gibbons wrote that the officers of his London battalion tended to go to their billets at the end of a march, leaving the men to the care of an NCO.[177] Similarly, Pte Tilsley, who served in a TF battalion in 55th Division, complained about poor officer–man relations in his unit – although he did praise one officer.[178] The differences that existed between individual units, and the problems of generalisation, are demonstrated by the fact that an officer of one TF unit, 1/5 Lancashire Fusiliers, spoke of the 'tradition'

> that an officer should look after his men, [and] see to their billets and food before anything else. Treat them on Parade with correct discipline and off Parade as human beings. Above all, one had to be fair. If one had to dish out punishment and it was fair, there was no ill feeling.[179]

Many of the points made above about Territorial units are also applicable to the New Armies, although unlike first-line TF units, Service battalions lacked a longstanding tradition of informal discipline, and were perhaps more vulnerable to Regular discipline creeping in. The 22/R. Fusiliers' enlightened disciplinary system survived until the battalion's disbandment in February 1918. Barnett-Barker took the 22/R. Fusiliers to France in November 1915 as its commander, and was succeeded in late 1917 by Maj. W.J.T.P. Phythian-Adams, who had joined the battalion as a temporary subaltern in 1914. Maj. C.R. Stone, who had joined the unit early in 1915, became second-in-command. Barnett-Barker expressed his fears as to what would have happened if the command of the battalion should fall to an outsider, who, he implied, would not have understood the ethos and peculiarities of the unit. A hard core of original rankers also seems to have remained with the 22/RF.[180] By contrast, the discipline of another Pals unit, 11/Welsh (Cardiff City) was became more 'Regular' following the arrival of a new colonel in November 1915. This event seems to have ended the use of auxiliary-style discipline in this unit.[181] One private complained in late 1916 that 'discipline is getting harder to stick every day'.[182]

Keith Grieves' research into the Southdown battalions (11, 12 and 13/R. Sussex), raised by Claude Lowther MP in 1914, suggests that these units had a 'greater sense of independent will and resistance to ill-considered military procedures' than was normal in the Regular army. The distinctive character of the battalions was diluted by non-Sussex drafts which arrived from September 1916 onwards, and the tradition of looser discipline probably declined as a consequence. However, it is noteworthy that 11/R. Sussex enjoyed a measure of continuity of command, which had obvious implications for the survival of the Pals ethos of the battalion.[183]

The extent to which a surviving cadre of rankers and commanders committed to a non-Regular ethos could preserve auxiliary discipline is illustrated by a formation akin to a Kitchener division, 63rd (Royal Naval) Division (RND). The RND clung to its distinct naval identity, which helped to create and nurture a high level of *esprit de corps* and inter-rank solidarity.[184] It also had a non-Regular approach to discipline, placing, for example, little emphasis on parade-ground drill. Three crucial factors in the survival of the RND's unique ideology may be mentioned. First, throughout the war, with one brief exception, the RND's commanders fought tenaciously to retain its naval character against military attempts to make it conform to army ways. Secondly, large numbers of men were promoted from the 'Lower Deck' to become officers in the RND, ensuring that a substantial proportion of the officer corps would consist of men schooled in the peculiar ethos of the formation. Thirdly, the return of RND veterans to the Division as reinforcements was all-important. As an RND officer commented, 'A division filled up day by day with strange reinforcements would in a week have lost its identity.'[185]

A Kitchener unit which suffered this fate was 16/R. Scots (2nd Edinburgh Pals). It sustained heavy casualties and by the end of 1916, in the assessment of R.W.F. Johnston, a temporary ranker-officer, 'the coinage was becoming debased as few of the original volunteers remained and as the experienced personnel became casualties'. A draft of men with less than three months' service arrived in November 1917, along with men who had formerly served with the ASC and cavalry: 'They were disgruntled and unwilling, in the main, to acquire the regimental spirit which must be fostered if personnel are to become good infantrymen.' Generalising from his experience with 16/R. Scots, Johnston, who also served with Regular and TF battalions, claimed that in 1917 Kitchener units 'varied in quality' and lacked the 'regimental pride' of the military professional.[186]

Clearly, this judgement is too sweeping. Many units, 10/Lincolns for one, maintained their identity to the end of the war.[187] On being posted in October 1918 to 13/R. Fusiliers, a Kitchener battalion, an officer was surprised and pleased to find that 'something of the old friendly spirit' had survived despite the turnover of personnel.[188] Maj. R.S. Cockburn, a ranker-officer, gave an important insight into evolution of discipline in 10/KRRC. He argued that the battalion's discipline was founded on comradeship and 'mutual support' of officers and men, who both recognised the need 'to carry on as best we could'. Early in the war 10/KRRC 'made a pretence' of abiding by Regular discipline, while in reality such standards could not be achieved: 'we had not the time, nor had we the NCOs'. However, the officers' belief that the battalion 'as a body' was 'more intelligent' than prewar recruits 'militated against a desire to be ruled entirely by the rod'. Therefore the 'ideal' was recognised, even if Regular disciplinary methods were not always employed, but

> Even the ideal itself faded into the very far distance at a later stage, when any satisfactory sort of discipline became difficult; when, that is we were given untrained drafts to replace casualties.[189]

The CO of 8/E. Surreys, Lt Col A.P.B. Irwin, believed that replacements that arrived in late 1917 were distinctly inferior in quality to the volunteers of 1914. However, to his great surprise, these 'huge drafts', which included conscripts, 'all became 8th East Surreys in no time at all'. This survival of *esprit de corps* owed much to the continuity of command (Irwin had been the adjutant in 1914). Officers and NCOs made a practice of briefing new drafts on the regimental history of the East Surreys, and the past glories of the battalion. Finally, as Irwin himself said, the fighting reputation of the battalion (and indeed the 18th Division as a whole) inspired strong loyalty to the unit.[190]

As the example of 12/Manchesters, bled dry on the Somme but revitalised by the appointment of an inspirational CO, Lt Col Magnay, demonstrates, battalion spirit could be rekindled, given the right commanding officer.[191] Likewise, the usual process whereby informal discipline was replaced by Regular discipline could be reversed, as in the case of 17/R. Fusiliers[192] and 6/R. Scots Fusiliers. Winston Churchill, who commanded the latter in 1916, was a 'fire-eating', aggressive commander, who in some respects attempted to introduce Regular methods into the battalion by drilling the men, improving their appearance and providing 'more style & polish'.[193] However, he also modified the disciplinary code of the battalion, most obviously by reducing 'punishment both in quantity, & method'.[194] What has not been grasped by previous writers[195] is that Churchill, despite his service as a Regular officer in the 4th Hussars, applied to the Scots Fusiliers the style of discipline he had known as an officer in a Yeomanry regiment, the Queen's Own Oxfordshire Hussars. This was very different from that imposed by the battalion's previous (Regular) CO, and the unit's temporary officers, schooled in Regular ways, initially reacted with hostility to Churchill's innovations. There is also some evidence that the paternal role of the officer had been neglected, since Churchill took pains to organise sports and concerts for the men, who appreciated their colonel's efforts on their behalf.[196]

A recent study of the casualties sustained by one Regular unit, 2/Yorkshires (Green Howards), in 1914–15 concluded that the prewar Regular army 'had virtually ceased to exist after some fifteen months of fighting'.[197] For the most part, the replacements received by Regular units were wartime volunteers and conscripts. Yet Regular units were more likely to have retained a distinctive character than their New Army or Territorial counterparts. Rawlinson noted in December 1918 that, despite containing only 'a sprinkling of professional soldiers', 1st, 2nd and 4th Divisions clung to 'the traditions of the Regular army', possessing even at that late stage a different character to 46th and 50th (TF) Divisions.[198]

Continuity of command helps to explain the survival of a distinctive Regular ethos. Regular officers tended to be appointed to command Regular units, which also seem to have received many, if not most, of the wartime products of Sandhurst. In addition, temporary officers posted to Regular units generally quickly absorbed the ethos of their regiment.[199] Surviving prewar Regular and Special Reservist rankers would also pass on the traditions of the unit. Writing of early 1915, Cpl John Lucy (2/R. Irish Rifles), horrified by new NCOs calling privates by their Christian names, stated that the Regular army 'was finished', but went on to say that 'We remnants' clung together in a 'form of freemasonry', which 'preserved and passed on the diluted *esprit de corps* of our regiment.'[200] Much the same process seems to have occurred in 1/R. Scots. When R.W.F. Johnston joined this unit as an NCO in December 1915, that is, after it had served for 11 months on the Western Front, he found that the Regular ethos was very much alive. He felt that 'life was easier and more agreeable in the rather different atmosphere' of his previous unit, 1/9 R. Scots (TF); in the Regular battalion, discipline was stricter and 'everything was done in the regimental way'. Johnstone also felt that 'the professionalism of the Regular Army even in 1918 was definitely evident to everyone despite its serious losses in officers and men'.[201]

This is not to say that discipline in all wartime Regular units remained as strict as it had been before the war. Both prewar officers and NCOs of 2/R. Welsh Fusiliers believed that standards had declined on active service.[202] On joining 2/R. Berks from its sister Regular battalion in 1915, Capt. Hanbury-Sparrow was perturbed to find the second battalion's combat performance had suffered because, he felt, 'strength-giving discipline' had been rejected in favour of 'easy-going ... South African War' discipline.[203] Several sources suggest that the barriers between Regular soldiers and officers began to drop even in the first months of fighting in France.[204] The example of one Regular battalion, 2/W. Yorks, suggests that this could have a modifying effect on discipline.

Sydney Rogerson, a temporary officer, published a book in 1930 which described life in 2/W. Yorks in late 1916. Liddell Hart described it as the most accurate recreation of the '*normal* atmosphere of a Battalion' that he had read. Rogerson felt that it was ridiculous to try to enforce 'the conventional formalities of discipline' while in the trenches. Indeed, he believed that the best way for an officer to motivate his men was to 'treat them as friends'. This entailed 'a relaxation of pre-war codes of behaviour', but without allowing familiarity.[205]

There are several points worthy of note about this passage. First, Rogerson's attitudes were something of a halfway-house between those of a prewar Regular officer and those of TF and New Army officers, some of whom clearly did allow what Rogerson would have considered to be 'familiarity'. Second, the battalion was commanded by a Regular officer, Lt Col J.L. Jack, who clearly did not share the radical views of Philip Howell. Nonetheless, Jack quietly modified the Regular discipline of 2/W. Yorks. In periods of bad weather, for instance, he relaxed his rule that men should shave daily.[206] Jack had the typical Regular officer's paternal concern for his men and it appears that he came to believe that a slight modification of discipline contributed to the well-being of the ranks.[207] Clearly, Rogerson learned his approach to officer–man relations from his colonel, whom he greatly admired. Jack, wrote one of his subalterns, exuded an air of 'dignity and control', winning the 'respect' of the men by his courage and by occasionally 'lifting ... the barrier of rank' and engaging in 'personal ... man-to-man encouragement', without ever courting 'familiarity'.[208]

As ever, broad generalisations about officer–man relations and discipline are difficult. One conscript private of 2/Worcesters bluntly described the notions of front-line 'comradeship between officers and men', and that 'an officer always looks after his men' as 'legend[s]'.[209] By contrast, a middle-class ranker in another regular unit, 1/Glosters, resented discipline and bull but singled out one of his officers as 'a nice fellow'.[210] Discipline was modified in many Regular units during the war. An officer of 3/Grenadier Guards recalled, in words reminiscent of Rogerson's, that 'the ultra-strict and sometimes too impersonal enforcement of discipline' was 'gradually modified' and a more informal and friendlier relationship evolved. However, 'the essentials of discipline were very much retained'.[211] Changes in discipline were not so drastic as to bring Regular units to resemble prewar auxiliary regiments; yet changes there were.

According to Keith Simpson,

> By 1917 the distinctions between what had been regular, territorial or New Army units had blurred or disappeared. Although many soldiers clung to their distinctive identity, for all intents and purposes the British soldier on the Western Front had become a 'National Serviceman'.[212]

As a generalisation, this view has much to recommend it, although there were numerous exceptions to the rule. It does seem that unless a cadre of 'original' soldiers and officers survived to pass on the traditions of a unit, or a sympathetic outsider was appointed to command it, a Regular style of discipline tended to replace auxiliary discipline in New Army and Territorial units. However, as argued in earlier chapters, this did not affect the relationship of the officer and his men – indeed, it may even have enhanced it, as junior officers sought to protect their men from the excesses of the disciplinary system.

9
Comparisons

This chapter attempts to set the wartime British army's style of officer–man relations and discipline into a wider context by examining the forces of two Dominions, Australia and Canada, and the British army of the immediate postwar period. The chapter begins, however, with a discussion of the experience of a force from outside the British tradition: the French army.

French discipline of the Great War has gained a reputation for savagery in the Anglo-Saxon world, although the traditional view has been considerably modified by recent research.[1] However, soldiers' dissatisfaction with discipline was a factor in the growth of 'tension' between French officers and their men during the war, compounded by the French officer's lack of paternalism. Unlike the British army, the French army was not renowned for its provision of baths, canteens, sports and other amenities for its troops.[2] Officers of the French army 'did not, by tradition or by custom picked up in the service', share the British concept of *noblesse oblige* and the primacy of the welfare of the men.[3] This was despite, or possibly as a consequence of, the fact that some 50 per cent of officers in the prewar French army were former NCOs. Unlike the British ranker, the *poilu* did not receive, in exchange for deference, his officer's paternal care. Some soldiers admired the bravery of their regimental officers, but it was their 'humanity', not their courage, that was in question.[4]

These problems had their roots in the prewar period, when, for a variety of social and political reasons the French army was beset by indiscipline, and the authority of officers and NCOs was undermined. The officer's role increasingly became one of administration rather than leadership, not least because to concentrate on administration was to avoid the difficult job of commanding undisciplined troops.

Inevitably, officers and men became increasingly distant, a situation exacerbated by the heavy casualties sustained by the Regular officer corps in the opening months of the war.[5] Contemporary British writers portrayed the French army's officer–man relationship as friendly and paternal.[6] In reality, French replacement officers had even less idea of translating the ritualistic verbal paternalism of calling their soldiers '*mes enfants*' into practical concern for their men than their Regular predecessors.[7]

Poor inter-rank relations played an important part in the French army mutinies of spring 1917. In a report of June 1917, postal censors noted that middle-aged rankers resented being treated in a cavalier fashion by officers who were little more than boys. Confidential reports of 15 and 21 July emphasised the importance of the officers' attitudes, noting the damage done by officers who were contemptuous of, or haughty towards, their men. By contrast, soldiers liked friendly, paternal officers. A recent writer has stressed the 'the crucial mediating role' of regimental officers in 'restraining the mutinies'.[8]

Pétain certainly realised the importance of restoring confidence between leaders and led. He issued orders that attempted to correct faults in the officer corps, and followed this up by visiting units, where he would give practical demonstrations of paternalism by ordering improvements in food and checking on leave-rosters.[9] By the summer of 1917, official reports were indicating that French officers were changing their attitudes to the ranks, and the men appreciated the change.[10] By September, the immediate crisis was over.

While it is not suggested that poor officer–man relations were solely responsible for the mutinies of 1917, they did play a significant role.[11] The French experience throws into stark relief the importance of the paternalistic ethos of the British officer corps in maintaining morale, and its survival among temporary officers. The loss of large numbers of Regular officers in 1914–15 did not bring about the end of paternalism in the British army.

The discipline and inter-rank relations of Canadian and Australian troops also provide useful comparisons with the British army. These forces were part of the BEF, and many Dominion officers and men were born in Britain. However, Australian and Canadian disciplinary codes have traditionally been depicted as looser, and officer–man relations more informal, than the British variety. This was supposedly the result of a frontier ethos that engendered in 'colonial' troops the dash, élan and initiative that was allegedly lacking in British units.

C.E.W. Bean's influence on perceptions of the Australian Imperial

Force (AIF) has been immense, and only recently have historians begun to challenge it. Bean, the official historian, argued that the AIF was an egalitarian, democratic army, characterised by the 'bush values' of toughness and self-reliance which influenced urban dwellers (who formed the majority of the army) as well as soldiers from rural areas. Bean saw these values, and the quality of the AIF's officers, most of whom came originally from the ranks, as vital factors in its military effectiveness.[12] Furthermore, he argued that British 'feudal' discipline and class-based inter-rank relationship were inappropriate for products of a society characterised by democracy and 'mateship'. Therefore, self-discipline and informal officer–man relations prevailed within the AIF.[13]

Australian society in 1914 was not classless but it was less hierarchical than its British equivalent, and in some units Australian officer–man relations tended to be rather less formal than the British variety. Cpl J.F. Edey (5/AIF) long remained a friend of one of his officers. Lt Frank Corlett would come along the trench to supervise Stand-To:

> Frank would grasp me by the PH [gas] helmet laying on my chest and ... say 'And how's the Western Front this morning, Jimmie?'[14]

In 1916, an amazed British private heard an Australian ranker address an officer as 'Joe'[15] while a British ranker attached to 2nd Australian Tunnelling Coy in 1917 recorded that Anzac officer–man relations were too informal for his taste, but he 'never saw an Australian officer lose the respect of his men'.[16] The reputation of the Australians for informality even reached the pages of a Canadian trench journal.[17] However, Dale Blair's recent research into 1/AIF suggests that at the very least, assumptions of inter-rank informality throughout the AIF should be called into question.[18]

Writers have often repeated Bean's views, sometimes in a form simplified almost to the point of parody,[19] for 'Mocking the upper-class Englishness of English officers was part of the [Anzac] tradition, the obverse of its egalitarian element, and so was condescension towards English soldiers.'[20] A recent view by an academic redresses the balance somewhat but still overstates the differences between Australian and British officer–man relations and discipline, and links them to allegedly superior Australian tactical performance.[21] Part of Bean's purpose in writing was celebrate the 'Digger', the ordinary Australian soldier. Bean was not an analytical historian; he was 'a myth maker, or bard – the Homer of the AIF'.[22] The AIF was indeed raised from a society that was different from Britain's, and its officer–man relations and discipline reflected this fact. Sir Ian Hamilton was one of a number of Britons who wrote on the differences between British and Australian society. He commented on the unwritten Australian rule that there should be no displays of respect, and argued that privates did not understand for what the officer stood.[23] Differences should not, however, be exaggerated. The comments of an Australian chaplain about the need for fair play towards the men were just as applicable to British soldiers.[24]

In reality the AIF was less egalitarian than Bean would have the reader believe. It was not until January 1915, for example, that it became *de rigueur* for officers to come from the ranks. In addition, it appears that few officers of the AIF came from 'labouring backgrounds'.[25] Bean himself rather inconsistently admitted that educational and social factors did play a role in the selection of officers, although he was at pains to argue that socially elite officers were no more effective than products of state schools.[26] Lt F.M. Stirling (29/AIF), an Australian public-school and 'Varsity man, displayed impeccable sporting and paternal attitudes of which an old Etonian would have been proud,[27] while a ranker of 8th Australian Light Horse referred approvingly to his old CO as 'An officer and a gentleman'. Cpl. W.C. Gamble (25/MG Coy, AIF), who served in France in 1917 and 1918, stated that ex-public schoolboys became officers 'irrespective of whether they had the backbone or ability to lead men'.[28]

In practice, the Australian system of commissioning officers was virtually identical to the British. The major difference was that it was usual for Australian ranker-officers to return to their former units. However, Bean exaggerated when he wrote that this policy 'was entirely opposed to the practice in the British Army'.[29] Edmonds, the British official historian, in commenting on a draft of one of Bean's 1918 volumes, took pains to correct some of Bean's misapprehensions about the wartime British officer corps. Edmonds pointed out that the British army had commissioned large numbers of rankers from humble social backgrounds, but Bean does not seem to have let the facts spoil his argument.[30]

Bean, who was far from the only Australian who misunderstood the British system of commissioning officers,[31] also claimed that 'A beneficial result of the whole system was that the Australian officer was much closer to his men than was his British colleague.'[32] This overlooks the fact that Australian officers, like their British counterparts, enjoyed a significantly superior lifestyle to that of the men they commanded. In 1917, Lt S.P. Boulton commented that as an officer he was 'appreciating to the fullest all the extra comforts etc. compared with all former experiences'.[33] Moreover, a close reading of Bean's own text reveals that inter-rank relations in the AIF were in some ways similar to those in the British army. Even on Gallipoli, a 'clear distinction had to be preserved between officers and men'.[34] By 1916 the ranker-officer had, for the sake of 'good discipline' to 'break with his old associates'. After a farewell dinner 'their relations were formal'.[35] But although Bean's claim that the 'character and competence' that men demanded of their leaders 'came to be the sole criteria' for the granting of commissions ignores the importance of social factors in this process,[36] it was in essence true enough.

However, much the same was true of the British army from mid-1915 onwards. Other virtues claimed by Bean for the AIF officer, such as battlefield leadership and paternalism, also characterised the British officer.[37] Finally, it should be noted that the fact that Australian inter-rank relations tended to be less formal than British does not imply that there was any less mutual respect, admiration and affection between the ranks in the BEF than there was in the AIF; each relationship reflected the nature of the home society.

Senior British, and indeed some Australian officers tended to blame AIF officers for the indiscipline of their troops.[38] After the war, General Sir John Monash, commander of the Australian Corps in 1918, proudly proclaimed that the AIF used the individuality of the Australian soldier as a basis on which to build collective 'battle discipline',[39] implicitly comparing Australian with British discipline, to the disadvantage of the latter. In the words of an AIF ranker, 'Australian discipline does not permit of unthinking obedience to senseless orders....'[40]

Australian discipline has attracted many myths. Self-discipline, as opposed to 'imposed' discipline was not a uniquely Australian trait. The popular image of the indisciplined Australian is based largely on the 1914–15 period, when the AIF experienced many disciplinary teething troubles. The British New Armies, another citizen force raised at the same time, faced similar, although less acute, disciplinary problems. Both forces of citizen soldiers shared the heritage of British 'auxiliary' discipline.[41] By the time the AIF reached the Western Front in 1916, its discipline was considerably tighter, and was to further improve by the end of the war. Paradoxically, 'discipline', in the sense of low numbers of military crimes, was generally good on Gallipoli, but grew steadily worse after the AIF's arrival on the Western Front, growing to crisis proportions by the end of the war.[42]

Battle-discipline is the obverse side of the coin from more general discipline. While the failure of Australian soldiers to salute officers was perhaps inconsequential, the same is not true of the 1915 riots in the Wasser, Cairo's brothel district.[43] *Pace* the postwar writings of Bean and Monash, Australian discipline had a great deal in common with British. A private of 59/AIF wrote in 1916 in terms that might have been used by a middle-class ranker in a British unit:

> One is practically a prisoner. You have to have lights out at certain hours and cannot talk after lights out. Even going on swimming parade you are marched by an officer or NCO in the strictest manner. I think they begrudge you your brains to think with at times.[44]

Monash's postwar views were at variance with his wartime ones.[45] Senior Anzac officers saw the tightening of discipline as essential to the maintenance of the reputation of their divisions as the elite 'shock-troops' of the BEF. Thus, although some historians believe that Australians 'would not stand for' Field Punishment No. 1,[46] in reality, 'The Anzacs accepted the standards in force in France and pursued them rigorously.'[47] However, for reasons of Australian domestic politics, the 121 soldiers sentenced to death were all reprieved. The view of the Anzac Provost Corps, that discipline in Palestine was 'good' in July 1918, with the few incidents being ascribed to small numbers of 'men of bad character' who 'have done their best to ruin the good name of the AIF', or the similar views of their counterparts in England,[48] contrasts sharply with the widespread indiscipline of 1915.

Until the appearance of Ashley Ekins' major study, any assessment of Australian discipline must remain provisional, but there does appear to have been a paradox: in the second half of the war, Australians were better disciplined, but committed more offences.

Lacking a Bean, the tradition of a Canadian frontier ethos producing informal discipline and officer–man relations is not as well developed, but it nonetheless exists.[49] This school has been attacked by scholarly historians, one arguing that

> [Canadian] battlefield excellence derived not from any innate superiority born of the North American frontier ... but primarily from British tutelage and the hard crucible of war.[50]

The myth of Canadian indiscipline is, like its Australian counterpart, rooted in the earliest months of the Canadian Expeditionary Force (CEF)'s existence. The problems of raising a citizen army virtually from scratch were exacerbated by politically appointed officers.[51] During this period the Canadians acquired a reputation for indiscipline[52] but in time, discipline developed, aided by tactful but firm handling by Byng, the British commander of the Canadian Corps in 1916-17.[53] In some cases Canadian units seem to have relied on 'self' rather than 'imposed' discipline, and liked to compare their informality with the British (or 'Imperial') approach. Canadians were, however, quite capable of insisting on a measure of 'spit and polish' behind the lines even if it was neglected in the trenches.[54] Jack Seely, the British commander of the Canadian Cavalry Brigade, concluded that Canadian discipline was informal but good:

> the Canadian Army was very flexible. It found room for everybody, and managed with great success to put people to their own jobs. But let nobody think that these eccentricities relaxed real discipline. I can truly say that ... I never had a rebellious word or look, nor once was an order disobeyed.[55]

Berton claims that 'an easiness' existed between Canadian officers and men that was 'foreign' to British forces.[56] Isabella D. Losinger's authoritative work on Canadian officer–man relations throws considerable doubt on the first part of this statement, as does the work of Desmond Morton,[57] and the memoir, based on diaries, of Pte D. Fraser (31/CEF and 6th Can. Bde MG Coy.) Fraser's journal could almost pass for an account by a soldier in a British Service or Territorial battalion. Fraser was an immigrant Scots clerk with no previous military service. Like his contemporaries in the BEF, he judged his officers largely by their treatment of the men and their performance as leaders.[58] When he wrote of 'the officers fraternizing with the men' on Christmas Day, 1915, the obvious implication was that normally officers did not fraternise with the rankers. However, the CEF often posted ranker-officers back to their old units and this probably enhanced the informality of the relationship. Thus Fraser wrote of one of his officers, a fellow Scot who had served as a private in Fraser's company: '[he] was a great friend of the writer. He confided in me greatly ...'. This situation does not seem to have presented any great problems of familiarity.[59]

The ranker who unfavourably compared inter-rank relations in his old unit, 3rd Canadian Division DAC, with those in his current unit, a TMB of the same division, is a salutary reminder of the difficulty of generalising about officer–man relations, which differed from unit to unit,[60] but in one case at least, Canadian indiscipline was linked by an observer to poor leadership. A prewar Canadian Regular officer, Capt. H.R. Hammond, steeped in the paternalistic ethos of the British Regular army, was unimpressed by the officers of 47/CEF, although he judged them to be little worse than officers of other Canadian battalions. He criticised a major for failing to set an example to his men, and the colonel for lacking paternalism and man-management skills, which led to indiscipline. 'It is not really the men's fault', commented Hammond, 'they are like children and have not been handled properly'.[61]

One can conclude that Canadian and Australian discipline tended to be looser and officer–man relations a little more informal than was usually the case in many British units, but the differences between 'Imperial' and 'Dominion' troops should not be overstated. The Dominion approach was rather more, and the British rather less, formal than is commonly believed, with the Canadian style perhaps more closely resembling the British than the Australian. Moreover, the idea that enlightened Dominion commanders deliberately tailored the disciplinary system to match the needs of their men, who were imbued with a frontier ethos, is at very least open to question.

On hearing the news of the Armistice on 11 November 1918 a battery sergeant-major exclaimed '"Now that the war is over, we can get down to real soldiering"': and 'everyone knew what that meant'.[62] In reality, another four years passed before the British army returned to a pre-1914 pattern of life. 1919–22 marked a period of transition from a mass citizen-army to a small Regular colonial gendarmerie. The army, 3.75 million strong at the Armistice, was rapidly demobilised. Most war-raised units were disbanded at a time when Britain had greater military commitments than ever before.[63] The army of this period, consisting of re-enlisted serving soldiers, young volunteers, and men conscripted in the latter part of the war and compulsorarily retained until the end of April 1920, was a very different force from the wartime BEF.[64]

After the Armistice, many soldiers changed their attitudes to authority and discipline. Many perceived the demobilisation procedure as unfair. Moreover, many wartime volunteers and conscripts seem to have believed that, in the words of one of them, 'Our contract is finished and we want our ticket', for in defeating Germany they had completed the job for which they had enlisted.[65] An NCO stated that there was 'a spirit of revolt against the system which had held the individual for so long'.[66] Men who had uncomplainingly accepted military discipline while hostilities were in progress now began to resent it, particularly if the military authorities attempted to impose prewar standards. A 1914 volunteer serving with 2/6 Sherwood Foresters in the Rhine Army spoke for many when he wrote: 'Spit & Polish we dont (*sic*) like that after Active Service'.[67]

While the most dramatic manifestations of indiscipline were the large-scale strikes in rear areas and in England, mainly involving lines of communication troops, there were also numerous minor incidents in front-line units.[68] A Regular staff officer noted that temporary officers as well as rankers adopted an attitude of 'We'll soldier no more.'[69] Certainly, many temporary officers had considerable sympathy with their men. Alan Thomas of 1/6 RWK recalled a company refusing to parade: 'They could scarcely be blamed. They were still, like most of us, not soldiers but civilians in uniform.'[70] According to a private of l/Devons, the battalion's officers pragmatically accepted the inevitable and eased discipline. This was one of many units in which this process occurred.[71] Some men defied their officers in a rather half-hearted fashion,[72] but there is no evidence of widespread hostility to regimental officers *per se*. Indeed, given the nature of the wartime officer–man relations, it would be surprising if there had been. Even Pte Alf Killick (AOC), a revolutionary socialist involved in the Calais mutiny of January 1919, recalled Capt. Rees, a 'fair man', and mentioned that the mutineers 'felt quite sorry' that the revolutionary struggle should begin while he was in command.[73]

The regimental officer had a vital role in managing the crisis of discipline that followed the armistice. As the postal censor commented, much depended on the character of individual officers;[74] while some soldiers complained about the 'indifference' of officers to the men's 'comfort', not infrequently soldiers praised the 'thoughtfulness' of their officers.[75] Rankers, those content with their lot and those who were not, frequently referred to the interest or otherwise displayed by their officers. The '"real officer" who acts like a gentleman towards his men' was often contrasted with those officers who in civil life were 'office-boys at 25/- a week'.[76]

In 1919 clumsy attempts to enforce 'long hours of tedious and unnecessary drill' provoked the men of 13 Siege Battery RGA into striking. However, as an officer remarked, 'they are still the same men who won the war and if treated in the right way and told the reason for things they are ready to do anything required of them'.[77] Tactful handling by sympathetic officers, who continued in their wartime role of defending their men against capricious authority, and the residual trust of men for their regimental officers, helped many units to maintain a measure of cohesion.[78] The role of the good regimental officer *vis à vis* his men had not changed. Maj.-Gen. Talbot Hobbs believed tactless officers who neglected their men were at the root of disturbances in 3rd Australian Division Artillery in January 1919.[79] However, some rankers' perceptions of the relationship had changed. In 1919 W.R. Bion (Tank Corps), imbued with the paternalistic ethos of the junior officer and accustomed to the close relationships existing within a combat unit, felt himself to be, at 21 years old, 'an antiquity, a survival from a remote past'.[80]

Four specific cases of postwar indiscipline suggest that good officer–man relations were as important in the army of 1919–22 as they had been in the BEF. 'Wully' Robertson, the commander of the Rhine Army in 1919, believed that a major factor in the unsatisfactory morale and discipline of this formation was that officers and men, serving in what amounted to 'new units', were to a large extent 'utter strangers to each other'.[81] Referring specifically to two cases of indiscipline, Robertson denounced officers who were 'out of touch with their men' and who had shown 'lack of consideration and efficiency in handling their units and in managing their interior economy'.[82]

Similarly, Maj.-Gen. Childs, the Director of Personnel Services, believed that the strike of 3/Coldstream Guards on 10 June 1919 resulted from an over-zealous training regime inflicted upon demob-happy soldiers who held a number of not unreasonable grievances.[83] The two most recent studies of the politically motivated mutiny of 1/Connaught Rangers in July 1920, which differ considerably in their interpretations, agree that a subsidiary cause was a failure of leadership by officers.[84]

The indiscipline of the months immediately following the Armistice were not typical of the interwar period as a whole. Anthony Clayton has pointed out that after the initial turbulence of 1919–20 subsided, the all-volunteer Regular army of the interwar years experienced only two, minor, outbreaks of indiscipline. Clayton highlights the role of the officer–man relationship in ensuring that 'loyalty to regiment took pride of place in men's minds'.[85]

The overall character of the officer class did change a little in the years following the war. The percentage of sons of 'Gentlemen' entering Sandhurst fell from 20.5 per cent in 1910 to 9.1 per cent in 1930.

Conversely, the percentage of sons of military professionals rose from 43.8 to 50.8 per cent.[86] Two other groups may have also contributed to this change in social profile. The first were wartime officers who converted their temporary commissions into Regular commissions; the second, officers of lower-middle-class origin who graduated from Sandhurst during the war, when the usual fees were waived.[87] Against this must be set the surprisingly large number of surviving prewar officers.[88]

The findings of the 1923 Haldane Committee on officer training and education would appear to show that there was an official wish to institutionalise the rough meritocracy of 1915–18. It argued that 'In these days it is neither necessary nor desirable to confine the selection of officers to any one class of the community.' The reality was rather different. Rather than lowering fees, the War Office asked county councils to make scholarships tenable at the two officer-training establishments as well as universities. This was not a policy likely radically to alter the social profile of the officer class. Moreover, the recommendation that the Territorials, should, as in 1914–15, be used as a source of officers prompted the comment that 'it is essential to review the exact conditions under which commissions were formerly given. Otherwise the door might prove inconveniently wide.'[89]

Between 1922 and 1930 some 189 rankers received commissions under the Y Cadet scheme, which was supposed to ensure that ex-rankers comprised 13.5 per cent of the officer corps, but a large proportion joined the unfashionable RASC. This threatened to create a social chasm with the rest of the army, and led to the halting of direct recruitment of officers into this corps. In the mid-1930s, Sandhurst intakes consisted of only about 5 per cent of ex-rankers.[90] In sum, although the social profile of the average British officer changed a little, the meritocratic officer corps of 1915–18 quietly vanished in the 1920s.

The ethos of the officer class of the interwar period, with its emphasis on sport,[91] social life, horses and the regimental family was very similar to that of the prewar period. As before, the average ranker was poorly educated, of a low medical standard, and possibly drawn from the ranks of the unemployed, and his life-style would have been instantly recognisable to his predecessor.[92] Not surprisingly, the officer–man relationship in the Regular army of the early 1920s was also very similar to that of twenty years before. Spike Mays, who enlisted as a trooper in the Royal Dragoons in this period, was not uncritical of military life, but he nonetheless wrote that officers

> always inspired confidence and respect in their men.... There was a deep division of status between commissioned and non-commissioned, which did not interfere with the good relationship between officers and men, and the Other Ranks in no way resented the money and splendour of their seniors. On the contrary, they admired them for it because it gave them a bit of cavalry dash and importance. There was friendship as well as discipline, and both were sure and certain.[93]

The evidence of rankers who served in other units in the early 1920s confirms that this picture of officer–man relations was not unique to the Royal Dragoons.[94] There were popular and unpopular officers: Bdr H.L. Horsfield, who enlisted in the Royal Artillery in 1920, recorded the general dislike of an unpopular commanding officer, but his admiration for one of the majors.[95] Pte Bradshaw of the King's testified to the continuation of the 'on parade, off parade' principle in the postwar army.[96] Officers testified to the importance of paternalism, although some believed that postwar rankers showed less automatic deference, the inter-rank relationship relying more on the professionalism and character of the officer.[97]

The character and ethos of the postwar Territorial Army (as the TF was renamed one year after its reconstitution in 1920) also had much continuity with the prewar period.[98] TA officers seem to have been of broadly middle-class origin, with many professional men taking commissions. According to Liddell Hart, the 'quality' of TA subalterns improved considerably by 1931. As before, Yeomanry officers tended to be of a higher social class. The expense incurred by officers ensured that there was a constant shortage of officers, and this fact would seem to preclude working-class men from becoming officers, although paradoxically service as a TA officer was popular with university students, who welcomed the modest stipend. Territorial rankers also seem to have been mainly drawn from the working class, including the unemployed. Liddell Hart commented that most TA rankers were of a similar class to those who had served in the prewar Militia. As before, in some Yeomanry units rankers were of a higher social class.[99]

A wartime officer of 1/5 York and Lancs believed that informal, friendly officer–man relations remained 'the great asset' of the postwar Territorials.[100] As late as 1938 a book on the TA devoted several pages to comparing the informal discipline of Territorial units with the more formal approach of the Regulars. It claimed that, impressive as the 'mutual understanding and cameraderie' that existed between the Regular officer and soldier was, officer–man relations in a TA unit were 'far deeper and more intimate'. The reasons given for the peculiar nature of discipline in the TA – 'the class of men concerned, the spirit which animates them, and the leadership which knows how to handle them and make use of that spirit'[101] – are very reminiscent of descriptions of pre-1914 auxiliary forces. Indeed, for many interwar Territorials, their annual camp was also their annual holiday, just had been the case with their prewar predecessors.[102] An officer who had served with the 1/4 Gordons believed that the atmosphere of the postwar 4/Gordons was even more democratic and egalitarian than it had been before the war.[103] Given the drawbacks of service in the TA of the interwar years – official indifference, hostility in some circles, inadequate rewards – the 'club' spirit was probably more essential than ever.[104]

10
Conclusion

In this book I have argued that officer–man relations in the British army of 1902–22 were generally good. The relationship was a reciprocal one: the ranker gave deference in exchange for the officer's paternalism. The relationship had its roots in the nature of British society, and can in retrospect be seen as a source both of strength and of weakness.

The morale of the BEF remained essentially sound throughout the 1914–18 war. The ethos of the prewar officer corps, which stressed the need for the officer to exercise paternal care for his men, was a major factor in the maintenance of wartime morale. Wartime temporary officers, many of whom came from outside the traditional officer-providing classes, imbibed this code of paternalism at officer-training units. The army's bureaucracy of paternalism reinforced this code in units of the field army. Thus the lower-middle-class and even working-class officers of 1918 were as paternal as their public school-educated predecessors of 1914. The fate of the French army in spring 1917 illustrates the problems of morale that could befall an army bereft of a paternal officer-class in this period.

Paradoxically, in many units good officer–man relations and harsh discipline existed side-by-side. Two disciplinary traditions were represented in the wartime army: that of the Regular army, and that of the Auxiliary forces. The latter was less strict and formal and characterised by a greater reliance on self-discipline. However, as the war progressed, Regular discipline tended to replace the less formal variety in a majority of units. Regimental officers helped to protect their men from some of the excesses of military discipline. The army appeared to many Other Ranks, especially wartime volunteers and conscripts, as a vast, impersonal, arbitrary, coercive machine. Sympathetic and paternal officers helped to off-set the impression that ordinary soldiers were helpless and friendless in the face of an all-powerful authority. Good officer–man relations in some cases enhanced the combat-effectiveness of units, while an important side-effect was the creation of a British war generation, which helped unite ex-soldiers of all ranks.

It is important to note the limitations of good officer–man relations. They could not prevent some cases of 'rough justice'. Neither could they do more than limit the damage wreaked by the widespread indiscipline that broke out in the army after the Armistice. Moreover, good officer–man relations did not necessarily lead to combat-effectiveness. If officers identified too closely with their men, this could lead to a reluctance to take aggressive action and thus put the group at risk. It might be added that there is little evidence to suggest that the regimental officers of 21st and 24th Divisions were any less paternal than those of any other New Army formation, yet the failure of battlefield-leadership was one of the reasons for the poor performance of these formations at Loos in September 1915.

The most important weakness of the officer–man relationship was that it created a culture of dependency. This reduced the scope for independent thought or initiative among the lower ranks because men relied so heavily on their officers, although this tendency should not be exaggerated. Moreover, it was not always appropriate to apply the paternal/deferential relationship and rigid discipline to educated, middle-class rankers. While officers sometimes modified the inter-rank relationship and discipline at unit level, the army neglected the opportunity to make good use of intelligent rankers. However, this fact should be weighed against the relative ease with which men could pass from the ranks to the officers' mess.

When one comes to draw up a balance sheet, the advantages of the officer–man relationship greatly outweighed the disadvantages. Above all, it played a crucial role in sustaining the morale of the British army through four years of gruelling attritional warfare.

Appendix 1

The Morale of the British Army on the Western Front, 1914–18

NB: for a more thorough discussion of this topic, see G.D. Sheffield, *The Morale of the British Army on the Western Front, 1914–18*, Occasional Paper No.2, Institute for the Study of War and Society, De Montfort University Bedford (1995).

One of the most useful definitions of morale, which links that of the individual with that of the group, is that of Irvin L. Child: 'morale pertains to [the individual's] efforts to enhance the effectiveness of the group in accomplishing the task in hand'.[1] The relationship between individual and collective morale can be described as follows: unless the individual is reasonably content he will not willingly contribute to the unit. High group morale is the product in large part of good morale experienced by the members of that unit; and the state of morale of a larger formation such as an army is the product of the cohesion of the units that compose that army. The possession of individual morale sufficiently high that a soldier is willing to engage in combat might be described as 'fighting spirit'.

The nineteenth-century military philosopher Clausewitz divided morale into two components: 'mood' and 'spirit'. The mood of an army [or indeed an individual] is a transient thing, which can change quickly; but an army with 'true military spirit' keeps 'its cohesion under the most murderous fire' and in defeat resists fears, both real and imaginary. Military spirit, Clausewitz argued, is created in two ways: by the waging of victorious wars and the testing of an army to the very limits of its strength.[2]

While the morale of the BEF remained fundamentally sound throughout the war, it is not being suggested that individual soldiers or indeed entire units were ecstatically happy all the time. Rather, the combat performance of British soldiers reflected their commitment to winning. The first winter of the war was a terrible trial for the soldiers in the trenches, and this experience affected morale. However, as the events of 1915 proved, the 'depression' of the winter of 1914–15 did not permanently erode the military spirit of the British soldier. Morale also suffered during the following winter, but the BEF's military spirit remained essentially intact.

The attritional battles of 1916–17 placed the men of the BEF under tremendous strain. The ultimate test of morale is willingness to engage in combat, and the BEF's divisions continued to fight throughout the campaigns. During the Battle of the Somme the BEF advanced about seven miles at the cost of 420 000 British casualties. Yet in November 1916 a report based on the censorship of soldiers' letters stated, 'the spirit of the men, their conception of duty, their Moral (*sic*), has never been higher than at the present moment'. Enthusiasm had been replaced by a 'dogged determination [,] to see the thing through at any cost'.[3] In 1917 further heavy losses came at Arras and at Ypres. Yet censorship reports and other evidence suggest that in the latter part of 1917 the BEF's morale was lower than in 1916 but remained sound.[4]

In 1918, the BEF had to relearn how to conduct mobile operations, first in a major retreat, and then advancing to victory. The initial success of the German Spring Offensive that began on 21 March 1918 has been traditionally attributed, in part at least, to a breakdown of British morale. A censorship report of July 1918, and the work of modern historians, suggests that this view is wrong.[5] Had the morale of the British Army indeed collapsed, the Germans would probably have won the war, for the autumn of 1918 was to demonstrate the serious consequences of a genuine weakening in an army's morale.

The events of the spring of 1918 cannot be divorced from the victories of the summer and autumn, when the BEF 'was at last obtaining a just reward for all its dogged and patient fighting'.[6] In the words of one temporary subaltern, the BEF fought the enemy to a standstill and then counterattacked

> and remain[ed] continuously on the offensive from August to November. The infantry at least had no doubt that they were winning, and their faith was justified when the greatest military Power of modern times finally collapsed in disordered retreat.[7]

Appendix 2

British Army Conscripts

Fifty-three point three per cent of all wartime enlistments occurred after the introduction of a form of conscription in January 1916. However, I have made no attempt to discuss relations between officers and conscripts *per se*. Assessing the conscripts' war experience is not easy. Relatively few soldiers who wrote memoirs or left diaries or letters admitted to the stigma of conscription.[1] Conscripts were drawn from a variety of social backgrounds, with a disproportionate number coming from white-collar, middle-class occupations.

Moreover, conscripts had widely differing wartime experiences. Men were conscripted into every arm, and served in Regular, New Army and TF units, which themselves may have had differing approaches to discipline and officer–man relations. Not surprisingly, conscripts' attitudes towards enlistment varied greatly. Some men attested under the Derby scheme, a 'half-way house' between voluntarism and compulsion. Some deliberately avoided the army during the voluntary phase, while others who were unable to volunteer welcomed conscription. Some men, who probably would have volunteered, only reached the age of enlistment after conscription was introduced.[2] Other men, such as Fred Dixon (Surrey Yeomanry) volunteered rather than await conscription, so he could join the regiment of his choice.[3] The writer Stephen Graham, conscripted in 1917, ended up in the ranks of the Scots Guards because he declined the chance of a commission.[4] In 1917, an 'old sweat' of 55th Division praised the keenness of 19-year-old conscripts, but this contrasted with the lack of enthusiasm evinced by conscripts in their mid-30s.[5] Given these variables, it would be a meaningless exercise to treat 'the conscripts' as a discrete group when discussing their attitudes towards officer–man relations and discipline.

However, some general remarks about conscripts are not out of place. Wartime volunteer officers and rankers often denigrated the discipline, motivation and military effectiveness of conscripted soldiers.[6] Thus the commander of 19th Division lamented that the replacements that arrived at the end of the Somme campaign in 1916 'lacked the cheerful eager look of the volunteer. We never had the same gallant adventurers in the ranks again.'[7] To be set against this subjective view is the fact that, whatever their facial expressions, in the last week of the war the men of 19th Division advanced 18 miles over difficult terrain against a tough enemy, taking 'fairly severe' casualties in the process.[8] Judged by the yardstick of military success, 19th Division's conscripts were effective soldiers. The same was true of the men of 9/DLI who were praised by their commander for their performance in September 1918:

> [They] showed the highest form of discipline while under the enemy barrage, never flinching although caught like rats in a trap. The conduct of the men was worthy of the highest traditions of the British army.[9]

In 1918, as in earlier years, some units were more effective than others, and a host of reasons determined military effectiveness. Leadership, morale, training and tactical ability were among the most important. In 1917–18 new men 'blended with surviving natural leaders to keep the show going'.[10] The survival of distinctive traditions and ethos in some units supports this contention. Conscription did not in itself affect combat performance to any great degree.

David Fraser argues that in the Second World War, British conscripts accepted the demands of military life if they experienced 'competent, understanding leadership and a well-run and pride-filled unit'.[11] Much the same might be said of the conscripts joining up in 1916–18, and indeed of the volunteers of 1914–15. Conscripts seem to have differed little in their attitudes to comparable volunteers. Alfred M. Hale, a 41-year-old artist and composer who loathed army life, 18-year-old E.C. Barraclough, who approached his service with enthusiasm and excitement, and F.A. Voigt, whose attitudes fell between those of the other two, are paralleled by those of volunteer soldiers.[12] The conscript's attitudes and assumptions mirrored those of other soldiers of his age, social class, education, and civilian and military experience. In this respect, the method by which a man joined the army was relatively unimportant.

Appendix 3

Discipline and Continuity in Small Units

Trench-mortar batteries, Royal Engineer field companies, machine-gun companies and similar units were relatively small, perhaps one hundred men strong, and commanded by fairly junior officers such as captains or even subalterns. Such units were often used as dumping grounds for officers and men who were not wanted by their parent units.[1] Paradoxically, members of TMBs and the like regarded themselves as an elite, as craftsmen with a skilled task to perform, unlike the 'general purpose' infantry. It is not surprising that officers and men who regarded themselves as part of an elite team should develop a working relationship that set aside the niceties of military etiquette.[2] They also shared a common identity as 'outsiders', disliked by the infantry. This antagonism arose partly because of their privileged, independent existence, but also because TMBs disturbed the live-and-let-live system, and drew fire on the infantry, while sappers often required working parties to be furnished by the infantry.[3]

Away from the supervision of more senior, perhaps Regular officers, commanders of small units often operated informal disciplinary systems. Lt P.G. Heath transferred from 8/E. Surreys to 55th TMB in 1916, and enjoyed an excellent relationship with his men: 'their discipline, according to Army standards, was deplorable ... but I preferred it that way ...'. He argued that in a TMB there was more scope for initiative, and many unorthodox soldiers and officers thrown out of infantry battalions revelled in the atmosphere of a TMB. A junior officer of 1/17 Londons volunteered for duty with a TMB to escape from formal discipline. He enjoyed the added responsibility, reduced fatigues, and gave his men all sorts of privileges. As a ranker-officer, he was well aware of the private's point of view.[4] Not every small unit operated a system of relaxed discipline,[5] but many did.

Falling between the trench-mortar battery or machine-gun company on one hand and the infantry battalion on the other was the artillery battery. Something of a 'craft union' attitude emerged among gunners of all ranks. One ranker argued that good officer–man relations in artillery batteries stemmed from their need to co-operate as part of a team.[6] One unit, 309 Siege Battery HAC, combined this attitude with the informality of a Territorial class corps.[7] The pattern of casualties that an artillery battery could expect to sustain also helped to maintain continuity of command and personnel. An officer of 62nd Divisional artillery calculated that in 22 months of action the unit had a casualty rate of 250 per cent for officers and 100 per cent for rankers. While these losses occurred in a 'steady drain' over the period, it was not uncommon for infantry battalions to take losses that crippled it 'in an hour of battle'. 62nd Division was a TF formation originally recruited in Yorkshire. Although individual replacements came from all over Britain, non-Yorkshire soldiers

> assumed a Yorkshire defiant and stubborn quality that characterised the 62nd Division from its earliest days, and were proud of it. D/310 Battery, never with more than twenty Yorkshire born men, had by now become Yorkshire to the core.[8]

An even smaller unit was the tank crew. The very special circumstances of a small group of soldiers, including an officer, sharing the danger and discomfort of a confined space led to the emergence of their own peculiar discipline and officer–man relations.[9] An officer wrote that the tank man was a specialist, 'not a military automaton.... The discipline of a Guards regiment would have been thoroughly out of place in a tank.'[10] The tank crew had an 'unwritten rule' that wounded were always evacuated by their luckier fellows. Within this pragmatic context, tank officers performed the roles of leading by example, risk-taking, and caring for their men.[11]

Appendix 4

Published Guides to Officership

A large number of articles and books on the nature of officership appeared during the Great War. Their influence is difficult to gauge. Many were widely distributed,[1] some being used by OCB instructors.[2] Given the conscientiousness displayed by most temporary officers, it is reasonable to assume that such guides were read by at least some of their intended audience.

Both official and unofficial guides offered much the same advice. They assumed that officers could be created out of men from non-traditional backgrounds, if they were taught leadership skills.[3] Fundamental was the winning of the men's respect, through the acquisition of 'character', defined by one writer as 'resolution, self-confidence, self-sacrifice'. The potential leader was enjoined to 'inspire your men by your example, sustain their courage in danger by your example, and their endurance in hardship by your example'.[4]

The young officer was urged to develop a relationship with his men that balanced paternal care and discipline. He needed to remember that 'the soldiers under you are individual human beings and not sheep or cattle; they have their individual feelings, tempers and temperaments'.[5] Moreover, the officer had to become aware of the Other Ranks' prejudices and thought processes, and become a kindly but firm father to his men, not a 'dry-nurse': '*we are all comrades* ... make the men feel that you realise this relationship *and love it*'.[6] A balance also had to be maintained between tact and discipline, with officers well aware of the limitations on their power: 'Never give an order which you cannot enforce. It is better to cancel an order than to allow it to be disregarded.'[7]

While published guides paid curiously little attention to the relationship between the NCO and the officer, advice on the welfare of the soldier was plentifully given. The officially produced *Notes for Young Officers* stated unequivocally that 'the care of his men must take precedence over every consideration of [the officer's] personal comfort', and through paternal care 'the trials and hardships of a campaign [will be] materially reduced ... [and] the goodwill of the men will be gained'. This pamphlet devoted an entire section to 'Relations between officers and men off duty' which included a list of ways in which the officer might brighten the lives of his soldiers.[8] Very similar advice appeared in other guides.

In sum, guides both officially and unofficially sponsored were produced in response to the widening of the social base of the wartime officer corps. Aimed at officers without a public school education, they reinforced the teachings of officer-training units, and were an additional means by which such men were inculcated with paternalistic concern for their soldiers.

Notes

Introduction

1 A.D. Thorburn, *Amateur Gunners* (Liverpool: Potter, 1933) p. 5.
2 B. Gardner, *The Big Push* (London: Quality: 1961) p. 5.
3 For this school, see A. Danchev, '"Bunking" and Debunking: the Controversies of the 1960s', in B. Bond (ed.), *The First World War and British Military History* (Oxford: Clarendon, 1991) pp. 263–88.
4 See, for instance, Maj. the Hon. J.J. Astor's introduction to R.A. Lloyd, *A Trooper in the 'Tins'* (London: Hurst and Blackett, 1938) p. 6.
5 J. Baynes, *Morale* (London: Cassell, 1967). See also idem, 'The Officer–Other Rank Relationship in the British Army in the First World War', *Quarterly Review* (Oct. 1966) pp. 442–52, which covers much the same ground. For officer–man relations in the prewar army, see also E.M. Spiers, *The Army and Society 1815–1914* (London: Longman, 1980) pp. 26–9; idem, 'The Regular Army in 1914', in I.F.W. Beckett and K. Simpson, *A Nation in Arms* (Manchester: Manchester University Press, 1985) pp. 43, 55, 57; idem, *The Late Victorian Army 1868–1902* (Manchester: Manchester University Press, 1992) pp. 112–14, and B. Farwell, *For Queen and Country* (London: Allen Lane, 1983) pp. 132–8.
   Some useful comparisons can be drawn with the Indian army of the Raj: J. Greenhut, 'Sahib and Sepoy: an Inquiry into the Relationship between the British Officers and Native Soldiers of the British Indian Army', *MA* XLVIII, (1984) 15–18. See also T.A. Heathcote, *The Indian Army* (Newton Abbot: David and Charles 1974), P. Mason, *A Matter of Honour* (London, Cape, 1974; D. Omissi, *The Sepoy and the Raj* (London: Macmillan, 1996). See also J.A. Crang, 'A Social History of the British Army 1939–45' (PhD, University of Edinburgh, 1992) pp. 121–59.
6 G.D. Sheffield, 'The Effect of War Service on the 22nd (Service) Battalion Royal Fusiliers (Kensington), with Special Reference to Morale, Discipline and the Officer–Man Relationship' (MA thesis, University of Leeds, 1984).
7 See, for instance, K. Simpson, 'The Officers', in Beckett and Simpson, *Nation in Arms* p. 85; D. Winter, *Death's Men* (Harmondsworth: Penguin, 1979) pp. 59–62; T. Denman, 'The Catholic Irish Soldier in the First World War: the "racial environment"', IHS XXVII, 108 (1991) pp. 352–65; A. Simpson, *Hot Blood and Cold Steel* (London: Tom Donovan, 1993) pp. 118–29. K.W. Mitchinson, *Gentlemen and Officers* (London: Imperial War Museum, 1995) pp. 98–100.
8 I.F.W. Beckett, 'The British Army, 1914–18: the Illusion of Change', in J. Turner (ed.), *Britain and the First World War* (London, Unwin Hyman, 1988) pp. 106–8: P. H. Liddle, *The Soldier's War 1914–1918* (London: 1988) pp. 80–81.
9 J.M. Bourne, *Britain and the Great War 1914–1918* (London: Arnold, 1989) p. 221.
10 I.D. Losinger, 'Officer–Man Relations in the Canadian Expeditionary Force,

1914–1919' (MA thesis, Carleton University, 1990).

11 C.E.W. Bean (ed.), *Official History of Australia in the War of 1914–1918* (Sydney: Angus and Robertson, 12 vols, 1921–42); B. Gammage, *The Broken Years* (Ringwood, Victoria: Penguin) pp. 85–88, 230–63; but see also Chapter 9 below.

12 C. Pugsley, *Gallipoli: the New Zealand Story* (Auckland: Hodder, 1984) pp. 13–14, 78, 258–60; C. Pugsley, *On the Fringe of Hell* (Auckland: Hodder and Stoughton, 1991), passim.

13 A.J. Peacock, 'Crucifixion No.2', *Gunfire* 4, pp. 211–21 is a typical example. See also P. Scott, 'Law and Orders: Discipline and Morale in the British Armies in France, 1917', in P. H. Liddle, *Passchendaele in Perspective* (London: Leo Cooper, 1997) pp. 349–68.

14 A. Babington, *For the Sake of Example* (London: Leo Cooper, 1983); J. Putkowski and J. Sykes, *Shot at Dawn* (Barnsley: Wharncliffe, 1989).

15 G.D. Sheffield, *The Redcaps* (London: Brassey's, 1994) pp. 49–96; idem, 'The Operational Role of British Military Police on the Western Front, 1914–18' in P. Griffith, (ed.), *British Fighting Methods in the Great War* (London: Cass, 1996) pp. 70–86.

16 G. Dallas and D. Gill, *The Unknown Army* (London, 1985); L. James, *Mutiny* (London: Buchan and Enright, 1987); J. Putkowski, 'Toplis, Etaples & "The Monocled Mutineer"', *ST* 18 (1986) pp. 6–11.

17 J.B. Wilson, 'Morale and Discipline in the British Expeditionary Force, 1914–18' (MA thesis, University of New Brunswick, 1978; J.G. Fuller, *Troop Morale and Popular Culture in the British and Dominion Armies 1914–1918* (Oxford: Clarendon, 1991); S.P. MacKenzie, 'Morale and the Cause: the Campaign to Shape the Outlook of Soldiers of the British Expeditionary Force, 1914–18', CJH XXV (1990) pp. 215–32; idem, *Politics and Military Morale* (Oxford, Clarendon, 1992) pp. 3–39; D.M. Simpson, 'Morale and Sexual Morality among British Troops in the First World War', unpublished paper, 1996; G.D. Sheffield, *The Morale of the British Army on the Western Front, 1914–1918* (Occasional Paper 2, Institute for the Study of War and Society, De Montfort University Bedford, 1995).

18 J. Bourke, *Dismembering the Male: Men's Bodies, Britain and the Great War* (London: Reaktion, 1996) pp. 27, 144–53.

19 M. Smith, 'The War and British Culture', in S. Constantine, M.W. Kirby and M.B. Rose, *The First World War in British History* (London: Arnold, 1995) p. 177.

20 Leonard V. Smith, *Between Mutiny and Obedience* (Princeton: Princeton University Press, 1994). See also C. Bryant, 'Obedience and Power: Changing Power Relations in a Habsburg Army Regiment, 1914–15', unpublished paper.

21 See the feature review by M.L. Buck in *WIH* 3, 1996, 107–19.

22 T. Ashworth, *Trench Warfare 1914–1918* (London: Macmillan, 1980) p. 8.

23 V.F. Eberle, *My Sapper Venture* (London: Pitman, 1973) p. 102.

24 Unpublished account, notebook IV, B.D. Parkin papers, 86/57/1, IWM.

25 Unpublished account, p. 116, R.C. Foot papers, IWM.

26 See, however, A. Thompson's controversial view of the construction of wartime memories in *Anzac Memories* (Melbourne, Oxford University Press, 1994). This should be compared with Nigel de Lee's view: 'Oral History and

British Soldiers' Experience of Battle in the Second World War', in P. Addison and A. Calder, *Time to Kill* (London: Pimlico, 1996) pp. 359–68.
27 H. Cecil, 'The Literary Legacy of the War: the Post-war British War Novel, a Select Bibliography' in P. H. Liddle, *Home Fires and Foreign Fields* (London: Brassey's, 1985) p. 205.

1 Officer–Man Relations and Discipline in the Regular Army, 1902–14

1 A.R. Skelly, *The Victorian Army at Home* (London: Croom Helm, 1977) p. 199.
2 J.E. Edmonds, *The Occupation of the Rhineland 1918–29* (London: HMSO, 1987) p. 168. See also FO comments included in Bridges to Edmonds, 22 July 1944, CAB 45/81, PRO.
3 A.J.A Wright, 'The Probable Effects of Compulsory Military Training on Recruiting for the Regular Army', *JRUSI* LV, (1911) 1590; Spiers, 'Regular Army', p. 44.
4 'Trades of men offering for enlistment, year ending September 30th 1913', *PP*. 1921, XX, CMD 1193.
5 Spiers, 'Regular Army' pp. 44–45, 53; Baynes, *Morale* p. 137; Skelly, *Victorian Army* pp. 289–90.
6 *PP*. 1921, XX, CMD 1193.
7 Unpublished account, p. 2, W.J. Nicholson papers, 78/31/1, IWM.
8 But see [W.E. Cairnes] *The Army from Within. By the Author of 'An Absent Minded War'* (London: Sands, 1901) p. 65.
9 A.H. Lee, 'The Recruiting Question: a Postscript to the Army Debate', *Nineteenth Century* 49, (1901) 1058.
10 J. Stevenson, *British Society 1914–45* (Harmondsworth: Penguin, 1984) p. 43.
11 Unpublished account, p. 3, H.J. Coombes papers, PP/MCR/119, IWM.
12 J.A. Balck, 'Recruiting in the German Army', *JRAMC* XV (1910) 567, 571–2.
13 Spiers, 'Regular Army', p. 39; P.E. Razzell, 'The Social Origins of British Officers in the Indian and British Home Army', *BJS* 14, (1963) 249–53.
14 G. Harries-Jenkins, *The Army in Victorian Society* (London: RKP, 1977) pp. 24–5.
15 Spiers, *Army and Society* p. 11.
16 [W.E. Cairnes] *'A British Officer': Social Life in the British Army* (London: Long, 1900) pp. xviii–xix.
17 See N. Newham-Davis, *Military Dialogues* (London: Long, n.d.) pp. 30–7; Cairnes, *Social Life* pp. XVIII, p. 61.
18 Simpson, 'The Officers', pp. 64–5; Harries-Jenkins, *Army* p. 96.
19 Quoted in C. Dalton, 'Commissions from the British Ranks 1706–1855', *JRUSI* XLIV (1900) 167.
20 R. Hargreaves, 'Promotion from the Ranks', *AQ* LXXXVI (1963) 200–10; Simpson, 'The Officers' p. 65.
21 '6197/R.G.A. (S.W.)', 'The Ranker Officer', *JRUSI* LXIII, (1918) 489.
22 D.M. Henderson, *Highland Soldier* (Edinburgh: Donald, 1989) p. 126. For the problems of ranker-officers, see W.R. Robertson, *From Private to Field Marshal* (London: Constable, 1921) pp. 29–31; V. Bonham-Carter, *Soldier True* (London: Muller, 1963) pp. 29–32.
23 Cairnes, *Army from Within* p. 197; Baynes, *Morale* pp. 29–30; letter, 8 May

1912, RMC Register of Letters File index 1910–12, 4843, RMASA.

24 E.A. Muenger, 'The British Army in Ireland, 1886–1914' (PhD thesis, University of Michigan, 1981) p. 93.

25 J.F.C. Fuller, *The Army in My Time* (London: Rich and Cowan, 1935) p. 6.

26 E.C. Vivian, *The British Army from Within* (London: Hodder, 1914) p. 52. See also R. Blatchford, *My Life in the Army* (London: Clarion, 1910) pp. 177–8.

27 F. Richards, *Old Soldier Sahib* (London: Faber, 1965) pp. 155–6.

28 Vivian, *British Army* p. 54.

29 J. Lucy, *There's A Devil in the Drum* (London: Naval and Military Press, 1992) p. 63.

30 Blatchford, *My Life*, pp. 177, 181; unpublished account, p. 20, J.W. Riddell papers, 77/73/1, IWM.

31 J. Hawke, *From Private to Major* (London: Hutchinson, 1938) p. 83.

32 F.H. Maitland, *Hussar of the Line* (London: Hurst and Blackett, 1951) p. 21.

33 Unpublished account, pp. 10, 12, 21, 44, J.W. Riddell papers, 77/73/1, IWM.

34 V.G. Kiernan, *European Empires from Conquest to Collapse 1815–1960* (London: Fontana, 1982) p. 137.

35 Unpublished account, p. 14, R. Garrod papers, IWM.

36 Unpublished account, p. 8, H.J. Coombes papers, IWM.

37 H. Wyndham, *The Queen's Service* (London: Heinemann, 1899) p. 151; idem, 'Officers and Men', *USM* XXI, (1902) 184–5.

38 D. Roberts, *Paternalism in Early Victorian England* (New Brunswick, NJ: Rutgers University Press, 1979) pp. 2–6.

39 P. Joyce, *Work, Society and Politics* (London: Methuen, 1980) p. 99. For one industrialist as a paternalist, see K. Grieves, *Sir Eric Geddes* (Manchester: Manchester University Press, 1989) pp. 129–31.

40 *Hansard*, 1st Series, Lords, XV (30 Mar. 1914), col. 795.

41 See N. Nicholson, *Alex* (London: Weidenfeld and Nicholson, 1973) p. 26; W.N. Nicholson, *Behind the Lines* (London: Cape, 1939) p. 105.

42 A.W. Taylor, *How to Organise and Administer a Battalion* (London: Rees, 1915) pp. 9–10.

43 Spiers, *Army and Society* pp. 26–9. On 'gentlemanliness', P. Mason, *The English Gentleman* (London: Pimlico, 1993) is suggestive.

44 N. Hamilton, *Monty – the Making of a General* (London: Hamish Hamilton, 1981) p. 72. See also the comments of a contemporary of Montgomery: B. Horrocks, *Corps Commander* (London: Sidgwick & Jackson, 1977) p. 154.

45 E.A.H. Alderson, *Pink and Scarlet, or Hunting as a School for Soldiering* (London: Hodder, 1913) pp. 198–9. See also E.A.H. Alderson, *Lessons from 100 Notes Made in Peace and War* (Aldershot: Gale & Polden, 1908) pp. 28, 89.

46 Blatchford, *My Life* pp. 205–8.

47 H.J.D. Clark, KRS Q.

48 A. Bryant, *Jackets of Green* (London: Collins, 1972) pp. 168–9.

49 T.W. Metcalfe, *Memorials of the Military Life* (London: Ivor Nicholson and Watson, 1936) p. 121.

50 Cairnes, *Social Life* p. 26.

51 *Standing Orders of the 2nd Battalion Essex Regiment (Pompadours)* (1903) p. 13.

52 Harries-Jenkins, *Army in Victorian Society* p. 148.
53 *RMCR* 1,(1912), p. 9; Diary, K.A. Garratt papers, especially 25, 30 Nov. 1913, 76-251 (i), RMASA.
54 Letter, 16 January 1913, WO/152/1913/5108, RMASA.
55 Montgomery of Alamein, *Memoirs* (London: Reader's Union, 1960) p. 20.
56 Metcalfe, *Memorials* p. 48.
57 H.D. Thwaytes, W.J. Jervois, KRS Q; for Woolwich, see unpublished account, pp. 16–20, L.A. Hawes papers, 87/41/1, IWM.
58 H.J.D. Clarke, KRS Q.
59 W. Childs, *Episodes and Reflections* (London: Cassell, 1930) p. 15.
60 J.A. Halstead, KRS Q.
61 C. Kernahan, *An Author in the Territorials* (London: Pearson, 1908) pp. 17–23, 45–6.
62 Officers have made similar comments to me about the value of training at present-day RMA Sandhurst.
63 W. Nasson, 'Tommy Atkins in South Africa', in P. Warwick (ed.), *The South African War* (London: Longman, 1980) pp. 127–8.
64 'Notes by Col. J.M. Grierson RA on Return from South Africa', pp. 75–7, WO 108/184, PRO; *Report on Office of Provost Marshal*, Pretoria, 15 July 1900, WO 108/259, PRO.
65 O. Mosley, *My Life* (London: Nelson, 1968) pp. 47–8.
66 Lucy, *Devil* p. 57. See also the comments of J. Espiner, quoted in P. F. Stewart, *The History of the XII Lancers (Prince of Wales)* (London: Oxford University Press, 1951) p. 227.
67 Lloyd, *Trooper* pp. 13, 17; unpublished account, p. 4, J.W. Riddell papers, 77/73/1, IWM.
68 Unpublished account, pp. 1–2, F.M. Packham papers, p. 316, IWM.
69 R. Edmondson, *John Bull's Army from Within* (London: Griffiths, 1907) p. 16.
70 *Regimental Standing Orders of His Majesty's Irish Regiment of Footguards* (1911).
71 See Orders of CMP, 8 May to 9 Oct 1914, Acn. 680, RMPA.
72 Lloyd, *Trooper* pp. 48–9.
73 Orders of CMP, 12 May 1914, Acc. 680, RMPA.
74 Diary, 6, 9 Apr. 1903, T. Grainger papers, 7104-31, NAM.
75 Unpublished account, p. 17, W.J. Nicholson papers, IWM.
76 Hawke, *From Private* p. 63.
77 P. Kennedy, 'Britain in the First World War', in A.R. Millett and W. Murray (eds), *Military Effectiveness* I (London: Unwin Hyman, 1988) p. 67.
78 Stewart, *XII Lancers* p. 227.
79 Maitland, *Hussar* pp. 84–5; H. Meyers, interview; H. Bolitho, *The Galloping Third* (London: Murray, 1963) p. 190.
80 F. Richards, *Old Soldier Sahib* (London: Faber, 1965) pp. 75, 82; Heathcote, *Indian Army* p. 166; letter, 26 Dec. [1912?], F. Williams papers, 80/23/1, IWM.
81 Unpublished account, pp. 2, 11, H.J. Coombes papers, PP/MCR/119, IWM; unpublished, unpaginated account, T.A. Silver papers, 74/108/1, IWM.
82 Unpublished account, p. 43, J.W. Riddell papers, 77/73/1, IWM.
83 A.P. Wavell to father, 23 Mar. 1914, in I.F.W. Beckett, *The Army and the*

Curragh Incident, 1914 (London: Bodley Head, 1986) p. 282.

84 Anon. to MacDonald, n.d., J. Ramsay MacDonald papers, PRO/30/69/1158, PRO. This letter may have been bogus, sent in an attempt to discredit the Labour Party.

85 F.A. Forster to family, Mar. 1914, in Beckett, *Curragh Incident* p. 132.

86 H. Benyon, *Working for Ford* (Harmondsworth: Penguin, 1984) p. 36.

87 Diary, 5, 9 Apr. 1903, T. Grainger papers, 7104-31, NAM.

88 A.F. Corbett, *Service through Six Reigns* (Norwich: privately published, 1953) pp. 47–8.

89 Unpublished account, p. 7, W. Fanton papers, 7802-78, NAM.

90 Cairnes, *Social Life* pp. 26, 116, 120–2.

91 Richards, *Sahib* p. 157; J.M. Brereton, *The British Soldier* (London: Bodley Head, 1986) pp. 110–11.

92 Wyndham, 'Officers and Men' 182.

93 Lucy, *Devil* p. 94.

94 Wyndham, 'Officers and Men' 183–6

95 Richards, *Sahib* p. 274.

96 Wyndham, 'Officers and Men' 183–90.

97 *Ibid.* p. 190; Lucy, *Devil* p. 94; Richards, *Sahib* pp. 155–6.

98 See, for instance, Maitland, *Hussar*, passim.

99 R. van Emden (ed.), *Tickled to Death to Go* (Staplehurst: Spellmount, 1996) p. 26.

100 Unpublished account, p. 10, R.G. Garrod papers, IWM.

101 Unpublished account, C. King papers, DS/Misc/91, IWM.

2 The Prewar Army: the Auxiliary Forces and Debates on Discipline

1 I.F.W. Beckett, *Riflemen Form* (Aldershot: Ogilby Trusts, 1982) p. 253. In general, militia and Special Reserve units, which mostly did not see active service in the Great War, are not discussed.

2 G. Fellows and B. Freeman, *Historical Records of the South Nottinghamshire Hussars Yeomanry* (Aldershot: Gale & Polden, 1928) p. XV.

3 For more details, see my forthcoming article 'Cavalry of the Territorial Force: the Yeomanry in Peace and War, 1902–18'.

4 *The LRB Record* V, (1907), 10.

5 H. Cunningham, *The Volunteer Force* (London: Croom Helm, 1975) pp. 33–4; Beckett, *Riflemen Form* pp. 82–3 has slightly different, but broadly comparable figures.

6 Evidence of Col. E.H. Bailey, 5 Nov. 1903, PP, 1904, XXXI, Cmd. 2062, 'Minutes of Evidence taken before the Royal Commission on the Militia and Volunteers', p. 296.

7 'Reports ... General Officers Commanding-in-Chief on the Progress made by the Territorial Force ...' [hereafter GOC Report] (1911) p. 9, W[ar] O[ffice] L[ibrary], now held in the PRO.

8 'Reports ... on the Physical Capacity of Territorial Force Troops ...' (1910) (hereafter Physical Capacity Report) p. 2, WOL/PRO. See also GOC Report (1909) p. 17, WOL/PRO.

9 GOC Report (1911) p. 20, WOL/PRO.

10 Bethune to PS of Secretary of State, 27 Nov. 1910 WO32/11242, PRO.

11 W. Richards, *His Majesty's Territorial Army* (London: Virtue, n.d.) III, p. 115;

H.D. Myers, KRS Q.
12 E. Riddell and M.C. Clayton, *The Cambridgeshires 1914–18* (Cambridge: Bowkes, 1934) pp. 1–2; E.C. Matthews, *With the Cornwall Territorials on the Western Front* (Cambridge: Spalding, 1921) p. 4.
13 Sir J.H.A. Macdonald, 'The Volunteers in 1905', *JRUSI*, XLIX, (1905) 917.
14 A.M. McGilchrist, *The Liverpool Scottish 1900–19* (Liverpool: Young, 1930) p. 4; *LRB Record* V, (1911) 32.
15 H. Morrey-Salmon, KRS Q; GOC Report (1909) p. 67, WOL/PRO.
16 B. Latham, *A Territorial Soldier's War* (Aldershot: Gale & Polden, 1967) p. 1. For the social history of the prewar LRB, see K.W. Mitchinson, *Gentlemen and Officers* (London: IWM, 1995) pp. 13–26.
17 'A Lieutenant Colonel in the British Army', *The British Army* (London: Sampson Low, 1899) p. 189.
18 Viscount Mersey, *Journals and Memories* (London: Murray, 1952) p. 134.
19 Cunningham, *Volunteer Force* p. 34.
20 Ibid., p. 58.
21 GOC Report (1911) p. 54, WOL/PRO (see also comments of Maj. Marker, 'Report on a Conference of General Staff officers at the Staff College' (1909) p. 22, SCL); A.H. Maude (ed.), *The 47th (London) Division, 1914–19* (London: Amalgamated Press, 1922) pp. 6–7.
22 Col. L. Banon, 'Report on the Supply of Officers for the Special Reserve and Territorial Force', 23 Dec. 1912, pp. 2, 3–5, WOL/PRO. See also Macdonald, 'Volunteers', 915–16.
23 Sir H. Roberts, 'The Auxiliary Forces Commission', *USM* XXIX, (1905) 504.
24 D. Wheatley, *Officer and Temporary Gentleman* (London: Hutchinson, 1978) pp. 53–4.
25 Comments of Col. R.H. Davies, 'Report of a Conference of General Staff Officers at the Staff College' (1910) p. 55, and Gen. Hon. Sir N.G. Lyttelton, 'Report on a Conference of General Staff Officers at the Staff College' (1908) p. 39 (both SCL).
26 Col. L. Banon, 'Report on the Supply of Officers for the Special Reserve and Territorial Force'. 23 Dec. 1912, pp. 5–6. WOL/PRO.
27 Cunningham, *Volunteer Force* p. 64.
28 GOC Report (1910) p. 63, WOL/PRO.
29 R.S.S. Baden-Powell, 'Training for Territorials', *JRUSI* LII, (1908) p. 1480.
30 R. Kipling, *The New Armies in Training* (London: Macmillan, 1915) p. 59.
31 C. Ponsonby, *West Kent (QO) Yeomanry and 10th (Yeomanry) Battalion the Buffs 1914–19* (London: Melrose, 1920) pp. 5–6. See also unpublished account, p. 5, A.S. Benbow papers, PP/MCR/146, IWM; unpublished account, p. 4, A.W. Bradbury papers, IWM.
32 GOC Report (1909), pp. 9, 45, 65, WOL/PRO.
33 GOC Report (1909) pp. 17, 54, WOL/PRO.
34 GOC Report (1911) p. 15, WOL/PRO.
35 E.R. Gladstone, *The Shropshire Yeomanry 1795–1945* (Manchester: Whitethorn Press, 1953) p. 191.
36 C.J. Blomfield, *Once an Artist Always an Artist* (London: Page 1921) pp. 20–1; 'An Adjutant', 'The Volunteer Company Officer', *USM* XXVII (1903) 313.
37 G.A. Steppler, *Britons, to Arms!* (Stroud: Sutton, 1992) pp. 43, 111.

38 I.F.W. Beckett, 'The Problem of Military Discipline in the Volunteer Force, 1859–1899', *JSAHR* LVI, (1978) 66–78.
39 Order Book, 5th VB Rifle Brigade 6 May 1896, WO 70/14, PRO.
40 Anon., *The 23rd London Regiment 1798–1919* (London: Times, 1936) pp. 6–7.
41 A. Keith-Falconer, *The Oxfordshire Hussars in the Great War* (London: Murray, 1927) p. 26.
42 *Hansard*, 4th Series, Commons, CLXXI, (22 Mar. 1907) col. 1289.
43 Quoted in M. Moynihan, *A Place Called Armageddon* (Newton Abbot: David and Charles, 1974) p. 120.
44 'Trainband', 'The Dearth of Volunteer Officers', *USM*, XXX, (1905) 649.
45 Anon., *23rd Londons* p. 6; see also I.F.W. Beckett, *The Amateur Military Tradition* (Manchester: Manchester University Press, 1992) p. 104.
46 Anon., *The History of the London Rifle Brigade* (London: Constable, 1921) p. 60.
47 J.H. Lindsay (ed.), *The London Scottish in the Great War* (London: privately published, 1925) p. 14.
48 Anon., *London Rifle Brigade* p. 60.
49 Latham, *Territorial Soldier* p. 2.
50 Lindsay, *London Scottish* p. 13.
51 E.V. Tempest, *History of 6th Battalion West Yorkshire Regiment* (Bradford: Percy Lund, 1921) pp. 1–2, 12.
52 G.B. Hurst, *With the Manchesters in the East* (Manchester: Manchester University Press, 1917) pp. 1,7.
53 Tempest, *6th West Yorks* p. 4.
54 Hurst, *Manchesters* p. 6.
55 Tempest, *6th West Yorks* p. 3.
56 Ibid. p. 10.
57 Hurst, p. 2; G.B. Hurst, in S.J. Wilson, *The Seventh Manchesters* (Manchester: Manchester University Press, 1920) p. xiii.
58 Tempest, *6th West Yorks* p. 10.
59 Hurst, *Manchesters* p. 8.
60 WO 32/11236, PRO.
61 See Kernahan, *Author*, passim.
62 GOC Report (1910) p. 14.
63 Unsigned letters, 10, 13, 14 Sept. 1914, H.E. Politzer papers, P.430, IWM.
64 Two notable exceptions were French and Baden-Powell. See R. Holmes, *The Little Field Marshal* (London: Cape, 1981) p. 131; Baden-Powell, 'Training for Territorials', 1474–7.
65 GOC Reports (1909) p. 20. See also minute by Bethune, 2 July 1912, WO 32/9192, PRO; GOC Reports (1909) p. 68.
66 GOC Reports (1911) p. 10.
67 W.C. Bridge in *JRUSI*, L (1906) 418–19; E.A. Loftus, KRS Q.
68 H.D. Meyers, KRS Q; letter, 13 Sept. 1914, S. Blagg papers, PP/MCR/220, IWM.
69 M. Howard, 'Men against Fire: the Doctrine of the Offensive in 1914', in P. Paret (ed.), *The Makers of Modern Strategy* (Oxford: Clarendon, 1986) pp. 521–2.
70 T. Travers, *The Killing Ground* (London: Unwin Hyman, 1987) p. 47.

71 A.F. Duguid to Edmonds, 27 Mar. 1926, CAB 45/155. PRO.
72 On Henderson and his influence, see Lord Roberts, 'Memoir' in G.F.R. Henderson (N. Malcolm, ed.), *The Science of War* (London: Longmans, 1906) pp. XIV–XXXVIII; J. Luvaas, *The Education of an Army* (London: Cassell, 1964) pp. 216–47.
73 G.F.R. Henderson, *Stonewall Jackson and the American Civil War* (London: Longmans, 1898), II, pp. 417, 429, 431, 455; idem, *Science* pp. 191, 208, 243.
74 Henderson, *Science* pp. 191–6, 244, 311; idem, *Jackson* II, pp. 438–42.
75 Henderson, *Jackson* II, pp. 431, 448–52; idem, *Science* p. 243.
76 Henderson, *Science of War* p. 310.
77 'Report of a Conference of General Staff Officers at the Staff College' (1909) p. 9, SCL.
78 Henderson, *Jackson* II, pp. 444–7, but see his views on Wellington's Light Division in *Science* pp. 200–1.
79 Idem, *Science* pp. 410, 412.
80 Ibid. pp. 393–5.
81 Sir J.F. Maurice, 'The Army Corps Scheme and Mr Dawkins's Committee', *Nineteenth Century* (1901) 145, 148.
82 E.L. Spears, *The Picnic Basket* (London: Secker, 1974) p. 73.
83 Sir R. Harrison, 'Thoughts on the Organisation of the British Army', *JRUSI* L (1906) 17–23.
84 [Baden-Powell], 'Manmastership', *CJ*, I, (1906) 410–21.
85 Henderson, *Science* pp. 392–3.
86 Maurice, 'Army Corps' p. 148.
87 See 'Colonel', 'How to Get Recruits', *USM* XXIX (1909) 611–19.
88 Minutes of 'Recruiting Conference', 5 May 1914, pp. 8, 11, 19, WO 106/364, PRO; C. Bonham-Carter, 'Suggestions to Instructors of Recruits', *AR* VII (1914) 119–27.
89 'Report of the Committee on Supply of Officers on Mobilization, 21 Oct. 1909', Precis No. 453 and 122nd Meeting of Army Council, 21 Mar. 1910, both in WO 163/15, PRO; E.M. Spiers, *Haldane: an Army Reformer* (Edinburgh: Edinburgh University Press 1980), pp. 141–2 and J. James, *The Paladins* (London: Macdonald, 1990) p. 112.
90 Minute by CIGS, 6 Mar. 1914, and terms of reference of report, WO 32/8386, PRO.
91 *Soldier's Small Books* of G.A. Balaam and V.G. Ellis, ERM.
92 Beckett, *Curragh Incident* p. 277; P. Verney, *The Micks* (London: Pan, 1973) p. 28. See also N. Nicolson, *Alex* (London: Pan, 1976) p. 58.
93 L. Cowper (ed.), *The King's Own – the Story of a Royal Regiment*, II, (Oxford: Oxford University Press, 1939) p. 304. This innovation would have pleased one ex-sergeant; Edmondson, *John Bull's Army* pp. 92–3.
94 'Extension of Service of the Footguards 1903–04', WO 32/6901, PRO.
95 J. Hall, *The Coldstream Guards 1885–1914* (Oxford: Oxford University Press, 1920) pp. 313–4.
96 F.P. Horsworthy to Edmonds, 27 July 1931, CAB 45/134, PRO.
97 Biographical details have been drawn from [R. Howell], *Philip Howell: a Memoir by his Wife* (London: Allen & Unwin, 1942) and Beckett, *Curragh Incident* p. 428.

98 Memo. from Haig, 12 Feb. 1906, P. Howell papers, IV/C/2/2, LHCMA; Haig to Howell, 8 Mar. 1907, P. Howell papers IV/C/2/17, LHCMA.
99 B. Bond, *The Victorian Army and the Staff College 1854–1914* (London: Eyre Methuen, 1972) p. 285; Howell, *Philip Howell* p. 42; Howell to C. Wigram, 22 Mar. 1914, in Beckett, *Curragh Incident* p. 104.
100 'Some Notes for Mr Asquith', n.d. (but 1908) P. Howell papers, IV/C/2/43, LHCMA.
101 Heathcote, *Indian Army* pp. 39–40.
102 G.J. Younghusband, *The Story of the Guides* (London: Macmillan, 1908) p. 194. For a description of a *durbar*, see F. Yeates-Brown, *Bengal Lancer* (London: Mott, 1984) pp. 22–6.
103 Draft of letter, 16 Dec. 1913, P. Howell papers, IV/C/2/41, LHCMA; Howell, *Philip Howell* p. 47.
104 Ibid. pp. 45–7.
105 Ibid. pp. 45, 47. See also draft of letter, 16 December 1913, p. Howell papers, IV/C/2/41, LHCMA.
106 'Some Notes for Mr Asquith', n.d. (but 1908), P. Howell papers, IV/C/2/43, LHCMA.
107 Howell to MacDonald, 3 Apr. 1914, J.R. MacDonald papers, PRO 30/69/1158, PRO. See also MacDonald's reply, 17 Apr. 1914, P. Howell papers, IV/C/2/53, LCHMA.
108 That British morale remained sound throughout the war is taken as a 'given' throughout this book. See Appendix 1.

3 The British Officer Corps, 1914–18

1 *The Times* 18 Sept. 1914, p. 10.
2 J.M. Winter, *The Great War and the British People* (London: Macmillan, 1987) pp. 83–99.
3 The latter figure excludes Indian, Dominion, and 'native' troops; with these included the total reaches 5 389 607. P. Simkins, *Kitchener's Army* (Manchester, Manchester University Press, 1988) p. XIV; *Statistics of the Military Effort of the British Empire During the Great War 1914–20* [hereafter SME] (London: HMSO, 1922) pp. 29–30.
4 Excluding Royal Defence Corps, Dominion and Indian Army commissions. Ibid. pp. 234–6.
5 M.A. Wingfield, 'The Supply and Training of Officers for the Army', *JRUSI*, LXII, (1924) 433.
6 Ibid. 434.
7 See Lord Derby's comments in G. Moorhouse, *Hell's Foundations* (London: Sceptre, 1992) p. 177.
8 *SME* p. 234; letter, 4 Nov. 1915, WPN1/95, W.P. Nevill papers, IWM.
9 K. Pearce Smith, *Adventures of an Ancient Warrior* (privately published, 1984), pp. 1–2.
10 *SME* p. 707.
11 Winter, *Great War* p. 83.
12 Book 7, p. 16, G.G.A. Egerton papers, 73/51/1, IWM.
13 *SME* p. 707.
14 R.C. Sherriff, *Journey's End* (London: Heinemann, 1958).
15 H. Quigley, *Passchendaele and the Somme* (London: Methuen, 1928) p. 28.

16 W. Hooper (ed.), *Letters of C.S. Lewis* (London: Fontana, 1988) p. 67.
17 Sheffield, 'Effect of War Service' pp. 21–6. The PRO's recent release of officers' personnel files, while too late for material to be included in this book, makes it possible for detailed analyses of officers' social and educational backgrounds to be undertaken.
18 PP, 1917, Cmd. 8642, 'Report of Committee on Promotion of Officers in the Special Reserve, New Army and Territorial Force', p. 608.
19 *SME* p. 253.
20 Kitchener to French, 16 Sept. 1914, quoted in G. Arthur, *Life of Lord Kitchener* III (London: Macmillan, 1920) p. 65.
21 Wilson to Kiggell, 15 Sept. 1914, III/1, L. Kiggell papers, LHCMA.
22 Blomfield, *Once an Artist* pp. 66–7; A.R. Haig-Brown, *The OTC and the Great War* (London: Country Life, 1915) p. 77.
23 Diary, 14 Nov., 3 Dec. 1914, B.O. Dewes papers, 84/22/1, IWM.
24 Sir J. French, *Dispatch* 2 Feb. 1915; E. Wallace, *Kitchener's Army and the Territorial Forces* (London, Newnes, n.d., c.1915) p. 175.
25 Rawlinson to Kitchener, 25 Nov. 1914, WB/7, PRO 30/57/51, Kitchener Papers, PRO.
26 J.Q. Henriques, *The War History of the 1st Battalion Queen's Westminster Rifles 1914–1918* (London: Medici Society, 1923) p. 73.
27 [A. Smith], *Four Years On the Western Front. By a Rifleman* (London: London Stamp Exchange, 1987) p. 69.
28 Diary, 14 Nov. 1914, B.O. Dewes papers, 84/22/1, IWM.
29 Simkins, *Kitchener's* pp. 212–25; G.H. Cassar, *Kitchener: Architect of Victory* (London: Kimber, 1977) pp. 202, 209–10.
30 'Instructions of Secretary of State', 6 Sept. 1914, WO 162/2, PRO.
31 *Daily Mirror* 10 Sept. 1914, p. 4.
32 Army Order 297, 1908. For the raising of the OTC, see E.M. Teagarden, 'Lord Haldane and the Origins of the Officer Training Corps', *JSAHR* XLV (1967) 91–6, and Spiers, *Haldane* pp. 135–41.
33 Richards, *Territorial Army* IV p. 165.
34 PP, 1907, XLVI, Cd 3294, 'Interim Report of the War Office Committee on the provision of officers (a) for service with with Regular Army in war, and (b) for the Auxiliary Forces'.
35 C.B. Otley, 'Militarism and Militarization in the Public Schools, 1900–1972', *BJS*, 29 (1978) p. 330.
36 Anon., 'Experiences of an O.T.C. Officer', *Blackwood's,* CXCIX, (1916) p. 399.
37 Haig-Brown, *OTC* pp. X, 97–9.
38 J. Gooch, 'The Armed Services' in Constantine *et al.*, *First World War* p. 186.
39 Simkins, *Kitchener's Army* p. 221.
40 A.H.H. MacLean, *Public Schools and the War in South Africa* (London, Stanford, 1902) p. 12.
41 Haig-Brown, *OTC* pp. 99–106.
42 See Chapter 1 above and I. Worthington, 'Antecedent Education and Officer Recruitment: the Origins and Early Development of the Public School–Army Relationship', *MA* XLI, (1977) pp. 187–8.
43 H.F.W. Deane and W.A. Evans, *Public Schools Year Book* (London: privately published, 1908).

44 D.W. Collett, *St Dunstan's College, Catford, Roll of Honour for the First World War 1914–19* (London: privately published, 1928).
45 *In Memoriam Royal Grammar School, Newcastle-upon-Tyne* (Newcastle, n.d.).
46 *Regulations for the Officers Training Corps, 1908.*
47 L.E.M. Savill, 'The Development of Military Training in Schools in the British Isles', MA thesis, University of London, 1937, pp. 15–16.
48 *Public Secondary Schools Cadet Association Camp Magazine* (August 1918) p. 11.
49 Hamilton to Haldane, 22 Feb. 1910, WO 32/6622, PRO.
50 R.C. Sherriff, 'The English Public Schools in the War' in G.A. Panichas (ed.), *Promise of Greatness* (London: Cassell, 1968) pp. 136–7.
51 Haig-Brown, *OTC* p. 99.
52 Unpublished account, p. 13, P. G. Heath papers, DS/Misc/60, IWM.
53 *The Times*, 23 Sept. 1914, p. 10; Simkins, *Kitchener's Army* pp. 224–5.
54 See correspondence in WO 32/11343, PRO; J.H. Hall to Edmonds, n.d., CAB 45/134, PRO.
55 Letter, 6 Nov. 1929, C. Howard, CAB 45/134, PRO.
56 Diary, 20 Nov. 1917, pp. 60–1, J.H. Dible papers, IWM.
57 Diary, 19 July 1915, H. Venables papers, LC.
58 A.O. 394, 16 Sept. 1914.
59 Circular of 16 Dec. 1914, WO 162/2, PRO.
60 J. Joll, '1914: The Unspoken Assumptions' in H. Koch (ed.), *The Origins of the First World War* (London: Macmillan, 1972) p. 309.
61 AROs, WD of Adjutant-General, GHQ, 22 Nov. 1914. WO 95/25. PRO.
62 R. Wilson, *Palestine 1917* (Tunbridge Wells: Costello, 1987) p. 29.
63 WD, First Army A and Q, 22 April 1915, WO 95/181, PRO.
64 Latham, p. 38; letters in P. G. Copson papers, IWM.
65 C.E. Jacomb, *Torment* (London: Melrose, 1920) p. 177.
66 Letter in 'General Aspects: Officers/Officer Cadets' file, LC.
67 See Travers, *Killing Ground*, Chapter 1.
68 'Wagger', *Battery Flashes* (London: Murray, 1916) p. 148.
69 J. Greenshield, KRS Q; B. Williams, *Raising and Training the New Armies* (London: Constable, 1918) p. 98.
70 J. Terraine (ed.), *General Jack's Diary 1914–1918* (London: Eyre & Spottiswoode, 1964) p. 195 W.R. Kingham, *London Gunners* (London: Methuen, 1919), foreword by Lord Denbigh, pp. XV–XVI.
71 P. W. Turner and R.H. Haigh, *Not for Glory* (London: Maxwell, 1969) p. 74.
72 Diary, 6 Sept. 1916, G. Powell papers, 76/214/1, IWM.

4 British Military Leadership, 1914–18: Influences and Training

1 W.D. Henderson, *Cohesion: the Human Element in Combat* (Washington, DC: National Defense University, 1985) p. 111.
2 F.C. Bartlett, *Psychology and the Soldier* (Cambridge: Cambridge University Press, 1927) pp. 138–9.
3 Quoted in R. Holmes, *Firing Line* (London: Cape, 1985) p. 340.
4 C.B. Handy, *Understanding Organisations* (Harmondsworth: Penguin: 1985) p. 118.
5 Ibid., pp. 121–7; Henderson, *Cohesion* pp. 111–14.
6 S.A. Stouffer *et al.*, *The American Soldier* (New York: Science Editions, 1965) II vols., passim.

7 Henderson, *Cohesion* pp. 11–16.
8 R.W. Seaton, 'Deterioration of Military Work Groups under Deprivation Stress', in M. Janowitz, *The New Military* (New York: Norton, 1964) pp. 225–49; G. Lewy, *America in Vietnam* (New York: Oxford University Press, 1978) pp. 155–6, 159.
9 J.A. Mangin, 'Athleticism: a Case Study of the Evolution of an Educational Ideology', in B. Simon and I. Bradley, *The Victorian Public School* (Dublin: Gill & Macmillan, 1975) p. 147.
10 *AQ* III, (1921) p. 106.
11 S. Foot, *Three Lives – And Now* (London: Heinemann, 1937) pp. 11–12.
12 See J. Morris, 'Richard Aldington and *Death of a Hero* – or Life of an Antihero?' in H. Klein (ed.), *The First World War in Fiction* (London: Macmillan, 1978) pp. 183–92; H. Cecil, *Flower of Battle* (London, Secker, 1995) pp. 11–41.
13 R. Aldington, *Death of a Hero* (London: Consul, 1965) pp. 285–6.
14 He also recognised that it could lead to snobbishness. Quoted in D. McCarthy, *Gallipoli to the Somme* (London: Leo Cooper, 1983) p. 39.
15 M. Stephen, *The Price of Pity* (London: Leo Cooper, 1996) pp. 20–1.
16 But see J. Lowerson, *Sport and the English Middle Classes 1870–1914* (Manchester: Manchester University Press, 1993) pp. 282–5.
17 Sir D. Haig, *A Rectorial Address Delivered to the Students in the University of St. Andrews, 14th May 1919* (St Andrews: Henderson, 1919) p. 16. For an excellent overview of Victorian sporting amateurism, see R. Holt, *Sport and the British* (Oxford: Clarendon, 1989) Ch. 2.
18 G. Best, 'Militarism and the Victorian Public School', in Simon and Bradley, *Victorian Public School* pp. 143–4. See also J. Gathorne-Hardy, *The Public School Phenomenon* (London: Hodder, 1977) pp. 144–56.
19 P. Parker, *The Old Lie* (London: Constable, 1987) pp. 83–4.
20 Memorial booklet pp. 4, 9, J.D. Burns papers, LC. The Australian public school system closely followed the British model.
21 Holt, *Sport* ch. 3, esp. pp. 152–6, 159–62, 168–9; R. McKibben, *The Ideologies of Class* (Oxford: Clarendon, 1994) pp. 146–8.
22 Diary, 15 Aug. 1914, T.S. Wollcombe papers, RMASL; *The Graphic* 5 June 1915, pp. 726–7.
23 J. Walvin, *Leisure and Society* (London: Longman, 1978) pp. 83–96 but see the qualifications in Holt, *Sport* pp. 162, 173–5; by 1914, rugby was more popular with the upper classes than 'soccer'.
24 *Yorkshire Post* 2 Sept. 1914.
25 Diary, 25 Apr. 1915, T.S. Wollcombe papers, RMASL.
26 Memorial booklet, p. 4., J.D. Burns papers, LC.
27 See, for example, P. Jones, *War Letters of a Public School Boy* (London, Cassell, 1918) passim.
28 Parker, *Old Lie* p. 84; Best, 'Militarism'.
29 S. Sillars, *Art and Survival in First World War Britain* (New York: St. Martin's Press, 1987) pp. 136–7. Writing late in life, Raymond criticised the 'naive romanticisms, the pieties, the too facile heroics and too uncritical patriotism' in *Tell England: The Story of My Days* (London: Quality Books, 1968) pp. 179–80.
30 Handy, *Understanding Organisations* pp. 60–2, 79.

31 A. Carton de Wiart, *Happy Odyssey* (London: Cape, 1950) p. 14.
32 Lt.Gen. Sir I. Maxse, 'Hints on Training. Preface', in *Notes for Commanding Officers* (Aldershot: 1918) pp. 12–15; *XV Corps. Notes on Minor Tactics Compiled from Lectures to Company Officers* p. 3, RMASL; 23 Oct. 1914, WD, 1/R. Irish Rifles, WO95/1730, PRO; 20 Mar. 1918, WD. 1/KRRC. WO95/1371. PRO.
33 E. Parker, *Into Battle 1914–18* (London, Longmans, 1964) p. 93.
34 Letter, 13 Sept. 1933, W. Braithwaite, CAB 45/126, PRO.
35 'Operations of 17th Division from 21 August 1918 to 11 November 1918' p. 1, Conf. 3355, SCL. For *ad hoc* training in this period, see Anon., *A Short History of the 19th (Western) Division 1914–1918* (London: Murray, 1919) p. 100.
36 C.N. Barclay, *Armistice 1918* (London: Dent, 1968) pp. 95–8.
37 *Assault Training, September 1917* (SS 185) pp. 5–6, 15.
38 *XV Corps Notes* pp. 5–6; *Notes and Hints on Training* (Jan. 1918) p. 6, RMASL.
39 C. Duffy, *Military Experience in the Age of Reason* (London: Routledge & Kegan Paul, 1987); Nicolson, *Alex* p. 53.
40 N. Fraser-Tytler, *With Lancashire Lads and Field Guns in France, 1915–18* (Manchester: Heywood, 1922) pp. 8, 140.
41 For a good description of a patrol, see E.C. Vaughan, *Some Desperate Glory* (London: Macmillan, 1985) pp. 120–1.
42 J.E. Edmonds, *Military Operations, France and Belgium, 1916*, I, Appendices, Appx. 6, pp. 43, 46.
43 *Report of the Committee on the Lessons of the Great War* p. 24, WO32/3116, PRO.
44 Parker, *Old Lie* pp. 213–15; C. Veitch, '"Play Up! Play Up! and Win the War": Football, the Nation and the First World War 1914–15', *JCH*, 20, (1985) 363–4.
45 G.H.F. Nichols, *The 18th Division in the Great War* (Edinburgh: Blackwood, 1922) p. 40; J.R. Ackerly, *My Father and Myself* (Harmondsworth: Penguin 1971) p. 50.
46 P. MacGill, *The Great Push* (London: Jenkins, 1916) p. 73 (London Irish at Loos, 25 Sept. 1915) R. Grieg to Edmonds, 22 Apr. 1930, CAB 45/134, PRO (16/R. Scots on Somme, 1 July 1916).
47 Veitch, 'Play up', 364.
48 M. Girouard, *The Return to Camelot: Chivalry and the English Gentleman* (London: Yale University Press, 1981) pp. 7, 16, 235.
49 Ibid. pp. 3–6, 8, 233–5.
50 Personal observation.
51 Eberle, *Sapper Venture* p. 22.
52 B. Gardner (ed.), *Up the Line to Death* (London: Magnum, 1976) pp. 141–2.
53 T.R. Glover, *The Ancient World* (Harmondsworth: Penguin, 1944) p. 28.
54 R.M. Ogilvie, *Latin and Greek* (London: Routledge, 1964) pp. 136, 143.
55 See, for example, C. Mackenzie, *Gallipoli Memories* (London: Panther, 1965 p. 61; A. Herbert, *Mons, Anzac and Kut* (London: Hutchinson, n.d.) p. 169.
56 S. Graham, *The Challenge of the Dead* (London: Cassell, 1921) p. 22.
57 T. King, 'The Influence of the Greek Heroic Tradition on the Victorian and Edwardian Military Ethos', unpublished paper.
58 Girouard, *Camelot* p. 66; R.W. Livingstone (ed.), *The Legacy of Greece*

(Oxford: Oxford University Press, 1921) p. 143.
59 See King, 'Influence'.
60 Mosley, *My Life* p. 28.
61 Raymond, *Tell England* pp. 273, 314; idem, *Story of My Life* pp. 122–3, 162; J. Richards, *Visions of Yesterday* (London: Routledge & Kegan Paul, 1973) pp. 153–4.
62 A.B. Cooper, 'A Friend of the Friendless', *Boy's Own Paper* 37 (1915) p. 474.
63 G. Arnold, *Held Fast for England* (London: Hamish Hamilton, 1979) pp. 17, 31–4; G.A. Henty, 'The Old Pit Shaft: a Tale of the Yorkshire Coalfields', in G.A. Henty *et al., Hazard and Heroism* (Edinburgh: Chambers, 1904) pp. 53–66. For a typical Henty book with a specifically military theme, see *In Times of Peril* (London: Griffith, Farran, 1881). Interestingly, the copy examined by the author came from the library of an Indian army officer who fought in the Great War.
64 Anon., 'A Boy's First Fight', in *Hazards and Heroism* pp. 317–42.
65 R. Kipling, 'The Army of a Dream' in *Traffics and Discoveries* (London: Macmillan, 1904) pp. 243–300.
66 M. Edwardes, 'Oh to meet an Army Man', in J. Gross (ed.), *Rudyard Kipling: the Man, His Work and His World* (London: Weidenfeld & Nicolson, 1972) p. 40.
67 R. Kipling, 'Only a Subaltern', in *Wee Willie Winkie* (London: Macmillan, 1920) pp. 101–20. See also the comments of Pte Mulvaney in Kipling's *Soldiers Three* (London: Macmillan, 1911) pp. 6, 9.
68 R. Kipling, *Puck of Pook's Hill* (Harmondsworth, Puffin, 1987) pp. 102–61.
69 C. Carrington, *Rudyard Kipling: His Life and Work* (Harmondsworth: Penguin 1970) pp. 446–7. See also Carrington to author, 10 Sep. 1983.
70 Carrington to author, 18 Aug. 1983; conversation with author 24 Feb. 1984.
71 Carrington, *Kipling* p. 152.
72 G. Orwell, 'My Country Left or Right', in S. Orwell and I. Angus (eds), *The Collected Essays, Journalism and Letters of George Orwell* I (Harmondsworth: Penguin, 1970) p. 589.
73 Parsons to War Office, 29 Nov. 1914, Sir L.W. Parsons papers, LHCMA.
74 T. Denman, *Ireland's Unknown Soldiers* (Dublin: Irish Academic Press, 1992) p. 43. See also pp. 42–6, 59–60.
75 Letters; 30 Nov. 1914, 14 Feb. 1916 W.P. Nevill papers, WPN/2, WPN 149, IWM.
76 H.E.L. Mellersh, *Schoolboy into War* (London: Kimber, 1978) p. 39.
77 E.A. James, *British Regiments, 1914–18* (London: Sampson, 1978) p. 119; B. Williams, *Raising and Training the New Armies* (London: Constable, 1918) pp. 96–7.
78 *SME* p. 235; Williams, *Raising and Training* p. 99.
79 *The Barncroft Magazine* (Berkhamstead, 1917) pp. 5–8.
80 Williams, *Raising and Training* p. 99.
81 Ibid. pp. 98–9.
82 Unpublished account, pp. 87–7a, A.J. Abrahams papers, P.101, IWM.
83 Smith, *Four Years* p. 74.
84 Unpublished account, p. 34, N.A. Pease papers, 86/9/1, IWM.
85 Unpublished account, p. 59, R.W.F. Johnston papers, 82/38/1, IWM.

86 'Ganpat', 'Fallen Angels', *Blackwood's*, CC, (1916) p. 5.
87 Unpublished account, p. 7, R.E. Barnwell papers, 85/7/1, IWM.
88 *The Times*, 5 Jan. 1916.
89 [D.H. Bell], *A Soldier's Diary of the Great War* (London: Faber, 1929) pp. 107–9.
90 E. Shephard, *A Sergeant-Major's War* (Ramsbury: Crowood, 1988) p. 137.
91 Williams, *Raising and Training* p. 101.
92 Hooper, *Letters of C.S. Lewis* p. 63.
93 M. Glover (ed.), *The Fateful Battle Line* (London: Leo Cooper, 1993) p. 157.
94 Letter, 19 Dec. 1915, L. Emmerson papers, PP/MCR/152, IWM.
95 Letter, 16 Apr. 1916, W.B. Medlicott papers, 87/56/1, IWM.
96 H. Strachan, *History of the Cambridge University Officers Training Corps* (Tunbridge Wells: Midas, 1976) pp. 145–6. See *Barncroft Magazine* passim for the undergraduate nature of OCB journals.
97 Glover, *Battle Line* p. 152; J.W.L. Napier, SR 7499, Dept. of Sound Records, IWM.
98 Notes of lecture of 3 Feb. 1916, C.R. Tobbit papers, 83/43/1, IWM.
99 Letter, 29 Apr. 1916, W.B. Medlicott papers, IWM.
100 *Blunderbus, being the Book of the 5th OCB, Trinity College*, 2 (Nov. 1916) p. 2.
101 See Hamilton's comments in WO 163/15, PRO.
102 Thorburn, *Amateur Gunners* pp. 7–11; C.M. Bowra, *Memories 1898–1939* (London: Weidenfeld and Nicolson, 1966) pp. 71–4.
103 Thorburn, *Amateur Gunners* p. 14.
104 Unpublished account, p. 34, N.A. Pease papers, 86/9/1, IWM.
105 J.B. Scrivenor, *Brigade Signals* (Oxford: Blackwell, 1937) pp. 18, 22–3. See also extract from privately printed memoir, pp. 69–70, J.C. MacDermott papers, LHCMA.
106 Letter, 4 Nov. 1916, T. Sherwood papers, IWM; Haigh and Turner, *Not for Glory* p. 81.
107 F. Hawkings (A. Taylor, ed.), *From Ypres to Cambrai* (Morley: Elmfield, 1974) p. 111.
108 R. Graves, *Goodbye to All That* (Harmondsworth: Penguin, 1960) p. 203.
109 J. Greenshields, KRS Q.
110 Shepherd, *Sergeant-Major's War* p. 140.
111 Henriques, *Queen's Westminster Rifles* p. 79. See also 'Ganpat', 'Fallen Angels' 172–3.
112 Blomfield, *Once an Artist* pp. 152–3.
113 See notes of lecture, 3 Feb. 1916, C.R. Tobbit papers, 83/43/1, IWM; notes of lecture by Major Shaw, item 1123, Misc. 74, IWM.
114 *Examination paper l, Military Law, Military Organisation and Interior Economy, Dec. 1916*, 'Officers: Officer Cadets' file, LC.
115 Unpublished notes, pp. 8–9, B.D. Parkin papers, 86/57/1, IWM.
116 Latham, *Territorial* p. 40. See also Henriques, *Queen's Westminster Rifles* p. 79.
117 For internalisation, see Handy, *Understanding* pp. 138, 140–2.
118 Hooper, *Letters of C.S. Lewis* p. 63.
119 Unpublished account, p. 34, N.A. Pease papers, 86/9/1, IWM.
120 Graves, *Goodbye to All That* p. 203. See also G.N. Wood, KRS Q.
121 See G.D. Sheffield, '"Oh! What a Futile War": Representations of the

Western Front in the Modern British Media and Popular Culture', in I. Stewart and S. Carruthers, *War, Culture and the Media* (Trowbridge: Flicks Books, 1996) pp. 54–74. For a different perspective, see D. Paget, 'Remembrance Play: *Oh! What a Lovely War* and History', in Tony Howard and J. Stokes, *Acts of War: the Representation of Military Conflict on the British Stage and Television since 1945* (Aldershot: Scolar, 1996) pp. 82–97.

5 Officer–Man Relations: the Disciplinary and Social Context

1 F. Moor, KRS Q.
2 J.N. Hall, *Kitchener's Mob* (London: Constable, 1916) p. 14. See also M.A. Mugge, *The War Diary of a Square Peg* (London: Routledge, 1920) pp. 24, 35.
3 N.D. Cliff, *To Hell and Back with the Guards* (Braunton: Merlin, 1988) p. 36.
4 Letters, 11 Feb. 1918, 24 Sept. 1916, H.S. Williamson papers, PP/MCR/333, IWM. The experience of battle was, however, common to regimental officers and Other Ranks.
5 Smith, *Four Years* pp. 120, 132.
6 Letter, 11 Aug. 1941, R.H. Tawney papers, BLPES.
7 F. Manning, *The Middle Parts of Fortune* (London: Granada, 1977) pp. 59, 142; G[eneral] R[outine] O[rders] 1388, in SS 309, *Extracts from General Routine Orders ... 1 Jan 1918* p. 67.
8 Letter, 29 Sept. 1917, L.B. Stanley papers, LC.
9 T.P. Marks, *The Laughter Goes from Life* (London: Kimber, 1977) p. 33. See also C.E. Montague, *Disenchantment* (London: Chatto and Windus, 1928) p. 84.
10 Graham, *Challenge* p. 121.
11 C. Duffy, *Russia's Military Way to the West* (London: Routledge & Kegan Paul, 1981) p. 132.
12 8 Oct., 13 Nov. 1914, and Appx. 74, 27 Mar. 1915, WD, Indian Corps, A & Q, WO 95/1091, PRO. For a flogging in France, see letter, 6 Oct. 1914, and diary, 4 Oct. 1914, G.F. Patterson papers, RMASA; for floggings in Aden, see 11 and 12 Nov. 1915, WD of APM Aden, WO 154/315, PRO. For an officer of 2/R. Irish Rifles who (quite illegally) struck British offenders, see C. Davison (ed.), *The Burgoyne Diaries* (London: Harmsworth, 1985) pp. 24–5.
13 J. Putkowski, 'A2 and the "Reds in Khaki"', *Lobster* No. 27, 19.
14 A. Babington, *For the Sake of Example* (London: Leo Cooper, 1983) pp. 3, 189; E.M. Spiers, *The Late Victorian Army* (Manchester, Manchester University Press 1992) p. 73; *Gunfire* No. 30, 43–4.
15 M. Messerschmidt, 'German Military Law in the Second World War', in W. Deist (ed), *The German Military in the Age of Total War* (Leamington Spa: Berg, 1985) p. 324.
16 W. Nicolai, *The German Secret Service* (London: Stanley Paul, 1924) p. 188; A.J. Fyfe, *Understanding the First World War – Illusions and Realities* (New York: Peter Long, 1988) p. 179.
17 Maj. W.F. Kelly, 'Report on Adjutant-General's Department ...' (Pretoria, 1900), pp. 1–2, WO 108/260, PRO.
18 '... Report of Committee on Punishments on Active Service', 5 Jan. 1904, p. 11, WO 32/4512, PRO.
19 For the discipline of the army in South Africa, see Maj. R.M. Poore, 'Report on Office of Provost Marshal' (Pretoria, 15 July 1900), WO 108/259, PRO;

Nasson, 'Tommy Atkins' pp. 125–37; Sheffield, *The Redcaps* pp. 42–6.

20 Diary, 4 July 1916, A. Surfleet papers, IWM. See also unpublished account, p. 26, E. Scullin papers, IWM.

21 'Report of Committee on Proposed Alterations in Military Penal System', WO 32/8734; Spiers, *Late Victorian Army*, pp. 74–5; R. Boyes, *In Glass Houses* (Colchester: MPSC Assoc., 1988) pp. 54–68.

22 Unpublished account, p. 33, P. Creek papers, 87/31/1, IWM.

23 A.G. Kingsmill, *The Silver Badge* (Ilfracombe: Stockwell, 1966) p. 30.

24 See the documents in 'Discipline – Vicarious Punishment of Troops' [June 1919] WO 32/9543. PRO.

25 W. Childs, *Episodes and Reflections* (London: Cassell, 1930) p. 142.

26 See Babington, *Sake of Example* passim; J. Putkowski and J. Sykes, *Shot at Dawn* (Barnsley: Wharncliffe, 1989) passim; A. Mackinlay, MP, 'The Pardons Campaign', *WFA Bulletin* No. 39 (June 1994) pp. 28–9; 'Condemned: Courage and Cowardice', six articles, including one by the author, in *JRUSI* 143 (1998) 51–69.

27 Unpublished account, pp. 14–15, R.L. Venables papers, 76/225/1, IWM.

28 J.C. Dunn, *The War The Infantry Knew* (London: Janes, 1987) p. 119. See also J. Nettleton, *The Anger of the Guns* (London: Kimber, 1979) pp. 166–7.

29 Diary, 12 Jan. 1917, A. Young papers, 76/101/1, IWM.

30 Unpublished account, A.W. Fenn papers, 75/16/1, IWM.

31 Unpublished account, p. 74, G. Buckeridge papers, p. 273, IWM.

32 Diary, 4 July 1916, A. Surfleet papers, IWM. For an interesting discussion, by an old soldier, of 'crucifixion', see R. Blatchford's articles in *Sunday Chronicle* 29 Oct. and 5, 12, 19, 26 Nov. 1916.

33 Bourke, *Dismembering* p. 81.

34 Steuart, *March, Kind Comrade* pp. 94–5. For a ranker's ambiguous reaction to a death sentence, see H.E. Harvey, *Battle-Line Narratives* (London: Brentano's, pp. 129–58. This is a fictionalised account of the events leading to the execution of Pte C.W.F. Skilton, 22/R. Fusiliers. See also Davison (ed.), *Burgoyne Diaries* p. 110.

35 Diary, 16 July 1915, A.H. Roberts papers, 81/23/1, IWM.

36 Lucy, *Devil* p. 204; unpublished account, p. 10, F.M. Packham papers, IWM.

37 G. Coppard, *With a Machine Gun to Cambrai* (London: Macmillan, 1986) p. 20; A. Rule, *Students under Arms* (Aberdeen: Aberdeen University Press, 1934) p. 52; letter, 15 Oct. 1963, R. Savage, BBC/IWM; R. Whipp, interview.

38 A.A. Hanbury-Sparrow, *The Land-Locked Lake* (London: Barker, 1932) pp. 217–18.

39 S. Graham, *A Private in the Guards* (London: Macmillan, 1919) pp. 1–2, 20.

40 A. French, *Gone for a Soldier* (Kineton: Roundwood Press, 1972) p. 23. See also unpublished account, pp. 33, 44, E. Partridge papers, LC.

41 Graham, *Private* p. 27. See also Cliff, *To Hell*, passim.

42 Unpublished account, pp. 22–3, W.H. Martin papers, PP/MCR/30, IWM; letter, 10 Oct. 1963, J.W. Binns, BBC/IWM.

43 J. Brophy and E. Partridge, *The Long Trail* (London: Deutsch, 1965) p. 13.

44 Babington, *Sake of Example* p. 131.

45 F.P. Crozier, *A Brasshat in No Man's Land* (London: Cape, 1930) p. 43.

46 C. Haworth, *March to Armistice 1918* (London: Kimber, 1968) p. 25.
47 Extracts from ROs 13 Aug. to 5 Sept. 1914, WO 95/3972, PRO.
48 ARO, 22 Dec. 1914, WO 95/25, PRO.
49 RO 29 Jan. 1916, WO 95/684, PRO. See also S.S. 309, *Extracts from General Routine Orders* p. 71.
50 R. Feilding, *War Letters to a Wife* (London: Medici Society, 1929) pp. 224–5; letter, 7 Oct. 1963, F. Blain, BBC/IWM.
51 S. Southwold, 'Rumours at the Front', in C. O'Riordan *et al.*, *A Martial Medley* (London: Scolartis, 1931) p. 106.
52 For *Auftragstaktik*, see M. van Creveld, *Fighting Power* (London: Arms & Armour, 1983) pp. 36–7; for criticisms, see C. Duffy, *Red Storm on the Reich* (New York: Atheneum, 1991) pp. 54–5, 363–4.
53 Manning, *Middle Parts* p. 71.
54 Winter, *Great War* pp. 37, 63; P. E. Dewey, 'Military Recruiting and the British Labour Force during the First World War', *HJ*, 27, (1984) 199–223.
55 J.M. Bourne, *Britain and the Great War 1914–1918* (London, Arnold, 1989) p. 205.
56 C. Denison [pseud.] 'From Two Angles', in O'Riordan, *Martial Medley* p. 82. See also Aldington, *Death of Hero* p. 243.
57 R. Roberts, *The Classic Slum* (Harmondsworth: Penguin, 1973) pp. 50–1; S. Meacham, *A Life Apart* (London: Thames & Hudson, 1977) pp. 161–2.
58 For instance, [D.S. Cairnes], *The Army and Religion* (London: Macmillan, 1919) pp. 101–3.
59 Quoted in J.H. Whitehouse, *Education* (London: Rich & Cowan, 1935) p. 170.
60 C. Barnett, *The Collapse of British Power* (Gloucester: Sutton, 1984) p. 104.
61 W.H. Beveridge, *Unemployment: a Problem of Industry* (London: Longmans, Green, 1930) p. 126.
62 Unpublished account, H. Warner papers, p. 462, IWM.
63 P. Thompson, *The Edwardians* (London: Paladin, 1977) p. 73: Meacham, *Life Apart* p. 171.
64 Quoted in Barnett, *Collapse* p. 104. See also J. Walvin, *A Child's World* (Harmondsworth: Penguin, 1982) pp. 54–6.
65 P. C. McIntosh, *Physical Education in England since 1800* (London: Bell, 1952) pp. 144. See also J.S. Hurt, 'Drill, Discipline, and the Elementary School Ethos' in P. McCann (ed.), *Popular Education and Socialisation in the Nineteenth Century* (London: Methuen, 1977); Bourke, *Dismembering* pp. 180–8.
66 J.O. Springhall, *Youth, Empire and Society* (London: Croom Helm, 1977) p. 16.
67 Ibid. p. 54; A. Summers, 'Edwardian Militarism' in R. Samuel (ed.), *Patriotism: the Making and Unmaking of the British National Identity*, I (London: Routledge, 1989) p. 254; G. DeGroot, *Blighty* (London: Longman, 1996) pp. 39–41.
68 Simkins, *Kitchener's Army* p. 20.
69 Physical training was also intended to produce a more rounded personality in the child. McIntosh, *Physical Education* pp. 141, 136, 149.
70 Hawke, *Private to Major* p. 43.
71 [F. Fox] 'GSO', *GHQ* (London: Philip Allan, 1920) pp. 72–3.

72 E.P. Thompson, 'Time, Work Discipline and Industrial Capitalism', *P&P*, 38, (1967) 60.
73 E.J. Hobsbawm, *Industry and Empire* (Harmondsworth: Penguin, 1968) p. 85. See also S. Pollard, 'Factory Discipline in the Industrial Revolution' *Economic History Review* XVI (1963) 254–71.
74 See R. McKibben, *The Ideologies of Class* (Oxford: Clarendon, 1994).
75 J. Melling, '"Non-Commissioned Officers": British Employers and Their Supervisory Workers, 1880–1920', *Social History* 5 (1980) 183–221.
76 G. Wagner, *The Chocolate Conscience* (London: Chatto & Windus, 1987) pp. 48–72.
77 J.M. Bourne, 'The British Working Man in Arms', in H. Cecil and P. Liddle, (eds), *Facing Armageddon* (London: Leo Cooper, 1996) p. 348.
78 R. Price, *An Imperial War and the British Working Class* (London: Routledge & Kegan Paul, 1972) pp. 7–10.
79 H. Macmillan, *Winds of Change* I (London: Macmillan, 1966) pp. 99–100; Thompson, *Edwardians* p. 147; unpublished account, p. 36, A. Simpson papers, Brotherton Library, University of Leeds.
80 H. Newby, *Country Life* (London: Cardinal, 1987) p. 58. For a discussion of the definition of the term, see idem, 'The Deferential Dialectic', *Comparative Studies in Society and History*, 17 (1975).
81 F.M.L. Thompson, *English Landed Society in the Nineteenth Century* (London: Routledge & Kegan Paul, 1971) pp. 184, 187; Newby, *Country Life* p. 69; Meacham, *Life Apart* p. 202; D. Winter, *Death's Men* (Harmondsworth: Penguin, 1979) p. 230.
82 C. Dakers, *The Countryside at War 1914–1918* (London: Constable, 1987) pp. 26–7; P. Horn, *Rural Life in England in the First World War* (New York, St. Martin's, 1984) pp. 28–9.
83 A. Howkins, *Poor Labouring Men* (London: Routledge & Kegan Paul, 1985) pp. 15–21.
84 Walvin, *Child's World* pp. 101–9.
85 Roberts, *Classic Slum* p. 168.
86 F.M.L. Thompson, *The Rise of Respectable Society* (London: Fontana, 1988) p. 198; R. McKenzie and A. Silver, *Angels in Marble* (London: Heinemann, 1968) pp. 197–8. Interestingly, one of the pieces of evidence offered by the latter is an ex-soldier's reminiscences of paternal officers of the Great War.
87 P. Joyce, *Work, Society and Politics* (London: Methuen, 1980) pp. 91, 94.
88 See B. Waites, *A Class Society at War* (Leamington Spa: Berg, 1987) p. 241; Joyce, *Work, Society & Politics* pp. 92–3; Roberts, *Classic Slum* pp. 4–6.
89 R. McKibbin, 'Why Was There No Marxism in Great Britain?', *EHR*, 99, (1984) 302–3; J. Harris, *Private Lives, Public Spirit: Britain 1870–1914* (Harmondsworth: Penguin, 1994) pp. 127, 135–8.
90 Thompson, *Respectable Society* pp. 58–9; B. Waites, 'The Government of the Home Front and the "Moral Economy" of the Working Class', in Liddle, *Home Fires* pp. 178–9.
91 W.J. Reader, *'At Duty's Call'* (Manchester: Manchester University Press, 1988); Spiers, *Late Victorian Army* pp. 180–203; but see S. Barlow, 'The Diffusion of "Rugby" Football in the Industrial Context of Rochdale, 1868–90: a Conflict of Ethical Values', *International Journal of the History of Sport*, X (1993) 49–67.

92 J.M. MacKenzie, *Propaganda and Empire* (Manchester: Manchester University Press, 1984) pp. 45, 181–3.
93 Mellersh, *Schoolboy* p. 72.
94 Cliff, *To Hell* p. 26; J.F. Tucker, *Johnny Get Your Gun* (London: Kimber, 1978) p. 12.
95 'Report of a Conference of General Staff Officers at the Staff College' (1910) pp. 74–6, SCL; Travers, *Killing Ground* pp. 39–40.
96 N. Macready, *Annals of an Active Life* I (London: Hutchinson, n.d.) p. 217.
97 F. Hammersley, 'Notes on 11th Division', CAB 45/237, PRO.
98 Simkins, *Kitchener's Army* pp. 237–44.
99 C.H. Potter and A.S.C. Fothergill, *The History of the 2/6 Lancashire Fusiliers* (Rochdale: published privately, 1927) p. 3.
100 V.W. Germains, *The Kitchener Armies* (London: Peter Davies, 1930) p. 111.
101 A.H. Atteridge, *History of the 17th (Northern) Division* (Glasgow: Maclehose, 1929) p. 14.
102 A. Jobson, *Via Ypres* (London: Westminster City, 1934) p. 2.
103 B. Cooper, *The Tenth (Irish) Division in Gallipoli* (London: Jenkins, 1918) pp. 14–15.
104 B.G. Buxton, KRS Q.
105 J. Ewing, *The History of the 9th (Scottish) Division 1914–1919* (London: Murray, 1921) p. 9.
106 W. Miles, *The Durham Forces in the Field 1914–18*, II (London: Cassell, 1920) p. 6.
107 Letter, 15 Sept. 1914, C. Jones papers, IWM.
108 See, for example, O. Rutter (ed.), *The History of the Seventh (Service) Battalion Royal Sussex Regiment* (London: Times, 1934) p. 2.
109 Anon., *Sixteenth, Seventeenth, Eighteenth, Nineteenth Battalions the Manchester Regiment: a Record 1914–1918* (Manchester: Sherrat & Hughes, 1923) p. 6.
110 Letter, 15 Sept. 1915, W.P. Nevill papers, WPN1/55, IWM.
111 [J.M. Mitchell] *The New Army in the Making by an Officer* (London: Kegan Paul ..., 1915) p. 14; A.P.B. Irwin, Oral History Interview, 000211/04, IWM.
112 Mitchell, *New Army* p. 17. For the composition of the battalion, see Nichols, *18th Division* p. 2.
113 Mitchell, *New Army* p. 21.
114 Ibid. pp. 21–2.
115 Unpublished account, pp. 25–6, P.G. Heath papers, DS/Misc/60, IWM.
116 Mitchell, *New Army* p. 22; unpublished account, p. 25, P.G. Heath papers, DS/Misc/60, IWM.
117 Mitchell, *New Army* pp. 24, 26.
118 Unpublished account, pp. 27–9, 51–4, P.G. Heath papers, DS/Misc/60, IWM.
119 Ts., history of 8/Norfolks, pp. 4, 15, C.F. Ashdown papers, IWM; [C.A. Maxwell], *I Am Ready* (London: Hazell, Watson, Viney, 1955) p. 128.
120 Sheffield, 'Effect of War Service' passim; letter, 18 Oct. 1914, W.J.T.P. Phythian-Adams papers.
121 E. Townshend (ed.), *Keeling Letters and Recollections* (London: Allen and Unwin, 1918) p. 219.
122 Unpublished account, S.W. Blythman papers, 80/40/1, IWM.

123 Unpublished account, pp. 29–30, C.F. Jones papers, LHCMA.
124 E. Renshaw, quoted in A.J. Peacock, 'A Rendezvous with Death', *Gunfire* No. 5, (1986) 256.
125 'Notes on the New Armies by a Divisional Commander', p. l, File 10, Sir I. Maxse papers, 69/53/5, IWM.
126 [J.H. Poett], 'A Dugout in War and Before', *GW* 1, (1989) 144, 146.
127 *Essex Countryside* (Dec. 1972) p. 41.
128 Diary 8 Sept.1914, 8–16 Sept. 1915, May 1916, R. Cude papers, IWM.

6 Officer–Man Relations: the Officer's Perspective

1 *The Times*, 7, 8 Sept. 1914.
2 *The Times*, 12 Sept. 1914. See also *Daily Sketch*, 8 Sept. 1914.
3 *The Times*, 9, 10, 22 Sept. 1914; *Daily Mirror*, 1 Aug. 1914.
4 *The War Illustrated*, 29 Aug. 1914, p. 54; ibid. 3 Oct. 191.
5 *Mr Punch's History of the Great War* (London: Cassell, 1919) pp. 54, 178; *The War Illustrated*, 11 Nov. 1916 p. 306.
6 Rawlinson to GHQ, 7 Nov. 1916, Fourth Army Papers, IWM.
7 J. Adair, *Developing Leaders* (Maidenhead: Talbot Adair, 1988) p. 13.
8 Montgomery, *Memoirs* pp. 486–7.
9 W. Slim, *Unofficial History* (London: Corgi, 1970) p. 94.
10 A. Thomas, *A Life Apart* (London: Gollancz, 1968) p. 33.
11 Letter, 24 Dec. 1914, W.B.P. Spencer papers, 87/56/1, IWM.
12 H.D. Myers, G.H. Cole, KRS Q.
13 Unpublished account, S.B. Abbott papers, 78/36/1, IWM.
14 Bell, *Soldier's Diary* p. 170.
15 Thomas, *Life Apart* p. 37.
16 R. Lewin, *Slim: the Standardbearer* (London: Pan, 1978) p. 9.
17 'Wisdom for Warts' p. 3, D. Hamond papers.
18 Document in C. Meadowcroft papers, LC.
19 Letter, May 1915, E.G. Venning, in L. Housman (ed.), *War Letters of Fallen Englishmen* (London: Gollancz, 1930) p. 282.
20 Unpublished account, Notebook III, B.D. Parkin papers, 86/57/1, IWM.
21 P. M. Campbell, *Letters from Gallipoli* (Edinburgh: published privately, 1916) p. 26.
22 *Divisional Trench Orders 1917* (AP&SS 1667 1500 1/17) pp. 3, 11.
23 *Supplement to SS 419 – Protection against Tear Gas Shells.*
24 'Entrainment and Move Orders 33 Division 26 Aug. 1916'. 71–210. RMASA.
25 ROs, 23 Nov. 1914, WO 95/25, PRO.
26 W.N. Nicholson, *Behind the Lines* (London: Cape 1939) pp. 33–4, 45. See also conference reports and memoranda in Maxse papers, 69/53/5, IWM.
27 Memorandum from AA & QMG, 66th Division, 28 Feb. 1918, P. Ingleson papers, LC.
28 W. Griffith, *Up to Mametz* (London: Severn House, 1981) pp. 180–1.
29 M.C. Morgan, *The Second World War 1939–45, Army, Army Welfare* (London: W.O. 1953) pp. 1–2. The padre also had an important paternal role: see below.
30 Unpublished account, p. 40, C.F. Jones papers, LHCMA.
31 C.E. Carrington, 'Some Soldiers', in Panichas, *Promise* p. 161. For a similar

incident in a Regular unit (2/Leinsters) see F.C. Hitchcock, *'Stand To': a Diary of the Trenches 1915–18* (Norwich: Gliddon, 1988) p. 211; for a New Army unit (10/Essex), see R.A. Chell, KRS Q.

32 Letter, 24 Dec. 1915, Sir W. Baynes papers, LC.
33 Douglas of Kirtleside, *Years of Combat* (London: Quality Books, 1963) p. 51.
34 Unpublished account, p. 30, H.D. Paviere papers, 81/19/1, IWM. See also letter, 5 Aug. 1915, G.E. Miall Smith papers, LC.
35 Letters from soldiers, 1915, in J. Pocock papers, LC; Anon., *Artillery and Trench Mortar Memories: 32nd Division* (London: Unwin, 1932) p. 662.
36 Letter, 10 July 1963, H. Boater, BBC/GW, IWM.
37 B. Martin, *Poor Bloody Infantry* (London: Murray, 1987) p. 115.
38 Letter, 9 Apr. 1916, WPN/177, W.P.Nevill papers, IWM. See also V.F. Eberle, *My Sapper Venture* (London: Pitman, 1973) p. 60.
39 Diary, 3 June 1915, H. Venables papers, LC.
40 Letters, 4 Feb. [Jan.] 1916, 3 Oct. 1915, H.A. Bowker papers, LC.
41 E.W. Flanagan to Edmonds, 17 Nov. 1935, CAB 45/133, PRO.
42 WD of 1 /R. Berks, 23 May 1916 & appx, WO 95/1371, PRO; diary, Sir H.H. Wilson, 24 May 1916, IWM; WD of IV Corps G.S., 23 May 1916, WO 95/713, PRO.
43 Notebook II, p. 50, 9 June 1916, G.G.A. Egerton papers, 73/51/1, IWM.
44 Extract from privately published memoir, p. 76, Lord MacDermott papers, LHCMA.
45 B. Adams, *Nothing of Importance* (Stevenage: Strong Oak, 1988) p. 211.
46 Comment by Gilbert Frankau, in P. Beaver (ed.), *The Wipers Times* (London: Macmillan, 1973) p. 79.
47 G. Macleod Ross, KRS Q.
48 See P. Fussell, *The Great War and Modern Memory* (London: Oxford University Press, 1975) pp. 270–309, and C.E. Carrington's hostile review in *Newsletter of the Friends of Amherst College Library* (Winter, 1976).
49 Poem and letters, 28 Apr. and 26 May 1916, G.S. Taylor papers, LHCMA.
50 Slim, *Unofficial History*, p. 32.
51 See also H. Read, 'My Company', in B. Gardner, *Up the Line to Death* (London: Magnum, 1976) pp. 87–90, and Fussell, *Great War* pp. 164–5.
52 'Wisdom for Warts' p. 3, D. Hamond papers.
53 Letter, T. Waterfield, 29 Sept. 1917, G.S. Taylor papers, LHCMA.
54 Letters, 12, 13 Dec. 1914, E. Taylor papers, LC. See also Adams, Nothing p. 29.
55 *The Times*, 5 Dec. 1916.
56 Nettleton, *Anger* pp. 68–9, 123–4.
57 R. Graves, *Goodbye to All That* (Harmondsworth: Penguin, 1960) p. 192.
58 Vaughan, *Some Desperate Glory* pp. 25–6. For a similar incident, see unpublished account, p. 43, H.D. Paviere papers. 81/19/1. IWM.
59 Note of 20 Nov. 1917, J.H. Dible papers, IWM.
60 Letter, 7 Oct. 1916, G.S. Taylor papers, LHCMA.
61 C.E. Carrington, *Soldier from the Wars Returning* (London: Hutchinson, 1965) p. 172. For an example of a court martial 'lean[ing] over backwards' to ensure a soldier had a fair trial, see Nettleton, *Anger* p. 166.
62 [A.M. Burrage], 'Ex-Pte.-X', *War is War* (London: Gollancz, 1930) p. 74; see also T.S. Hope, *The Winding Road Unfolds* (London: Tandem, 1965) p. 124.

63 G.H. Cole, KRS Q; Nettleton, *Anger* p. 200.
64 Diary, 9 Jan. 1917, W.R. Acklam papers, 83/23/1, IWM.
65 Letter, 16 Feb. 1916, H.A. Bowker papers, LC. For a similar example, see D. Williams, 'An Artilleryman's War 1914–19', *ST!* 29 (1990) p. 27.
66 J.L. McWilliams and R.J. Steel, *The Suicide Battalion* (St Catharine's, Ontario: Vanwell, n.d.) p. 186; Eberle, *Sapper Venture* p. 73.
67 Turner and Haigh, *Not for Glory* pp. 86–7; see also Carrington, *Soldier* p. 170.
68 D. Jones, *In Parenthesis* (London: Faber, 1963) p. 15.
69 'Wisdom for Warts', p. 9, D. Hamond papers. See also 'Mark VII', *Subaltern* p. 21.
70 R. Coldicott, *London Men in Palestine* (London: Edward Arnold, 1919) p. 125.
71 Ibid., pp. 66–70.
72 Housman, *Letters* p. 186.
73 D. Ascoli, *The Mons Star* (London: Harrap, 1981) pp. 8, 101.
74 Winter, *Great War* pp. 83–92; Hamilton to Kitchener, 1 Sept. 1915, WO 95/4266, PRO.
75 *AQ* III, (1921) p. 164.
76 Sheffield, 'Effect of War Service', pp. 21, 24a.
77 [C.E. Carrington] 'C. Edmonds', *A Subaltern's War* (London: Peter Davies, 1929) p. 18.
78 Unpublished account, p. 40, FAS/3, F.A. Shuffrey papers, PP/MCR/261, IWM.
79 Such a draft arrived at 6/Northants. early in 1918. Letter, 14 Feb. 1918, C.E.G. Parry Okeden papers, 90/7/1, IWM. For the experience of a young soldier, see F.J. Hodges, *Men of 18 in 1918* (Ilfracombe: Stockwell, 1988) passim.
80 E. Parker, *Into Battle* (London: Longman, 1964) pp. 36, 38.
81 Diary, 15 May 1916, J.W.B. Russell papers, LC.
82 Diary, 1 Apr. 1919, F.C. Shuffrey papers, FAS/7, PP/MCR/261, IWM.
83 V. Bartlett, *No Man's Land* (London: Allen & Unwin, 1930) p. 168.
84 W.R. Bion, *The Long Week-End 1897–1917* (London: Free Association, 1982) p. 119.
85 Letter, 12 Feb. 1916, C.R. Stone papers, IWM.
86 Unpublished account, pp. 69–70, C.J. Lodge Patch papers, IWM 66/304/1.
87 Note, 11 Nov. 1917, pp. 179–80, J.H. Dible papers, IWM. In December 1917, Haig commented on the disparity with the French, whose soldiers were given leave every four months. S. Fay, *The War Office at War* (Wakefield: EP, 1973) p. 106.
88 Circular from Military Secretary to C-in-C, 9 Jan. 1918, P. Ingleson papers, LC.
89 Until 1916, thousands of men were leaving the army on the expiration of their service. F.W. Perry, *The Commonwealth Armies* (Manchester: Manchester University Press, 1988) p. 18; *The Times*, 6 Apr. 1916.
90 Unpublished account, p. 4, J. Woollin papers, PP/MCR/110, IWM.
91 Notebook I, B.D. Parkin papers, 86/57/1, IWM.
92 Martin, *Poor Bloody Infantry* p. 131.
93 I.F.W. Beckett, 'The British Army, 1914–18: the Illusion of Change' in J.

Turner (ed.), *Britain and the First World War* (London: Unwin Hyman, 1988) p. 107.

94 Sheffield, 'Effect of War Service', pp. 57–8.

95 'Mark VII', *Subaltern* p. 95.

96 B.G. Buxton, KRS Q.

97 R.H. Mottram, 'A Personal Record' in R.H. Mottram *et al.*, *Three Personal Records of the War* (London: Scolartis, 1929) pp. 62–3.

98 I. Hay, *The First Hundred Thousand* (Edinburgh: Blackwood's, 1916) p. 297.

99 Diary, 16 Aug. 1917, P. Fraser papers, 85/32/1, IWM.

100 L. Humphreys, KRS Q.

101 Letter, 9 July 1916, R.E. Wilson, KRS Q.

102 A.H. Cope, KRS Q.

103 G.A. Studdert-Kennedy, 'Religious Difficulties' in F.B. MacNutt (ed.), *The Church in the Furnace* (London: Macmillan, 1918) p. 391.

104 Diary, 7 Feb. 1917, R.H. Sims papers, 77/130/1, IWM.

105 Diary, 16 Dec. 1916, J.M. Thomas papers, 88/56/1, IWM.

106 N. Gladden, *Somme 1916* (London: Kimber, 1974) p. 73; Coppard, *Machine Gun* p. 13.

107 Pearce-Smith, *Ancient Warrior* p. 9.

108 Notebook I, pp. 31–3, G.G.A. Egerton papers, 73/51/1, IWM.

109 Diary, 25 Nov., 31 Dec. 1917, R.L. Mackay papers, IWM. For similar examples, see P.J. Campbell, 'The Ebb and Flow of Battle', in *In the Cannon's Mouth* (new edition, London: Hamish Hamilton, 1986) p. 203; 'F.O.O.', *The Making of a Gunner* (London: Eveleigh Nash, 1916) p. 101.

110 Report on operations, 3/4 Aug. 1916, in Sir. G. McCrae papers, LHCMA.

111 Unpublished account, p. 38, A.J. Abraham papers, P.191, IWM.

112 Diary, 2 May 1917, R. Cude papers, IWM.

113 This judgement has recently been confirmed by Ian Whitehead, 'Not a Doctor's Work? The Role of the British Regimental Medical Officer in the Field', in Cecil and Liddle, *Facing Armageddon* pp. 470–1. See also K. Simpson, 'Dr. James Dunn and Shell-Shock', in ibid. pp. 502–20.

114 M. Linton-Smith, 'Fellowship in the Church' in MacNutt, p. 110. For some perceptive comments on the difficulties faced by padres, see Cecil, *Flower* pp. 164–5.

115 M. Brown, *The Imperial War Museum Book of the First World War* (London: Sidgwick and Jackson, 1991) p. 247.

116 E. Thompson, *These Men Thy Friends* (London: Knopf, 1927) p. 19.

117 Linton-Smith, 'Fellowship', pp. 109–113.

118 J. Horne, *The Best of Good Fellows* (London: J. Horne, 1995) p. 154.

119 Brown, *Imperial War Museum* p. 247; *Artillery and Trench Mortar Memories* p. 672. For the paternal padre of 4/R. Fusiliers, see A. Brown, *Destiny* (Bognor: New Horizon, 1979) pp. 5–6.

120 Letters, 21 May 1917, 13 Jan. 1918, C.E. Raven papers, LC. See also a novel by a gunner, C.R. Benstead's *Retreat* (London: Methuen, 1930), which deals with the breakdown of a padre who becomes totally alienated from his flock.

121 For a recent corrective, see J.M. Bourne, 'British Generals in the First World War', in G.D. Sheffield (ed.), *Leadership and Command: the Anglo-American Military Experience since 1861* (London: Brassey's, 1997) pp. 93–112.

122 G. Powell, *Plumer: the Soldier's General* (London: Leo Cooper, 1990) p. 281; R.H. Mottram, *Journey to the Western Front* (London: Bell, 1936) p. 27.
123 D. Clarkson, *Memoirs of a Company Runner* (Edinburgh: Scottish National Institute for War Blinded, 1972) p. 18. For Kellett, see Sheffield, 'Effect of War Service' pp. 68–71.
124 P. R. Munday, KRS Q; unpublished account, p. 1, S. Ragget papers, 90/1/1, IWM.
125 Carrington, *Soldier* p. 99.
126 Smith, *Four Years* p. 187.
127 Unpublished account, Chapter 10, p. 8, J. Woolin papers, PP/MCR/110, IWM.
128 Col. M.G.N. Stopford, lecture, p. 13, Conf. 3898, SCL.
129 Carrington, *Soldier* p. 100.
130 For a disparaging cartoon, see *Mr. Punch's History* p. 123.
131 D. Wheatley, *Officer and Temporary Gentleman* (London: Hutchinson, 1978).
132 Unpublished account, p. 22, W.G. Wallace papers, LHCMA; Parsons to Secretary, WO, 29 Nov. 1914, Sir L.W. Parsons papers, LHCMA.
133 J.E.H. Neville, *The War Letters of a Light Infantryman* (London: Sifton Praed, 1930) p. 6.
134 R.T. Rees, *A Schoolmaster at War* (London: Haycock, n.d.) p. 79.
135 12 May 1915, WD, Adjutant-General, GHQ, WO 95/25, PRO.
136 A. Calder, *The People's War* (London: Panther, 1971) p. 285; Crang, 'British Army' pp. 132–4; *The Times*, 29 Apr. 1963.
137 Sir E. Swinton, in *Twenty Years After*, II (London: Newnes, 1938) p. 1188.
138 H. Essame, *The Battle for Europe, 1918* (London: Batsford, 1972) pp. 12–13, 107–8, 111, 179–81. See also Barclay, *Armistice 1918* pp. 100–1.
139 Edmonds, *OH 1918* IV pp. 183, 292, 515; V pp. 575–6, 577, 578, 592.
140 P. Simkins, 'The War Experience of a Typical Kitchener Division: the 18th Division, 1914–18' in Cecil and Liddle, *Facing Armageddon* pp. 304–5. Simkins' findings are supported by the present author's research.
141 F. Moor, KRS Q.
142 'F.O.O.', *Making* p. 121.
143 Crozier, F.P. *Impressions and Recollections* (London: T. Werner Laurie, 1930).
144 Sir A. Bishop, KRS Q.
145 Dunn, *War the Infantry Knew* p. 357.
146 L.H.M. Westropp, KRS Q.
147 Lucy, *Devil* p. 353; G.H. Cole, KRS Q; Sheffield, 'Effect of War Service', p. 61.
148 W. Robertson to Edmonds, c. 1936, CAB 45/137, PRO. For a discussion of this question, see [B.H. Liddell Hart] 'Bardell', 'Study and Reflection v. Practical Experience', *AQ*, VI, (1923) 318–31; I.F.W. Beckett, 'The Territorial Force', in Beckett and Simpson, *Nation* pp. 140–3.
149 Letter, 14 Feb., 29 (?) Mar., 8 May 1918, C.E.G. Parry-Okenden papers, 90/7/1, IWM; letter, 7 May 1917, H.M. Dillon papers, IWM.
150 'Rationing: Food Queues ...' MAF 60/243, PRO. See also Waites, 'Government of the Home Front' pp. 188–9.
151 Letters, n.d. and 12 Oct. 1918, A.H. Swettenham papers, 83/31/1, IWM.
152 T. O'Toole, *The Way They Have in the Army* (London: John Lane, 1916) p. 163.

153 These figures are drawn from items in H.F. Bowser papers, 88/56/1, IWM.
154 Nettleton, *Anger* p. 57.
155 Simpson, 'The Officers' p. 77; Mellersh, *Schoolboy* p. 56.
156 R.H. Arhrenfeldt, *Psychiatry in the British Army in the Second World War* (London: Routledge & Kegan Paul, 1958) p. 51.
157 Simpson, 'The Officers' p. 84.
158 H. Williamson, *The Golden Virgin* (London: Macdonald, 1984) pp. 225–6.
159 For the value of his novels as historical evidence, see H. Cecil, 'Henry Williamson: Witness of the Great War' in B. Sewell (ed.), *Henry Williamson, the Man and His Works: a Symposium* (Padstow: Tabb House, 1980) pp. 69–82.

7 Officer–Man Relations: the Other Ranks' Perspective

1 V.W. Tilsley, *Other Ranks* (London: Cobden-Sanderson, 1931) pp. 126–7.
2 F. Manning, *The Middle Parts of Fortune* (London: Cardinal, 1977) p. 75.
3 These comments were recorded by a gentleman ranker: R.K.R. Thornton (ed.), *Ivor Gurney War Letters* (London: Hogarth, 1984) p. 217.
4 'Mark VII' [M. Plowman], *A Subaltern on the Somme* (London: Dent, 1928) pp. 101–2. See also E.M. Spiers, 'The Scottish Soldier at War', in Cecil and Liddle, *Facing Armageddon* p. 319.
5 Privately published article, L.B. Stanley papers, LC.
6 See the views of a gunner of 61st Divisional Artillery on his CO, Maj. Harris, expressed in letters and diary entries; unpublished account, pp. 47, 61, 78, S.L.C. Edwards papers, IWM.
7 Bion, *Long Week-End* p. 133.
8 Clarkson, *Memoirs* p. 24.
9 Jobson, *Via Ypres* p. 186.
10 Diary,14, 20 Jan. 1915, J.W. Gower papers, 88/56/1, IWM.
11 Unpublished account, S.B. Abbot papers, 78/36/1, IWM; 28–30 Apr. 1917, WD, 86 Bde. MG Coy, WO 95/2302, PRO.
12 Unpublished memoir, p. 9, F.M. Packham papers, p. 316, IWM.
13 Nettleton, *Anger* p. 191.
14 J. Gibbons, *Roll On, Next War!* (London: Muller, 1935) p. 69.
15 J.R. Tucker, *Johnny Get Your Gun* (London: Kimber, 1978) p. 152.
16 Sheffield, 'War Service', p. 76.
17 Coppard, *Machine Gun* pp. 117–18. See also Gibbons, *Roll On* p. 70.
18 Anon., 'Memories V. Concerning Officers and NCOs', *Twenty Years After*, supplementary vol. p. 371.
19 Moran also argued, however, that no man has an inexhaustible supply of courage. Lord Moran, *The Anatomy of Courage* (London: Constable, 1945) pp. X, 10, 19–20, 26, 67.
20 E. Linklater, *Fanfare for a Tin Hat* (London: Macmillan, 1970) p. 61.
21 Anon., 'Mesopotamian Diary: with the 5th Buffs along the Tigris 1915–16', *GW*, 2 (1989) 24–5; ibid., 2, No. 3 (1990) pp. 65–6.
22 Unpublished account, pp. 8a–9, 54, 84, A.J. Abraham papers, p. 191, IWM.
23 F. Dunham (R.H. Haig and P. W. Turner, eds), *The Long Carry* (Oxford: Pergamon, 1970) pp. 10, 31, 36, 42, 68, 82.
24 Unpublished account, G.S. Chaplin papers, IWM.
25 Diary, 3. Oct. 1915, J. Griffiths papers, IWM.
26 Diary, 16 Feb. 1916, A. Moffat papers, LC.

27 Letter, 5 Sept. 1918, H. Innes papers, LC.
28 Personal observation of Sandham Memorial Chapel, Burghclere, Berkshire. The official guide leaflet misses the significance of this figure. See also J.M. Winter, *Sites of Memory, Sites of Mourning* (Cambridge, Cambridge University Press, 1995) pp. 167–71.
29 Coppard, *Machine Gun* pp. 17, 69.
30 Diary, 10, 12 Aug. 1915, G. Brown papers, 85/11/1, IWM.
31 D. Hankey, *A Student in Arms* (London: Melrose, 1916) pp. 59–70.
32 R. Davies, 'Donald Hankey, a Student in Arms' *ST!* 47, Sept. 1996, 11–14.
33 Coppard, *Machine Gun* p. 5.
34 Shephard, *Sergeant-Major* pp. 82, 97.
35 French, *Gone for a Soldier* p. 29.
36 G.E.M. Eyre, *Somme Harvest* (London: Jarrolds, 1938) p. 168.
37 Anon., *George Elton Sedding: the Life and Work of an Artist Soldier* (published privately, n.d.) pp. 152, 154.
38 Unpublished account, p. 54, J.W. Riddell papers, 77/73/1, IWM.
39 Anon, 'Memories V', *Twenty Years After*, supp. vol., pp. 369–72.
40 Spiers, *Late Victorian Army* p. 103.
41 Tucker, *Johnny Get Your Gun* p. 41.
42 'Ex-Pte. X', *War is War* p. 217.
43 Diary, 6 Feb. 1917, 8–16 Sept. 1915, R. Cude papers, IWM.
44 G.W. Grossmith, KRS Q. See also the comments of a Regular officer of the Devons in A.H. Cope, KRS Q.
45 M.W. Parr, KRS Q.
46 Thomas, *Life Apart* p. 67.
47 C. Mizen, interview.
48 'Ex-Pte. X', *War is War* pp. 71–2.
49 [Bell,] *Soldier's Diary* p. 170
50 P. M. Morris, 'Leeds and the Amateur Military Tradition: the Leeds Rifles and its antecedents, 1815–1918', PhD, University of Leeds, 1983, p. 804.
51 Gibbons, *Roll On* p. 72.
52 M.L. Walkington, KRS Q. See also F.H. Keeling's letter, 4 May 1916, in Townshend, *Keeling Letters* p. 281.
53 G.S. Mitchell, *'Three Cheers for the Derrys'* (Derry: YES! Publications, 1991) pp. 141–2 (10/R. Innis. F.); Sheffield, 'Effect of War Service' p. 61 (22/RF); L. Milner, *Leeds Pals* (Barnsley: Wharncliffe, 1991) p. 51 (15/W. Yorks).
54 Diary, 19 Oct. 1914, T.S. Wollocombe papers, RMASL; E.S. Woods (ed.), *Andrew R. Buxton, the Rifle Brigade* (London, Robert Scott, 1918) p. 278.
55 G.H. Cole, KRS Q. See also A. Rule, *Students under Arms* (Aberdeen: Aberdeen University Press) 1934) p. 77.
56 See for, example, A. Bird (ed.), *Honour Satisfied* (Swindon: Crowood, 1990) p. 31; R.H. Mottram, *Journey to the Western Front* (London: Bell, 1936) p. 161.
57 'Ex-Pte. X', *War is War* p. 71.
58 Brophy and Partridge, *Long Trail* p. 225. Officers also used this phrase, or a variation on it, in the context of relations between senior and junior officers. *Notes for Young Officers* (London: HMSO, 1917) p. 2; Mosley, *My Life* p. 46.
59 Diary, December 1915, R. Cude papers, IWM.

60 C. Crutchley, *Shilling a Day Soldier* (Bognor Regis, New Horizon, 1980) p. 51.
61 'Notes on A. and Q. Conference, 29-5-16 at Headquarters XI Corps', WO 95/885, PRO.
62 Winter, *Death's Men* p. 67.
63 Letter, 22 Aug. 1915, W.P. Nevill papers, WPN1/34, IWM; 'Memo. on Trench to Trench Attacks ...' 31 Oct. 1916, p. 4. WO 158/344, PRO.
64 K Sykes to Edmonds, 9 May 1936, CAB 45/137, PRO.
65 Coppard, *Machine Gun* p. 69.
66 Quoted in C.H. Potter and A.S.C. Fothergill, *The History of the 2/6th Lancashire Fusiliers* (Rochdale: published privately, 1927) p. 49.
67 Coppard, *Machine Gun* p. 69.
68 Campbell, *Cannon's Mouth* pp. 59, 65.
69 Coppard, *Machine Gun* p. 69.
70 J. Hargrave, *The Suvla Bay Landing* (London: Macdonald, 1964) p. 116.
71 Letter, June 1916, T.M. Sibley papers, DS/Misc/44, IWM.
72 Letter, n.d. c.1963, K.J. Box, BBC/GW, IWM.
73 Unpublished account, R.D. Fisher papers, 76/54/1, IWM.
74 Crozier, *Impressions* p. 25.
75 For a good selection of such letters, see M. Brown, *Tommy Goes to War* (London: Dent, 1978), pp. 199–200; Brown, *IWM First World War* p. 234.
76 Winter, *Sites of Memory* pp. 51–2.
77 Letter; 24 June 1916, S. Brashier. I am grateful to Mr N. Lucas for sending me a copy of this letter.
78 B. Webb, *Edmund Blunden: a Biography* (New Haven: Yale University Press, 1990) pp. 52–3.
79 L.W. Griffith, 'The Pattern of One Man's Remembering', in Panachias, *Promise* p. 287.
80 Spiers, 'Scottish Soldier', in Cecil and Liddle, *Facing Armageddon* p. 322.
81 Letter of 6 Apr. 1918, T. Sherwood papers, IWM.
82 A. Eden, *Another World* (London: Allen Lane, 1976) pp. 80–1.
83 S. Graham, *Part of the Wonderful Scene* (London: Collins, 1964) pp. 163–5.
84 Martin, *Poor Bloody Infantry* pp. 156–7.
85 Unpublished memoir, p. 8, W.H. Davies papers, 8201–13, NAM. See also unpublished account, p. 32, D. Starrett papers, 79/35/1, IWM.
86 Letter, Mar. 1915, C.H. Rastall papers, IWM; see also letters of 17 and 21 Apr. 1916.
87 Caldicott, *London Men* pp. 45, 176.
88 Vaughan, *Desperate Glory* pp. 17–18.
89 Sherriff, *Journey's End* pp. 8–9, 16–17.
90 Letter, 7 July 1930, C.F. Hill, CAB 45/134, PRO.
91 Unpublished account, p. 12, H.L. Adams papers, 83/50/1, IWM.
92 Eberle, *Sapper Venture* p. 149.
93 Clarkson, *Memoirs* p. 13.
94 J. Gillam, *Gallipoli Diary* (Stevenage: Strong Oak, 1989) p. 283.
95 H. Williamson, *A Fox under My Cloak* (London: Macdonald, 1985) pp. 35–6.
96 O'Toole, *The Way They Have* pp. 148–9.
97 For one such NCO, Sgt. Ross of 32nd Division TMB, see *Artillery and Trench Mortar Memories* p. 677.

98 N. Hancock, 'War from the Ranks', in O'Riordan *et al.*, *Martial Medley* p. 183.
99 C.E. Montague, *Disenchantment* (London: Chatto & Windus, 1928) pp. 16–20, 24–5.
100 Simkins, *Kitchener's Army* p. 228; Middlebrook, *Somme* p. 18.
101 Jobson, *Via Ypres* pp. 177–8; Baynes, *Morale* p. 197. See also W.L. Andrews, *Haunting Years* (London: Hutchinson, n.d.) pp. 74–5.
102 H. Dalton, *With the British Guns in Italy* (London: Methuen, 1919) p. 70.
103 Letter, 26 Jan. 1917, O. Hopkin, quoted in A.J. Peacock, 'A Rendezvous with Death', *Gunfire* 5 (1986) p. 368.
104 Baynes, *Morale* pp. 190, 192.
105 'Wisdom for Warts' p. 6, D. Hamond papers.
106 Unpublished memoir, p. 30, P. Creek papers, 87/31/1, IWM.
107 Hooper, *Letters* p. 67.
108 For a list of platoon sergeant's duties, see B.C. Lake, *Knowledge for War – Every Officer's Handbook for the Front* (London: Harrison, n.d.) p. 149; for a description of the duties of the CSM, see Shephard, *Sergeant Major* p. 125.
109 Marks, *Laughter* p. 110; Shephard, *Sergeant Major* p. 126.
110 Unpublished account, J.W. Riddell papers, p. 50, 77/73/1, IWM; R. Whipp, interview.
111 Anon., 'Memories V', *Twenty Years After*, supp. vol., p. 374.
112 Kingsmill, *Silver Badge* pp. 75–6; Sheffield, 'Effect of War Service', pp. 46–7.
113 Letter, 5 May 1917, A. Young papers, 76/101/1, IWM.
114 Unpublished account, pp. 94–5, R.L. Venables papers, IWM.
115 Letter, 15 Nov. 1914, F.H. Keeling, in Townshend, *Keeling Letters* p. 199.
116 Unpublished account, pp. 86–7, 302, 348, C.F. Jones papers, LHCMH.
117 Adams, *Nothing of Importance* p. 62.
118 Manning, *Middle Parts* p. 237.
119 R.H. Tawney, *The Attack and Other Papers* (Nottingham: Spokesman, 1981) p. 12.
120 *OH, 1916*, I p. 26.
121 Griffith, *Mametz* pp. 135–6.
122 S. F. Hatton, *The Yarn of a Yeoman* (London: Hutchinson) pp. 87–8.
123 Aldington, *Death of a Hero* p. 338. See also Manning, *Middle Parts* p. 132.
124 Fraser-Tytler, *Field Guns* p. 21. See also see Martin, *Poor Bloody Infantry* p. 50 (infantry); A.J.L. Scott, *Sixty Squadron R.A.F. 1916–1919* (London: Greenhill, 1990) p. 4.
125 Letter, 16 July 1963, T. Boyce, BBC/GW, IWM.
126 *Notes for Young Officers* p. 6; H. Gordon, *The Unreturning Army* (London: Dent, 1967) pp. 60–1.
127 Campbell, *Cannon's Mouth* pp. 70–1.
128 F. Moor, KRS Q.
129 See, for instance, Shephard, pp. 68–9. See Simpson, 'The Officers', p. 83, for some points which could be used to construct a contrary argument.
130 R. Devonald-Lewis, (ed.), *From the Somme to the Armistice* (London: Kimber, 1986) p. 69.
131 Diary, 2 Aug. 1917, R.L. Mackay papers, P. 374, IWM.
132 Shephard, *Sergeant Major*, passim.
133 See, for example, B. Clarke, 'From "Cog" to Partnership', *The War*

Illustrated, 21 Dec. 1918, p. 311; but see also G. Philips, 'The Social Impact' in Constantine *et al. First World War* p. 133.

134 *The Times*, 23 Oct. 1916. For the sceptical reaction of an officer to a similar speech by the Archbishop of York, see letter, 28 July 1917, in Neville, *War Letters* p. 50.

135 M. Ferro, *The Great War 1914–1918* (London: Routledge, 1973) p. 145; B. Bond, *War and Society in Europe, 1870–1970* (London: Fontana, 1984) p. 120; M. Howard, introduction to Manning, *Middle Parts* p. vi.

136 G.H. Woolley, *Sometimes a Soldier* (London: Benn, 1963) p. 54.

137 Steuart, *March Kind Comrade* p. 69.

138 G.D. Sheffield, 'The Effect of The Great War on Class Relations in Britain: the Career of Major Christopher Stone DSO MC', *W&S*, 7 (1989) 87–105.

139 Tawney, *Attack* p. 22.

140 Webb, *Blunden* p. 94.

141 E.G.D. Liveing, 'Cut Off in a Cave', in J. Buchan, (ed.), *The Long Road to Victory* (London: Nelson, 1920) p. 225.

142 See R.W. Little, 'Buddy Relations and Combat Performance' in Janowitz, *The New Military* (New York: Norton, 1969) pp. 195–223; E.A. Shills and M. Janowitz, 'Cohesion and Disintegration in the Wehrmacht' in D. Lerner (ed.), *Propaganda in War and Crisis* (New York: Arno, 1972) pp. 367–415. O. Bartov, *Hitler's Army* (New York: Oxford University Press, 1991) challenges, somewhat unconvincingly, the role of the primary group in the Wehrmacht's unit cohesion.

143 Brophy and Partridge, *Long Trail* p. 151. See also letter 21 Mar. 1917, Pte B.F. Eccles papers, 82/22/1, IWM.

144 Letter, 22 Oct. 1917, J. Mudd papers, 82/3/1. IWM.

145 See for instance, unpublished story, p. 33, E. Partridge papers, LC; unpublished account, p. 88, R.D. Fisher papers, 76/54/1, IWM; P.G. Ackrell, *My Life in the Machine Gun Corps* (Ilfracombe: Stockwell, 1966) p. 14.

146 Steuart, *March, Kind Comrade* pp. 97–8; 'Ex-Pte. X', *War is War*, passim. For a critical analysis of the concept of disillusionment, see G.D. Sheffield, '"Disillusionment" and Other Myths of British Army Morale in the First World War', unpublished paper; K. Grieves, 'C.E. Montague and the Making of *Disenchantment* 1914–21', *WIH* 4 (1997) 35–54.

147 P. MacGill, *The Red Horizon* (London: Caliban, 1984) p. 94; E. Linklater, *The Man on My Back* (London: Macmillan, 1950) p. 36.

148 Letter, 11 Feb. 1918, H.S. Williamson papers, PP/MCR/3, IWM.

149 Undated extract from diary, quoted in Grey, *Confessions* pp. 35, 72.

150 A. Waugh, 'A Light Rain Falling', in Panachias, *Promise* pp. 342–3.

151 H. Carpenter, *J.R.R. Tolkien: a Biography* (London: Unwin Paperbacks, 1978) p. 89.

152 Unpublished account, pp. 123–5, R.C. Foot papers, IWM.

153 Letter, 24 June 1969, W.M. Jenner, P. Blagrove papers, LHCMA.

154 C.M. Bowra, *Memories 1898–1939* (London: Weidenfeld & Nicolson, 1966) pp. 89–91.

155 Crutchley, *Shilling a Day Soldier* pp. 73–4.

156 WO 32/5455, PRO.

157 I.F.W. Beckett, 'The Nation in Arms, 1914–18' in Beckett and Simpson, *Nation in Arms* p. 24.

158 Bourke, *Dismembering* p. 153–5.
159 *66th East Lancashire Division Dinner Club* (Manchester: Falkener, 1924).
160 *Artillery and Trench Mortar Memories 32nd Division* p. 687.
161 Moorhouse, *Hell's Foundations* p. 187; R. Whipp, interview; Hatton, *Yarn* p. 228.
162 While Haig believed that officers should play a paternal role in the British Legion, there was in fact a 'low level of officer participation'. (N.J.A. Barr, 'Service not Self – the British Legion 1921–1939', PhD, University of St Andrews, 1994, pp. 99–100.) This suggests that the Legion was a rather different type of organisation from many unit OCAs.
163 A. Gregory, *The Silence of Memory* (Oxford: Berg, 1994) p. 54.
164 Woolley, *Sometimes* p. 175; Sheffield, 'Effect of War Service', pp. 95–6; C.E. Crutchley, *Machine Gunner 1914–19* (London: Purnell, 1975).
165 See the embittered epilogue in Hatton, *Yarn* pp. 283–6 and M. Petter, '"Temporary Gentlemen" in the Aftermath of the Great War: Rank, Status and the Ex-Officer Problem', *HJ*, 37 (1994) 127–52.
166 W. Turner, *Pals: the 11th (Service) Battalion (Accrington) East Lancashire Regiment* (Barnsley: Wharncliffe, n.d.) p. 188.
167 For the process of commemoration of the dead of 1/4 KOSB, see G. Richardson, *For King and Country and Scottish Borderers* (Hawick: published privately, 1987) pp. 84–9.
168 K.R. Grieves, 'Making Sense of the Great War: Regimental Histories, 1918–23', *JSAHR* LXIX, (1991) 14.
169 Dunn, passim; C. Stone, *A History of the 22nd (Service) Battalion, Royal Fusiliers (Kensington)* (privately published, 1923) pp. 11–26; unusually, two NCOs, O.F. Bailey and H.M. Hollier, wrote *The Kensingtons: 13th London Regiment* (London: published privately, 1935). This was of course a 'class corps'.
170 Circular from Old Contemptibles Association, n.d., in letter, July 1963, O.Y. Smith, BBC/GW, IWM.
171 See, among many others, *City of London Rifles Quarterly Journal*; *Old Comrades Journal, 2/4 The Queens*; *Newsletter* of 19th Division RFA Old Comrades Association: *Mufti, the Peacetime Record of the Fighting 22nd* [22/RF OCA journal].
172 J. Roebuck, *The Making of Modern English Society from 1850* (London: Routledge & Kegan Paul, 1973) p. 101. See also A. Marwick, *The Deluge* (Harmondsworth: Penguin, 1967) p. 218.
173 Ferro, *Great War* pp. 156–7, 225.
174 Election address in Coll. Misc. 567, BLPES.
175 K. Harris, *Attlee* (London: Weidenfeld & Nicolson, 1982) p. 55; R. Rhodes James, *Anthony Eden* (London: Weidenfeld & Nicolson, 1986) plate between pp. 186 and 187.
176 S.R. Ward, 'Great Britain: Land Fit for Heroes Lost' and J.M. Diehl, 'Germany: Veterans' Politics under Three Flags', both in S.R. Ward (ed.), *The War Generation* (Port Washington, NY: Kennikat Press, 1975) pp. 31–5, 135–86. In 1918 Ernest Thurtle, a ranker-officer, stood unsuccessfully for parliament as the candidate of an-ex serviceman's group. Later he became a Labour MP. E. Thurtle, *Time's Winged Chariot* (London, Chaterson, 1945) pp. 59–60.

177 Letter, n.d., c.1963, S.A. Boyd, BBC/GW, IWM.
178 R. Bessel and D. Englander,'Up from the Trenches: Some Recent Writing on the Soldiers of the Great War', *European Studies Review* 11 (1981) p. 393.
179 Quoted in Wilson, *Myriad Faces* p. 802.
180 Keegan, *Face of Battle* p. 225.
181 G. DeGroot, *Blighty* (London: Longman, 1996) pp. 164–5, 297–8. He cites and makes use of material in Sheffield, 'Effect of the Great War' yet gives no indication of the case propounded in this article; indeed, his own argument effectively inverts it.
182 Wilson, *Myriad Faces* p. 827.
183 S. Andreski, *Military Organisation and Society* (London: Routledge & Kegan Paul, 1968).
184 Bessell and Englander, 'Up from the Trenches', p. 393.
185 For an interesting discussion that stresses the complexity of veterans' attitudes to their wartime experience, see Gregory, *Silence of Memory*, Chapter 2.

8 Officer–Man Relations: Morale and Discipline

1 Sherriff, 'English Public Schools', in Panachias, *Promise* pp. 134, 152.
2 For the problems of trench papers as a source, see D. Englander, 'Soldiering and Identity: Reflections on the Great War', *WIH* 1 (3) (1994) pp. 303–4.
3 Fuller, *Troop Morale* pp. 53–6.
4 R.C. Sherriff, 'My Diary' in *The Journal of the East Surrey Regiment* I (1937) 109–18. For some comments on ranker-officers by an officer of 9/E.Surreys, see L.C. Thomas, KRS Q.
5 R.M. Bracco, *Merchants of Hope* (Oxford: Berg, 1993) pp. 158–9, 169–70.
6 T.A.M. Nash (ed.), *The Diary of an Unprofessional Soldier* (Chippenham: Picton, 1991) pp. ix, 27.
7 F. Hawkings (A. Taylor, ed.), *From Ypres to Cambrai* (Morley: Elmfield, 1974) pp. 6, 133.
8 G.S. Hutchison, *Warrior* (London: Hodder, 1932) p. 13.
9 Eberle, *Sapper* p. 44.
10 F.O.O., *Making of a Gunner*, pp. 79–80; unpublished account, p. 4, S.H. Raggett papers, 90/1/1, IWM.
11 Douie, p. 42.
12 A. Horne, *Macmillan 1894–1956* (London: Macmillan, 1988) p. 36. See Lewin, *Slim* pp. 261–2 for a comparison with Attlee, who believed that industrial relations in the 1940s should be conducted on the lines of the officer–man relationship that he had experienced in the Great War.
13 Letter, 13 Nov. 1915, J.O. Coop papers, 87/56/1 IWM.
14 Waites, 'Government of Home Front' p. 188.
15 Anon., 'Memories III – 'The Minor Pleasures of War', in *Twenty Years After, Supp. Vol.* p. 239; Dunn, *War: the Infantry* p. 65.
16 See, for example, letter, 22 May 1917, W.T. King papers, 89/7/1, IWM in which a private comments on the incidence of shellshock among officers and men. See also Englander, *Soldiering and Identity* pp. 307–110; M. Brown, *Imperial War Museum Book of the Somme* (London: Sidgwick & Jackson, 1996) pp. 110, 201.
17 Letter, 6 Aug. 1917, W.E. Hoad papers, LC; 1 Nov. 1914, 23 Feb. 1915, WD,

Chief Censor, Advanced Base, WO 95/39877, PRO; GRO 602, 612, Extracts from General Routine Orders, 71–273, RMASA; Army Order 121, 24 July 1916, 71-225, RMASA.

18 1 Apr. 1915, WO 95/3987, PRO. For an individual officer's perfunctory censorship of letters, see letter, 10 Oct. 1915, H.A. Bowker papers, LC.

19 Unpublished account, notebook II, B.D. Parkin papers, 86/57/1, IWM.

20 For a comparison with 1939–45, see D. Butler, 'The British Soldier in Italy'; pp. 13–32, CAB 101/224 and S.R. McMichael, *An Historical Perspective on Light Infantry* (Ft Leavenworth, KS: Combat Studies Institute, 1987) pp. 33–5.

21 Tucker, *Johnny* p. 49.

22 C. Barnett, 'A Military Historian's View of the Great War', *Essays by Divers Hands, Being the Transactions of the Royal Society of Literature*, XXXVI, 1970, pp. 1–18; Barnett, *Collapse* pp. 428–35; Bourne, *Britain and the Great War 1914–1918* (London: Arnold, 1989) p. 220; M. Howard, 'The Art of the Tat', *The Times Literary Supplement*, 9–15 Feb. 1990, p. 138.

23 See Fussell, *Great War*; E.J. Leed, *No Man's Land* (Cambridge: Cambridge University Press, 1979); S. Hynes, *A War Imagined* (London: Bodley Head, 1990). Robin Prior and Trevor Wilson mount a splendid counter-offensive against the literary tendency in their 'Paul Fussell at War', *WIH* 1 (1994) 63–80.

24 Unpublished account, pp. 1–6, 41, P. Creek papers, 87/31/1, IWM.

25 Letter, 21 Nov. 1915, P.H. Jones papers, p. 246, IWM. For similar sentiments expressed by another middle-class ranker, see letter, 15 Mar. 1915, N.F. Ellison papers, DS/Misc/49, IWM.

26 Diary, 6 May 1917, R.L. Mackay papers, IWM. See also H. Dearden, *Medicine and Duty* (London: Heinemann, 1928) p. IX.

27 F. Majdaleny, *The Monastery* (London: Corgi, 1957) p. 103.

28 Unpublished account, p. 284, C.F. Jones papers, LHCMH.

29 Letter, Oct. 1963, A.R. Armfield, BBC/GW, IWM.

30 Letter, 28 Aug. 1917, W.L. Fisher papers, 85/32/1, IWM. See also H. Williamson, introduction to [Bell], *Soldier's Diary* p. xiv.

31 See Fuller, *Troop Morale*, passim. For the importance of hobbies to working-class males in this period, see McKibben, *Ideologies*, Chapter 5.

32 1–10 Feb. 1917, WD, 9/R. Irish Rifles, WO 95/2503, PRO.

33 C.F. Wurtzburg, *The History of the 2/6 (Rifle) Battalion 'The King's' (Liverpool Regiment) 1914–19* (Aldershot: Gale & Polden, 1920) pp. 93–4.

34 Henriques, *War History* p. 164–5.

35 C. Dudley Ward, *The Welsh Regiment of Foot Guards 1915–1918* (London: Murray, 1936) p. 52.

36 Col. M.G.N. Stopford, lecture on trench warfare, p. 13, Conf. 3898, SCL; G.F.R. Hirst, KRS Q.

37 Potter and Fothergill, *2/6th LF* p. l6.

38 E.T. Dean, '"We Will All Be Lost and Destroyed" Post-Traumatic Stress Disorder and the Civil War', *Civil War History* XXXVII (1991) 146.

39 Lord Moran, *The Anatomy of Courage* (London: Constable, 1945) p. X.

40 Holmes, *Firing Line* pp. 214–15; A. Kellett, *Combat Motivation* (Boston, MA: Kluwer-Nijhoff, 1982) pp. 276–8, 300; T. Copp and B. McAndrew, *Battle Exhaustion* (Montreal: McGill–Queens University Press, 1990) pp. 5, 81–2.

41 P.J. Leese, 'A Social and Cultural History of Shellshock, with Particular Reference to Experience of British Soldiers during and after the Great War', (PhD, Open University, 1989) pp. 22–4; B. Shepherd, 'Shell-shock on the Somme', *JRUSI* 141 (1996) pp. 51–6.
42 R.H. Arhrenfeldt, *Psychiatry in the British Army in the Second World War* (London: Routledge & Kegan Paul, 1958) p. 10; Kellett, *Combat Motivation* p. 274.
43 F. Fernadez-Armesto, *Millennium* (New York: Scribner, 1995) p. 492.
44 For the thesis that the hero of *Lord of the Rings*, Frodo Baggins, had shell-shock, see F. Spufford, 'The War that Never Stopped' *The Times Literary Supplement*, 23 Mar. 1996, p. 11. J.R.R. Tolkien was a Western Front veteran. The detective is of course Dorothy L. Sayers's Lord Peter Wimsey.
45 7 Aug. 1916, WD of 86th MG Coy, WO 95/2302, PRO.
46 Evidence of A.F. Hurst, 'Report of the War Office Committee of Enquiry into "Shell-Shock"' (1922) Cmd. 1734, p. 25.
47 Unpublished account, p. 95, E. Partridge papers, LC.
48 See, for instance, A. Macphail, *Official History of the Canadian Forces in the Great War 1914–19: the Medical Services* (Ottawa: Dept of National Defence, 1925) p. 278.
49 G. Sparrow and J.N. MacBean Ross, *On Four Fronts with the Royal Naval Division* (London: Hodder, 1918) p. 237; T. Bogacz, 'War Neurosis and Cultural Change in England 1914–22: the Work of the War Office Committee of Enquiry into "Shell Shock"', *JCH* 24, (1989) pp. 227–56.
50 See C.S. Myers, *Shell Shock in France, 1914–18* (Cambridge: Cambridge University Press, 1940) pp. 50–62, 106–7. For a modern view on the nature and treatment of battle-stress, see R. Gabriel, *The Painful Field* (New York: Greenwood, 1989) esp. pp. 25–45, 137–58.
51 'Shell-Shock Report' p. 93; see also pp. 13–14, 36–7, 50. For a less dogmatic view, see Myers *Shell Shock* pp. 38–9. See also K. Simpson, 'Dr. James Dunn and Shell-Shock', in Cecil and Liddle, *Facing Armageddon* pp. 84–5.
52 Copp and McAndrew, *Battle Exhaustion* pp. 5, 81–2; Holmes, *Firing Line* p. 259.
53 A.M. McGilchrist, *The Liverpool Scottish, 1900–19* (Liverpool: Young, 1930) p. 103.
54 C. Headlam, *History of the Guards Division in the Great War 1915–1918*, I (London: Murray, 1924) p. 193.
55 Anon., 'Memories XIV – the Minor Miseries of War', in *Twenty Years After*, II, p. 1121; C.G. McBride, diary, 22 Oct. 1918, Armed Forces Museum, Camp Shelby, MS.
56 Smith, *Four Years* p. 100.
57 Letter, 5 Sept 1915, G. Banks-Smith papers, LC.
58 Diary, 8 Nov. 1917, J.M. Thomas papers, 88/56/1 IWM.
59 J. Brophy, 'The Soldier's Nostrils', in O'Riordan, *Martial Medley* pp. 124–5. See also Dunn, *War the Infantry Knew* p. 18.
60 Hitchcock, *Stand To* p. 122; R.D. Fisher papers, p. 12, 76/54/1 IWM; Divisional Trench Orders 1917 p. 9; S.S. No.408, 'Some of the Many Questions a Platoon Commander Should Ask Himself on Taking over a Trench, and at Intervals Afterwards', p. 7; Steuart, *March Kind Comrade* p. 182.

61 Letter, 18 Sept. 1916, B.F. Eccles papers, 82/22/1, IWM.
62 Letter, 19 July 1915, H. Venables papers, LULLC.
63 Bowra, p. 81: R.J. Wyatt, 'The Major "Minor Horror of War"', *ST*, 23 (1988) pp. 29–30; Aldington, *Death of a Hero* p. 250.
64 R. Whipp, interview. See also J. Ellis, *Eye Deep in Hell* (London: Fontana, 1976) passim.
65 Unpublished account, pp. 19–20, R.D. Fisher papers, pp. 19–20, 76/54/1, IWM.
66 Quoted in Brown, *Tommy* p. 56.
67 G.H. Cole, KRS Q.
68 Diary, 4 Mar. 1916, R.H. Sims papers, 77/130/1, IWM.
69 Brown, *Tommy* p. 56.
70 Diary, 2, 5, 6, 8 Mar., 11 July 1917, J. Williams papers, 83/14/1, IWM.
71 Diary, 21 July 1917, J. Williams papers, 83/14/1, IWM.
72 Diary, 16 June 1917, J. Williams papers, 83/14/1, IWM.
73 Grey, *Confessions*, dedication and pp. 127–8.
74 Unpublished memoir, notebook 1, B.D. Parkin papers, 86/57/1, IWM.
75 Quoted in Brown, *IWM First World War* p. 55.
76 Unpublished account, pp. 89–90, R.D. Fisher papers, 76/54/1, IWM.
77 Aldington, *Death of a Hero* p. 237.
78 Manning, *Middle Parts* p. 229.
79 Ibid., pp. 2–4.
80 C.N. Smith, 'The Very Plain Song of It: Frederic Manning, Her Privates We', in H. Klein (ed.), *The First World War in Fiction* (London: Macmillan, 1978) pp. 176–8.
81 B. Gammage, *The Broken Years* (Ringwood, Victoria: Penguin, 1975) p. 240.
82 C.E. Jacomb, *Torment* (London: Melrose, 1920) p. 320. This statement conflicts with some of Jacomb's earlier comments about paternal NCOs and officers: ibid. pp. 73, 171.
83 C. Haworth, *March to Armistice 1918* (London: Kimber, 1968) p. 28.
84 Crozier, *Impressions* p. 214. See also criticisms of excessive risk-taking among officers: 'Memorandum on Trench to Trench Attacks', 31 Oct. 1916, WO 158/344, PRO.
85 R. Little, 'Buddy Relations and Combat-Performance', in Janowitz, *New Military* pp. 195–223. The view of a recent writer that British officers were 'enthusiastic but tactically incompetent schoolboys' can be safely dismissed: B.I. Gudmundsson, *Storm Troop Tactics* (Westport, Conn.: Praeger, 1995) p. 175.
86 See, for example, letter, 6 July 1916, Pte W.A. Hollings, quoted in L. Milner, *Leeds Pals* (Barnsley: Wharncliffe, 1991) p. 141.
87 H. Horne, KRS Q; Tilsley, *Other Ranks* pp. 30, 170; Thorburn, p. 183.
88 Diary, 22 (?) Nov., 13 Dec. 1915, J. Griffith papers, IWM.
89 Unpublished account, p. 2, H.L. Adams papers, 83/50/1, IWM. See also Stopford, lecture on trench warfare, Conf. 3898, SCL.
90 Letter, 22 Oct. 1917, J. Mudd papers, IWM; MAF 60/243, PRO.
91 Letter, 27 Apr. 1915, G.W. Durham papers, IWM.
92 H.R. Williams, *The Gallant Company* (Sydney: Angus & Robertson, 1933) pp. 62–3.
93 Manning, *Middle Parts* pp. 13–21; H. Munday, *No Heroes, No Cowards*

(Milton Keynes: People's Press, 1981) p. 17.
94 Letter, 12 June 1916, H.R. Hammond papers, LC.
95 Hanbury-Sparrow, *Land-Locked Lake* p. 19.
96 Nettleton, *Anger of the Guns* p. 113. See also Bion, *Long Week-End* p. 201; G. Brennan, *A Life of One's Own: Childhood and Youth* (Cambridge: Cambridge University Press, 1979) p. 226.
97 H. Boustead, *The Wind of Morning* (London: Chatto & Windus, 1974) p. 37. See also A. Behrend, *As from Kemmel Hill* (London: Eyre & Spottiswoode, 1963) p. 147; E. Foster-Hall, KRS Q.
98 Letter, 16 Apr. 1917, B.F. Eccles papers, 82/22/1, IWM.
99 Gibbons, *Roll On* p. 69.
100 Dunham, *Long Carry* pp. 68, 82.
101 Quoted in Peacock, 'Rendezvous with Death' p. 344.
102 Keegan, *Face of Battle* p. 332.
103 W.H. Binks, diary, Sept. 1916, in *The White Rose*, Dec. 1994, 66.
104 Citation for *Croix de Guerre*, quoted in P. Bryant, *Grimsby Chums* (Hull: Humberside Libraries & Arts, 1990) p. 92.
105 Letter, Oct. 1963, A.R. Armfield, BBC/GW, IWM.
106 Appx, Sept. 1918, WD of l/Gordons, WO95/1435, PRO.
107 'Memo on Trench to Trench Attacks' 31 Oct. 1916, pp. 12, 25, WO 158/344, PRO.
108 D. Lamb, *Mutinies* (Oxford and London: Solidarity, n.d.) p. 3.
109 T.S. Hope, *The Winding Road Unfolds* (London: Tandem, 1965) pp. 48–9.
110 Sheffield, 'Operational Role of Military Police' in Griffith, *British Fighting Methods* pp. 76–8.
111 S. Snelling, *VCs of the First World War – Gallipoli* (Sutton: Stroud, 1995) p. 122. See also J. Terraine (ed.), *General Jack's Diary 1914–1918* (London: Eyre & Spottiswoode, 1964) p. 277.
112 F.P. Crozier, *A Brass Hat in No Man's Land* (London: Cape, 1930) 109–10; idem, *The Men I Killed* (London: Joseph, 1937) pp. 54–6, 61–3.
113 Letter, n.d., Sir G. Jeffrys, CAB 45/114, PRO.
114 Lamb, *Mutinies* pp. 3–5; C. Denison, 'From Two Angles' in C. O'Riordan, *Martial Medley* p. 80; letter, 3 May 1930, W.B. Spender, CAB 45/137, PRO.
115 Gordon, *Unreturning Army* p. 116.
116 J. Putkowski and J. Sykes, *Shot at Dawn* (Barnsley: Wharncliffe, 1989) pp. 84–5, 130.
117 'Wisdom for Warts', p. 6, D. Hamond papers.
118 H. Myers, interview.
119 Putkowski and Sykes, *Shot at Dawn* p. 278.
120 'GHQ Great Britain, Weekly Intelligence Summaries, 22 Dec. 1917 to 29 Mar. 1918', AIR 1 538/16/15/55, PRO.
121 G. Ashurst, *My Bit: a Lancashire Fusilier at War 1914–18* (Ramsbury: Crowood, 1988) p. 61. For a similar incident, see unpublished account, p. 13, A.J. Abraham papers, p. 191, IWM.
122 Diary, 2 Nov. 1916, A. Young papers, 76/101/1. IWM
123 Unpublished account, pp. 61–4, C.F. Jones papers, LHCMA 7.
124 Unpublished account, p. 62 J.W. Riddell papers, 77/73/1, IWM.
125 *Manual of Military Law*, quoted in S.T. Banning, *Military Law Made Easy* (Aldershot: Gale & Polden, 1904) p. 16.

126 J. Putkowski, 'A2' *Lobster* 28 (1994) 19–20.
127 L. James, *Mutiny* (London: Buchan & Enright, 1987) pp. 9–15. For some interesting reflections on the nature of strikes, see H. Benyon, *Working for Ford* (Harmondsworth: Penguin, 1984) pp. 180–1.
128 Simkins, *Kitchener's Army* pp. 200–2, 238–89, 243–4.
129 Lewin, *Slim* p. 16; P. Croney, *Soldier's Luck* (Ilfracombe: Stockwell, 1965) pp. 10–11.
130 For general discussions of the mutiny, see G. Dallas and D. Gill, *The Unknown Army* (London: Verso, 1985) pp. 63–81; James, *Mutiny* pp. 89–98; J. Putkowski, 'Toplis, Etaples & "The Monocled Mutineer"', *ST* 18, (1986) pp. 6–11. As Putkowski convincingly demonstrates, W. Allison and J. Fairley, *The Monocled Mutineer* (London: Quartet, 1978) is unreliable.
131 J. Putkowski, 'British Army Mutinies in World War One', unpublished paper.
132 Letter, n.d., A.F. Sheppard, BBC/GW, IWM.
133 Letter, 3 Aug. 1915, P.H. Jones papers, p. 246, IWM.
134 Diary,11 Sept. 1917, pp. 178–9, J.H. Dible papers, IWM.
135 Babington, *Sake of Example* pp. 132–3, but see Liddle, *Soldier's War* pp. 79–80.
136 9–10 Sept. 1917, WD of Commandant, Etaples Base Camp, WO 95/4027, PRO.
137 Letter, G.F. Oppenshaw, *Observer*, 23 Feb. 1964.
138 E. Parker, *Into Battle 1914–1918* (London: Longman, 1964) p. 86. See also C.B. Brereton, *Tales of Three Campaigns* (London: Selwyn & Blount, 1926) p. 139 for the views of a NZ officer sympathetic towards the mutineers. Although the accounts of Oppenshaw, Parker and Brereton are inaccurate in details, they are very revealing of junior officers' attitudes towards the mutiny.
139 Dallas and Gill, *Unknown Army* p. 76.
140 Letter, J. Putkowski, *The Guardian*, 26 Sept. 1986.
141 Bourne, *Great War* p. 179.
142 See J. Stevenson, *Popular Disturbances in England 1700–1870* (London: Longman, 1979).
143 Diary, p. 182, 11 Sept. 1917, J.H. Dible papers, IWM.
144 Thomas, *Life Apart* pp. 114–17. See also E.L.G. Griffith-Williams, KRS Q.
145 Mosley, *My Life* p. 66.
146 [C. Maxwell, (ed.)], *Frank Maxwell, Brig.-General VC CSI DSO: a Memoir and Some Letters* (London: Murray, 1921) pp. 139, 140, 172.
147 Putkowski and Sykes, *Shot at Dawn* pp. 82–4.
148 Letter, n.d., G.W. Shepperd, CAB 45/137, PRO.
149 Letter, 29 Mar. 1930, W.F. Jeffries, CAB 45/135, PRO; unpublished account by F.H. Wallis, CAB 45/132, PRO. For an official explanation of the cross-posting policy, see N. Macready, *Annals of an Active Life* (London: Hutchinson, n.d.) I, pp. 255–7.
150 Letter, n.d., P. Alder, BBC/GW, IWM. See also K.W. Mitchinson, 'The Reconstitution of 169 Brigade: July–October 1916', *ST* No.29 (1990) 8–11.
151 Diary, 15 May 1918, R.L. Mackay papers, p. 374, IWM. See also J.P.W. Jamie, *The 177th Brigade 1914–18* (Leicester: Thornley, 1931) p. 34; Dunham, pp. 121–3.
152 L. Davidson, interview.

153 Bartlett, *No Man's Land* p. 165; F.A.J. Taylor, *The Bottom of the Barrel* (London: Regency, 1978) p. 99; French, *Gone for a Soldier* p. 32; B. Newman, *The Cavalry Went Through* (London: Gollancz, 1930) pp. 25, 35.
154 H. Colbourne, W. Gilbert, L. Davidson, interviews; Marks, *Laughter* pp. 69–70; Coppard, *Machine Gun* p. 29.
155 Beckett, 'Territorial Force', pp. 137, 147.
156 Letter, A.B. Hill, n.d. (c.1935) CAB 45/114, PRO.
157 Unpublished account, in letter, 24 Mar. 1935, E.J. King, CAB 45/135, PRO.
158 Tempest, *6th West Yorks* pp. 280–1.
159 'Mespotamian Diary' Part Two', *GW* 1, No. 4 (Aug. 1989) p. 152.
160 Unpublished account, G.H.W. Cruttwell papers, DERR Museum.
161 Anon., *The War History of the 1/4th Battalion the Loyal North Lancashire Regiment 1914–18* (published privately, 1921) pp. 80, 107.
162 Jamie, *177th Bde* p. 18.
163 C.A.C. Keeson, *The History and Records of Queen Victoria's Rifles* (London: Constable, 1923) p. 219.
164 Unpublished account, pp. 13–14, 32, W.J. Bradley papers, LC; J. Baynes, *The Forgotten Victor* (London: Brassey's, 1989) pp. 21–2.
165 G.H. Cole, KRS Q. For similar comments about 1/4 Gordons, see N.C.S. Down, KRS Q.
166 Glover, *Fateful Battle Line* p. 371.
167 W.H.A. Groom, *Poor Bloody Infantry* (New Malden: Picardy, 1983) pp. 45, 160. For changes in LRB personnel, see Mitchinson, *Gentlemen* pp. 85–96, 140.
168 Smith, *Four Years* p. 157.
169 Quoted in Mitchinson, 'Reconstruction' p. 10.
170 Hurst, introduction to Wilson, *Seventh Manchesters* pp. XI–XIII; Andrews, *Haunting Years* pp. 270, 286–8.
171 Maude, *47th Division* p. 211.
172 Wilson, *Palestine* p. 42.
173 V. de Sola Pinto, *The City that Shone* (London: Hutchinson, 1969) p. 219. For factors in the survival of the original ethos of another unit, see E.W. Gladstone, *The Shropshire Yeomanry* (Manchester: Whitethorn Press, 1953) pp. 216, 233.
174 Unpublished account, part 2, pp. 3–4, J.W. Wintringham papers, 78/9/1, IWM.
175 Nicholson, *Behind the Lines* p. 48.
176 'Note by Lt. Col. J.W. Simpson' May 1915, in WD of 1/6 Londons, WO 95/2729, PRO. For the paternalistic attitudes of another officer of this unit, see H.D. Meyers, KRS:Q.
177 Gibbons, *Roll On* p. 71.
178 Tilsley, *Other Ranks* pp. 202, 249.
179 G. Horridge, KRS Q.
180 Sheffield, 'Effect of War Service', pp. 64–6.
181 Diary, 18, 21, 26, 27 Nov. 1915, W.A. Rogers papers, 87/62/1, IWM.
182 V. Ireland, *The Story of Stokey Lewis VC* (Haverford West: published privately, n.d.) pp. 30, 52.
183 K. Grieves, '"Lowther's Lambs": Rural Paternalism and Voluntary Recruitment in the First World War', *Rural History* 4, No. l, (1993) pp. 65,

67, 69; idem, communications with author.

184 W.S. Churchill, introduction to D. Jerrold, *The Royal Naval Division* (London: Hutchinson, n.d.) pp. xiv–xv; H. Gough, *The Fifth Army* (London: Hodder 1931) p. 152.

185 Jerrold, *RND*, pp. 185–8, 208–9, 311, 322–3. See also L. Sellers, *The Hood Battalion* (London: Leo Cooper, 1995) pp. 171–2, 202, 207.

186 Unpublished account, pp. 88, 106, R.W.F. Johnston papers, 82/38/1, IWM.

187 Bryant, *Grimsby Chums* pp. 23, 29–30, 182–3.

188 Parker, *Into Battle* p. 93. He mistakenly states that 13/RF was a TF unit.

189 Unpublished account pp. 22–3, R.S. Cockburn papers, 78/4/1, IWM.

190 A.P.B. Irwin, Oral History Interview, 000211/04, IWM; regimental history lecture in WD, 8/E. Surreys, WO 95/2050, PRO.

191 Nicholson, *Behind the Lines* p. 152.

192 Unpublished account, S.V.P. Weston papers, LC; Sheffield, 'Effect of War Service' pp. 53–4.

193 Unpublished account, p. 16, C.E.L. Lyne papers, 80/14/1, IWM; letter, 17 Jan. 1916, W.S. Churchill, in M. Gilbert (ed.), *Winston S. Churchill: Companion* (London: Heinemann, 1972) Vol. III, Part 2, p. 1377.

194 Letter, 14 Jan. 1916, W.S. Churchill, in ibid. p. 1374; [A.D. Gibb] 'Capt. X', *With Winston Churchill at the Front* (London: Gowans and Grey, 1924), pp. 75–8.

195 For example, M. Gilbert, *The Challenge of War: Winston S. Churchill 1914–1916* (London: Minerva, 1990) pp. 637–8.

196 Letter, 17 Jan. 1916 W.S. Churchill, in Gilbert, *Companion* Vol. III, Part 2, p. 1377; idem, *Challenge of War* p. 640. The WD of 6/R.Scots Fusiliers contains little evidence of sports or baths organised for the men before Churchill's arrival: WO 95/1772, PRO.

197 J.S. Sly, 'The Men of 1914', *ST* No. 35, (1992) pp. 11–13.

198 F. Maurice, *The Life of General Lord Rawlinson of Trent* (London: Cassell, 1928) p. 252.

199 J. Greenshields, KRS Q.

200 Lucy, *Devil* pp. 293–4.

201 Unpublished account, R.W.F. Johnston papers, pp. 41, 51, 105–6.

202 Dunn, *War: the Infantry* pp. 46, 64; Richards, *Old Soldiers* p. 315.

203 Hanbury-Sparrow, *Land-Locked Lake* 147, 150. See also idem, 'Discipline or Enthusiasm?', *BAR*, 20 (1965) pp. 8–13.

204 Unpublished account, p. 48, L.A. Hawes papers, IWM; J. Charteris, *At GHQ* (London: Cassell, 1931) p. 41; T. Carew, *Wipers* (London: H. Hamilton, 1972) p. 220.

205 S. Rogerson, *Twelve Days* (Norwich: Glidden, 1988) pp. 86–8.

206 Terraine, *General Jack's Diary* p. 186.

207 Rogerson, pp. 97–8.

208 M. McConville, quoted in Terraine, *General Jack's Diary* p. 86.

209 Taylor, *Bottom of the Barrel* pp. 81, 107.

210 Marks, *Laughter* p. 41.

211 G.F.R. Hirst, KRS Q.

212 K. Simpson, 'The British Soldier on the Western Front', in Liddle, *Home Fires* p. 144.

9 Comparisons

1 A. Horne, *The Price of Glory* (Harmondsworth: Penguin, 1964) p. 72: L.V. Smith, 'The Disciplinary Dilemma of French Military Justice, September 1914–April 1917: the Case of 5e Division d'Infanterie', *J. Mil. Hist.* 55, (1991) 47–68.
2 D. Englander, 'The French Soldier, 1914–18', *French History* 1, (1987) p. 54.
3 A. Brett-James, 'Some Aspects of British and French Morale on the Western Front, 1914–1918', 7, unpublished paper, 1978; D. Porch, 'The French Army in the First World War', in A. Millett and W. Murray, *Military Effectiveness* I (London: Unwin Hyman, 1988) p. 200. For examples of failures of paternalism, see M. Bloch, *Memoirs of War 1914–15* (Cambridge: Cambridge University Press, 1980) pp. 161–2.
4 Englander, 'French Soldier' pp. 55, 59. See S. Audoin-Rouzeau, *Men at War 1914–1918* (Oxford: Berg, 1992) pp. 57–9, 63, for a slightly rosier view of French officer–man relations.
5 D. Porch, *The March to the Marne* (Cambridge: Cambridge University Press, 1981) pp. 131–2; idem, 'The French Army in the First World War' pp. 222–3.
6 'Ex-Trooper', *The French Army from Within* (London: Hodder, 1914) pp. 27–8; *The Times*, 8 Apr. 1916.
7 C. Dawbarn, *Joffre and His Army* (London: Mills & Boon, 1916) pp. 98–9; Brett-James, 'Some Aspects of ... Morale', 7.
8 G. Pedroncini, *Les Mutineries de 1917* (Paris: Presses Universitaires de France, 1967) pp. 246, 250; L.V. Smith, 'War and 'Politics': the French Army Mutinies of 1917', *WIH* 2 (1995) 190. This is the most recent analysis of the mutinies, which takes issue with some of Pedroncini's findings.
9 P. Pétain, 'A Crisis of Morale in the French Nation at War, 16 April–23 October-1917' (15 May 1926) in E. Spears, *Two Men Who Saved France* (London: Eyre & Spotiswoode, 1966) pp. 99–100, 104–7.
10 Pedroncini, *Les Mutinieries* p. 250.
11 For a useful comparison with the breakdown of German discipline in 1918, see W. Diest, 'The Military Collapse of the German Empire: the Reality of the Stab-in-the-Back Myth', *WIH* 3, (1996) 192–3.
12 J. Grey, *A Military History of Australia* (Cambridge: Cambridge University Press, 1990) pp. 91–2.
13 C.E.W. Bean, *Official History of Australia in War of 1914–1918* [hereafter *AOH*] I (Sydney: Angus & Robertson, 1921) pp. 7, 47, 607; C.E.W. Bean, *AOH*, VI (Sydney: Angus & Robertson, 1942) p. 1085.
14 Unpublished account, J.F. Edey papers, LC. For other examples of informal officer–man relations, see L.H. Harris, *Signal Venture* (Aldershot: Gale and Polden, 1951) p. 21; diary, 4 Aug. 1915, G.B. Edwards papers, LC.
15 Unpublished account, S.B. Abbott papers, 78/36/1, IWM.
16 Unpublished account, P. R. Hall papers, 87/55/1, IWM.
17 *The Listening Post*, 10 Dec. 1916, p. 129.
18 See Dale Blair's forthcoming Victoria University of Technology PhD.
19 See Peter Weir's 1981 film *Gallipoli*; P. Firkins, *The Australians in Nine Wars* (London: Pan, 1973) pp. 125–6, 136; J. Laffin, *Digger* (London: Cassell, 1959) pp. 39–40, 171–3, and the Bean-influenced memoir of an AIF veteran: W.D. Joynt, *Saving the Channel Ports 1918* (North Blackburn:

Wren, 1975) pp. 2–3.

20 K.S. Inglis, 'Anzac and the Australian Military Tradition', *RIHM* 72 (1990) 3.

21 E.M. Andrews, *The Anzac Illusion* (Cambridge: Cambridge University Press, 1993) pp. 152–4. For a corrective, see P. Simkins, 'Co-Stars or Supporting Cast? British Divisions in the "Hundred Days", 1918' in Griffith, *British Fighting Methods* pp. 53–8, 60–1, 65–6.

22 E.M. Andrews, 'Bean and Bullecourt: Weaknesses and Strengths of the Official History of Australia in the First World War', *RIHM*, 72, (1990) 47. See also J. Mordike, 'The Story of Anzac: a New Approach', *JAWM*, 16 (1990) pp. 5–17; T.H.E. Travers, 'From Surafend to Gough: Charles Bean, James Edmonds, and the Making of the Australian Official History', *JAWM*, 27 (1995) pp. 16, 22, 23 and, generally, J. Ross, *The Myth of the Digger* (Sydney: Hale & Iremonger, 1985).

23 *Report on an Inspection of the Military Forces of the Commonwealth of Australia, by General Sir I. Hamilton* (20 May 1914) p. 34. See also C.E. Jacomb, *God's Own Country* (London: Max Goschen, 1913) pp. 43–4. For modern views, see B. Kingston, *The Oxford History of Australia, III, 1860–1900* (Melbourne: Oxford University Press, 1988) p. 280; Andrews, *Anzac Illusion*, p. 151.

24 'Notes on Chaplaincy work with the A.I.F.', 10 Dec. 1921, RC Racklyeft papers, 1 DRL 642, AWM.

25 Grey, *Military History* p. 92.

26 Ibid. p. 92; C.E.W. Bean, *AOH*, III (Sydney: Angus & Robertson, 1929) p. 54. For Australian public schools, see Kingston, *Oxford History*, p. 202.

27 Diary, 12, 22 Sept. 1916, F.M. Stirling papers, LC.

28 P. Burness, *The Nek* (Kenthurst, NSW: Kangaroo Press, 1996) p. 44; unpublished account, p. 58, W.C. Gamble papers, LC. See also letter, 1916, T. Gardner, quoted in J. Morice, *Six-Bob-a-Day Tourist* (Ringwood, Victoria: Penguin, 1985) p. 59.

29 C.E.W. Bean, *AOH*, III, p. 51. In *AOH*, VI, p. 20, Bean repeats then contradicts this statement in the space of a paragraph.

30 Edmonds to Bean, 8 Dec. 1938, 3 DRL 7953, item 34, AWM.

31 See lectures (?) on 'Stopping German Offensive, 1918', and 'Leadership and Discipline', C.H. Brand papers, 3 DRL 2750, AWM.

32 Bean, *AOH*, VI, p. 21. See also idem, *AOH*, III, p. 125.

33 Letter, 27 Nov. 1917, S.p. Boulton papers, 1 DRL 138, AWM. For an Australian ranker-officer who recognised that commissioned status demanded a different code of behaviour, see diary, 13 Dec. 1916, TJ Richards papers, 2 DRL 794, AWM.

34 Bean, *AOH*, I, p. 530.

35 Bean, *AOH*, III, p. 52.

36 Bean, *AOH*, I, p. 530.

37 Bean, *AOH*, I, p. 550, VI, p. 21.

38 Sir A. Murray to Robertson, 18 Mar. 1916, Rob I/32/13/1; Birdwood to Murray, 25 Feb. 1916, Rob I/32/13/2, both Sir W.R. Robertson papers, LHCMA. For the weakness of officers in AIF in 1914–15, see D. Winter, *25 April 1915: the Inevitable Tragedy* (St Lucia: University of Quebec Press, 1994) pp. 38–40.

39 J. Monash, *The Australian Victories in France in 1918* (London: IWM, n.d.) pp. 292–3.
40 Harris, *Signal Venture* p. 56.
41 See P. Stanley, '"Soldiers and Fellow-Countrymen" in Colonial Australia' in Mc.Kernan and M. Browne, *Australia: Two Centuries of War & Peace* (Canberra: AWM/Allen & Unwin, 1988) pp. 80–2, 85–7, and C. Wilcox, *For Hearths and Homes: Citizen Soldiers in Australia, 1854–1945* (St Leonards, NSW: Allen & Unwin, 1998).
42 I am grateful to Ashley Ekins for discussing this point with me. See also G. Wahlert, 'Provost: Friend or Foe? The Development of an Australian Provost Service 1914–1945' (MA thesis, University of New South Wales, 1996) pp. 44, 48–58; Burness, *The Nek* p. 78.
43 Gammage, *Broken Years* pp. 38–40; S. Brugger, *Australians and Egypt 1914–1919* (Carlton, Victoria: Melbourne University Press, 1980) pp. 145–7.
44 Letter, T. Gardner, 1916, quoted in Morice, *Six Bob* p. 59.
45 For his imposition of strict discipline on 3rd Australian Division, see G. Serle, *John Monash* (Carlton, Victoria: Melbourne University Press, 1982) pp. 272, 280.
46 Fuller, *Troop Morale* p. 50.
47 Pugsley, *On the Fringe of Hell* pp. 101, 131–5. Pugsley has drawn upon Ekins' unpublished work.
48 Appx to July 1918, WD, Anzac Provost Corps, WO 154/129, PRO, letter, n.d. (c. June 1918) Capt. W.C. Fussell, quoted by Wahlert, 'Provost', p. 69.
49 P. Berton, *Vimy* (Markham, Ontario: Penguin, 1987) pp. 49–50.
50 J.A. English, *The Canadian Army and the Normandy Campaign* (New York: Praeger, 1991) p. 308. See also A.M.J. Hyatt, *General Sir Arthur Currie* (Toronto: Toronto University Press, 1987) pp. 114–15; D. Morton, 'The Canadian Military Experience in the First World War, 1914–18', in R.J.Q. Adams, (ed.), *The Great War, 1914–18* (College Station, TX; Texas A&M, 1990) pp. 88–9.
51 G.W.L. Nicholson, *Canadian Expeditionary Force 1914–1919* (Ottawa: Queen's Printer, 1962) pp. 24–38. For problems in 11/CEF see letters, 4 Nov., 28 Dec. 1914, G.W. Durham papers, IWM.
52 Letter, 21 Nov. 1914, G.F. Patterson papers, RMASA.
53 J. Williams, *Byng of Vimy* (London: Leo Cooper, 1983) p. 131.
54 *Listening Post* 10 Aug. 1916, p. 120; McWilliams and Steel, *Suicide Battalion* p. 49.
55 J.E.B. Seely, *Adventure* (London: Heinemann, 1931) p. 236.
56 Berton, *Vimy* p. 161.
57 I.D. Losinger, 'Officer–Man Relations in the Canadian Expeditionary Force, 1914–1919' (MA thesis, Carleton University, 1990); D. Morton, *When Your Number's Up* (Toronto: Random House, 1993) esp. p. 104.
58 R.H. Roy (ed.), *The Journal of Private Fraser* (Victoria, BC: Sono Nis Press, 1985) pp. 30–1, 81, 172, 241.
59 Ibid., pp. 70, 270. For an example of informal inter-rank relations in Canadian Corps Cyclist Battalion, see letter, Oct. 1916, G.W. Durham papers, IWM.
60 Letter, 22 Dec. 1917, W.E. Hoad papers, LC.

61 Letters, 27 Nov., 5, 11 Dec. 1915, 2 Feb. 1916 H.R. Hammond papers, LC. For the relatively poor quality of Canadian officers, see Morton, *Number's Up* pp. 108–9; S.J. Harris, *Canadian Brass* (Toronto: University of Toronto Press, 1988) pp. 98–100.
62 Bowra, *Memories* p. 87.
63 See K. Jeffery, *The British Army and the Crisis of Empire* (Manchester: Manchester University Press, 1984) passim.
64 K. Jeffery, 'The Post-War Army', in Beckett and Simpson, *Nation* pp. 214–15.
65 'Postal Censorship Report on Demobilisation etc.' [hereafter PCR] 23 December 1918, Haig papers, Acc.315/220, file d, N[ational] L[ibrary] of S[cotland].
66 Unpublished account, p. 90, G. Buckeridge papers, IWM.
67 Letter, 3 May 1919, A.E. Slack papers, IWM. See also PCR 20 Jan. 1919, Haig papers, NLS.
68 See A. Rothstein, *The Soldiers' Strikes of 1919* (London: Journeyman, 1985) passim.
69 Nicholson, *Behind Lines* p. 294.
70 Thomas, *Life Apart* p. 159. See also unpublished account, p. 35, A.S. Benbow papers, PP/MCR/146, IWM.
71 Unpublished account, G.R. Sullivan papers, IWM. See also unpublished account, p. 89, G. Buckeridge papers, IWM.
72 Unpublished account, p. 95, J.K. Stanford papers, DS/Misc/75, IWM.
73 A. Killick, *Mutiny!* (Brighton: Militant, n.d.) p. 6.
74 PCR, 23 Dec. 1918, Haig papers, NLS.
75 PCR, 20 Jan. 1919, Haig papers, NLS.
76 PCR, 6 Jan. 1919 Haig papers, NLS.
77 Unpublished account, L. Parrington papers, IWM.
78 Diary, 1 Dec. 1918, E.R. Hepper, PP/MCR/138, IWM.
79 Hobbs to Gellibrand, 11 Jan. 1919, 3 DRL 1473 item 58, AWM.
80 Bion, *Long Week-End* p. 286.
81 Robertson to WO, 11 May 1919, Robertson papers, Rob I/28/21C, LHCMA; D.G. Williamson, *The British in Germany 1918–1930* (Oxford: Clarendon, 1991) pp. 32–7.
82 'Notes of Conference …', 17 May 1919, Robertson papers, Rob I/28/8b, LHCMA.
83 Minute by Childs, WO 32/9543, PRO.
84 A. Babington, *The Devil to Pay* (London: Leo Cooper, 1991) pp. 86–8; J. Putkowski, 'The Mutiny of 1st Bn. (*sic*) Connaught Rangers', unpublished paper.
85 A. Clayton, *The British Empire as a Superpower* (Athens, GA: University of Georgia, 1986) pp. 510–11.
86 Simpson, 'The Officers', p. 91; T.A. Heathcote, *The Military in British India, 1600–1947* (Manchester: Manchester University Press, 1995) pp. 213–14.
87 Beckett, 'The British Army' p. 114; P. Simkins, 'The Four Armies 1914–18' in D. Chandler and I. Beckett, *The Oxford History of the British Army* (Oxford: Oxford University Press, 1994) p. 260.
88 *Report of Committee on the Education and Training of Officers*, p. 3, WO 32/4353, PRO.

89 'Precis for Army Council No.1152', p. 23, WO 32/4353, PRO.
90 G.F. Spillan, 'Manpower Problems in the British Army 1918–1939: the Balancing of Resources and Commitments' (DPhil thesis, University of Oxford, 1985) pp. 125–6.
91 *The Hoghunters Annual* II (Bombay: Times of India Press, 1929) p. 5, refers to the revival of the sport of pig-sticking in recent years.
92 B. Bond, *British Military Policy Between the Two World Wars* (Oxford: Clarendon, 1980) pp. 62–70, 143, 145; S. Mays, *Fall Out the Officers* (London: Eyre & Spottiswoode, 1969) pp. 4–5, 60, 66, 152–3; A. Dixon, *Tinned Soldier* (London: Cape, 1941) passim; Wilson to Plumer, 30 Mar. 1920, Wilson papers, HHW 2/56/1a. IWM.
93 Mays, *Fall Out* p. 92.
94 S. Finch (E. Yorks), 000943/03; L.P. Gaines (Dorsets) 000874/09; T.M. Stevens (RGA), 000776/07, Oral History Interviews [hereafter OHI], IWM; Dixon, *Tinned Soldier* pp. 176, 190, 227–8.
95 H.L. Horsfield, 0001780/05, OHI IWM.
96 A.H. Bradshaw, 000811/11, OHI, IWM.
97 J. Prendergast, *Prender's Progress* (London: Cassell, 1979) p. 42; J.G. Halstead, KRS Q.
98 P. Dennis, *The Territorial Army 1907–1940* (Woodbridge: Boydell, 1987) pp. 154–5; I.F.W. Beckett, *The Amateur Military Tradtion 1558–1945* (Manchester: Manchester University Press, 1991) p. 249; B. Peacock, KRS Q; Mitchinson, *Gentlemen* p. 248.
99 Dennis, *Territorial Army* pp. 161, 164; B.H. Liddell Hart, 'This Territorial Year', *AQ*, 21 (1930–1), 243–4; Gladstone, *Shropshire Yeomanry* pp. 263–4.
100 C.D. Fox, KRS Q.
101 G.R. Codrington, *The Territorial Army* (London: Sifteon Praed, 1938) pp. 66–8.
102 Verdin, *Cheshire Yeomanry* p. 194.
103 R.W.F. Johnson, KRS Q. See also Verdin, *Cheshire Yeomanry* p. 185.
104 Dennis, *Territorial Army* p. 181; Steppler, *Britons, to Arms!* p. 137.

Appendix 1 The Morale of the British Army on the Western Front, 1914–18

1 Quoted in I. McLaine, *Ministry of Morale* (London: Allen & Unwin, 1979) p. 8.
2 C. von Clausewitz, (M. Howard, and P. Paret, eds) *On War* (Princeton, NJ: Princeton University Press, 1976) pp. 187–9.
3 'Report on Complaints, Moral etc.' [Nov. 1916] pp. 6, 12, and 'Summaries of Censorship Reports on General Conditions in British Forces in France', M. Hardie papers, 84/46/1, IWM, quoted in Sheffield, *Morale of the British Army* pp. 11, 13.
4 See the reports in the Hardie papers, IWM; 'The British Armies in France as gathered from Censorship' 18 Dec. 1917, GT 3044, CAB 24/36, PRO and 'Memorandum by General Smuts', 3 Jan. 1918, GT 3198, CAB 24/37, PRO.
5 'The British Armies in France as Gathered from Censorship', Appx to July 1918, WO 256/33, PRO; M. Middlebrook, *The Kaiser's Battle* (London: Allen Lane, 1978) pp. 105, 300–18, 341; Sheffield, *Redcaps* pp. 73.
6 R.E. Priestley, *Breaking the Hindenburg Line* (London: T. Fisher Unwin, 1919) p. 87.

7 C. Douie, *The Weary Road* (London: Murray, 1931) pp. 15–16.

Appendix 2 British Army Conscripts

1 I.F.W. Beckett, 'The Real Unknown Army: British Conscripts 1915–19', *GW* 2 (Nov. 1989) p. 4. The only full-length study gives an interesting overview, but its usefulness for this study is unfortunately limited: I.R. Bet-El, 'Experience into Identity: the Writings of British Conscript Soldiers, 1916–18' (PhD thesis, University of London, 1991).
2 Beckett, 'Real Unknown Army', p. 8.
3 T. Quinn, *Tales of the Old Soldiers* (Stroud: Sutton, 1993) p. 28.
4 S. Graham, *Part of the Wonderful Scene* (London: Collins, 1964) p. 153.
5 Hope, *Winding Road* p. 162. For a perceptive contemporary novel on the experiences of a conscript, see J. Spurr, *Conscript Titch* (London: Chambers, 1917).
6 Beckett, 'Real Unknown Army' pp. 4–5.
7 T. Bridges, *Alarms and Excursions* (London: Longman, 1938) p. 162.
8 Anon., *A Short History of the 19th (Western) Division (1914–1918)* (London: Murray, 1919) p. 108.
9 E.G. Crouch to Edmonds, 22 Aug. 1933, CAB 45/126, PRO.
10 S. Bidwell and D. Graham, *Fire-Power* (London: Allen & Unwin, 1982) p. 117.
11 D. Fraser, *And We Shall Shock Them* (London: Sceptre, 1988) p. 104.
12 P. Fussell (ed). *The Ordeal of Alfred M. Hale* (London: Leo Cooper, 1975) passim; unpublished account, E.C. Barraclough papers, 86/86/1/IWM; F.A. Voigt, *Combed Out* (London: Cape, 1929); Simkins, *Kitchener's Army* pp. 206–8.

Appendix 3 Discipline and Continuity in Small Units

1 Officer of 8/KRRC to Edmonds, 13 Apr. 1934, CAB 45/133, PRO.
2 Latham, *Territorial Soldier* pp. 49, 51.
3 T. Ashworth, *Trench Warfare 1914–1918* (London: Macmillan, 1980) pp. 163–8; L. Davidson, interview.
4 Unpublished account, pp. 92–3, 161–2, P. G. Heath papers, DS/Misc/60, IWM. For the relaxed atmosphere of 99th MG Coy, see Sheffield, 'Effect of War Service' pp. 66–7.
5 P.G. Ackrell, *My Life in the Machine Gun Corps* (Ilfracombe: Stockwell, 1966) p. 24; A. Russell, *With the Machine Gun Corps* (London: Dranes, 1923) p. 166.
6 A.J. Cummings, 'Books and Plays of the Great War', *Twenty Years After*, supp., vol., p. 10.
7 Kingham, *London Gunners* esp. pp. 76, 82.
8 Unpublished account, pp. 89, 121–2, R.C. Foot papers, IWM.
9 Liddle and Richardson, 'Voices from the Past', p. 657.
10 F. Mitchell, *Tank Warfare* (Stevenage: Spa Books, 1987) p. 289.
11 D. Fletcher, *Tanks and Trenches* (Stroud: Sutton, 1994) pp. 53, 103, 129, 153, 171.

Appendix 4 Published Guides to Officership

1 *The Times*, 14 Nov. 1916; A.H. Trappman, *Straight Tips for 'Subs'* (London, May 1915 edn.), p. 4.

2 *Blunderbuss* 2, 60; 'Regular', *Customs of the Army: a Guide for Cadets and Young Officers* (London, 1917) p. 3.

3 'C.N.W.', 'An Open Letter to the Very Young Officer', *JRUSI*, LXIII (1917) 69–70; [T.D. Pilcher] *A General's Letters to his Son on Obtaining His Commission* (London: Cassell, 1917) p. 9; 'Basilisk', *Talks on Leadership Addressed to Young Officers* (London: Hugh Rees, 1941) p. 1. This edition was produced during the Second World War – in itself an interesting fact – but the text is essentially similar to its Great War predecessor.

4 Anon., 'The Duties of an Officer', *The Times*, 1 April 1916. This article, based on an address by a senior officer to a tactical school in France, was subsequently published as an unofficial pamphlet and eventually as an official publication (SS 415). See also 'Esterel', 'To the Junior Subaltern', *Journal of the Royal Artillery XLIII* (1916–17) 231; B.C. Lake, *Knowledge for War – Every Officer's Handbook for the Front* (London: Harrison, n.d.) p. 18.

5 Trapmann, *Straight Tips* pp. 48–9.

6 Ibid., p. 50; Pilcher, *General's Letters* p. 24; 'Basilisk', *Talks* p. 56; *The Times*, 1 Apr. 1916.

7 Pilcher, *General's Letters* p. 56.

8 *Notes for Young Officers* (London: HMSO, 1917) pp. 6–9, 68–70.

Sources and Select Bibliography

Primary sources

(Key: O = Officer; R = Ranker; G = General)

Unpublished sources

Australian War Memorial, Canberra

Papers of:

- C.E.W. Bean (O)
- S.P. Boulton (R/O)
- C.H. Brand (O)
- R.C. Racklyeft (O)
- T.J. Richards (R/O)

British Library of Political and Economic Science, London School of Economics

Papers of:

- H. Dalton (O)
- G. McCrae (O)
- R.H. Tawney (R)

Brotherton Library, University of Leeds

Papers of:

- A. Simpson (R)

Camp Shelby, Mississippi: Armed Forces Museum

Papers of:

- C.G. McBride

Duke of Edinburgh's Royal Regiment Museum

Papers of:

- G.H.W. Cruttwell (O)

Essex Regiment Museum

Miscellaneous papers

Imperial War Museum

Papers of:

- A.J. Abrahams (R)
- A.E. Abrey (R)
- S.B. Abbott (R)
- W.R. Acklam (R)
- H.L. Adams (R)
- J. Allison (R)

C.F. Ashdown	(O)
R.E. Barnwell	(O)
E C Barraclough	(R)
A.S. Benbow	(R)
S. Blagg	(R)
S.W. Blythman	(R)
H.F. Bowser	(O)
A.W. Bradbury	(R)
G. Brown	(R)
G. Buckeridge	(R)
G.S. Chaplin	(R)
A.G. Clark	(R)
R.S. Cockburn	(O)
H.J. Coombes	(R)
J.O. Coop	(O)
P.G. Copson	(R)
P. Creek	(R)
R. Cude	(R)
B.O. Dewes	(R/O)
J.H. Dible	(O)
H.M. Dillon	(O)
G.W. Durham	(R)
B.F. Eccles	(R)
S.L.C. Edwards	(R)
G.G.A. Egerton	(G)
L. Emmerson	(R/O)
R. Farley	(R)
A.W. Fenn	(R)
R.D. Fisher	(R)
W.L. Fisher	(R)
R.C. Foot	(O)
F.A. Foster	(O)
P. Fraser	(R)
R. Garrod	(R)
J. Griffiths	(R)
J.W. Gower	(R)
P.R. Hall	(R)
M. Hardie	(O)
L.A. Hawes	(O)
P.G. Heath	(R/O)
E.R. Hepper	(O)
R.W.F. Johnston	(R/O)
P.H. Jones	(R)
C. King	(R)
W.T. King	(R)
C.J. Lodge Patch	(O)
C.E.L. Lyne	(O)
R.L. Mackay	(O)
W.H Martin	(R)

F.I. Maxse	(G)
W.B. Medlicott	(R/O)
A. Moffat	(R)
J. Mudd	(R)
P.D. Munday	(R)
W.P. Nevill	(O)
W.J. Nicholson	(R)
F.M. Packham	(R)
B.D. Parkin	(O)
L. Parrington	(O)
C.E.G. Parry Okenden	(O)
H.D. Paviere	(O)
N.A. Pease	(O)
H.E. Politzer	(R)
G. Powell	(R)
S.H. Ragget	(R)
C.H. Rastall	(R)
J.W. Riddell	(R)
A. Roberts	(R)
W.A. Rogers	(R)
R.H. Sims	(R)
E. Scullin	(R)
T. Sherwood	(R/O)
F.A. Shuffrey	(O)
T.M. Sibley	(O)
T.A. Silver	(R)
A.E. Slack	(R)
W.B.P. Spencer	(O)
J.K Stanford	(O)
D. Starrett	(R)
C.R. Stone	(R/O)
G.W. Sullivan	(R)
A. Surfleet	(R)
A.H. Swettenham	(R)
J.M. Thomas	(R)
C.R. Tobbit	(R/O)
R.L. Venables	(R)
F. Williams	(R)
J. Williams	(R)
H.S. Williamson	(R)
H.H. Wilson	(G)
J.W. Wintringham	(O)
J. Woolin	(R)
A. Young	(R)

Film, *Battle of the Somme*
Correspondence concerning the BBC television series *The Great War.*
'Lecture delivered by Brig. Gen. R.A.K. Montgomery … on 30 October 1914'
Misc. 74, Item 1123
Fourth Army Papers

Oral History Interviews with

E. Finch	(R)
L.P. Gaines	(R)
H.L. Horsfield	(R)
A.P.B. Irwin	(O)
T.M. Stevens	(R)

Liddell Hart Centre for Military Archives, King's College London

Papers of:

P. Blagrove	(O)
P. Howell	(O/G)
C. F. Jones	(R)
L. Kiggell	(G)
G. McCrae	(O)
Lord McDermott	(O)
L.P. Parsons	(G)
W. Robertson	(G)
G.S. Taylor	(O)
W.G. Wallace	(O)

Liddle Collection, University of Leeds

Papers of:

A.B. Ashby	(O)
G. Banks-Smith	(R)
W. Baynes	(O)
H.A. Bowker	(O)
J.D. Burns	(O)
W.J. Bradley	(R)
D.W. Croft	(O)
G.B. Edwards	(R)
R. Gillespie	(O)
J.F. Edey	(R)
W.C. Gamble	(O)
E. Partridge	(R)
J.E. Phesey	(O)
J. Pocock	(O)
C.E. Raven	(O)
J.W.B. Russell	(O)
L.B. Stanley	(R)
F.M. Stirling	(O)
E. Taylor	(O)
H. Venables	(R)
W. Watson	(R)
S.V.P. Weston	(O/G)

'General Aspects: Duties of an Officer' file
General Aspects: Officer/Officer Cadets' file

Old War Office Library, Whitehall (now located at Public Records Office)

Territorial Army papers

National Army Museum

Papers of:

W.H. Davies	(R)
W. Fanton	(R)
T. Grainger	(R)
Surrey Yeomanry papers	
Westminster Dragoons papers	

National Library of Scotland

Haig papers	(G)

Privately held material

Papers of:

G.D.A. Black	(O) (in possession of N. Lucas)
W.J.T. Phythian-Adams	(O) (in possession of H. Phythian-Adams)
D. Hamond	(O) (in possession of Dr. R. Hamond)

Author's collection of Oral History Interviews with:

L. Davidson	(R)
C. Mizen	(R)
H. Meyers	(R)
R. Whipp	(R)

First World War Officers Questionnaires (in possession of K.R. Simpson MP)

Public Record Office

Papers of:

Kitchener	(G)
Haig	(G)
Ramsay MacDonald	

Official papers, series:

AIR 1
CAB 24
CAB 45
CAB 101
WO 32
WO 70
WO 95
WO 106
WO 108
WO 154
WO 158
WO 162
WO 163

Royal Military Academy Sandhurst, Archives

Papers of:

'An old Gunner of 155 Brigade RA'	(R)
K.G. Garratt	(O)

G.F. Patterson (O)

RMC Register of Letters
Miscellaneous orders and papers

Royal Military Academy Sandhurst, Library

T.S. Wollocombe papers (O)

Royal Military Police Museum

Corps Order Books
Report on Straggler Posts

Staff College Library, Camberley

M.G.N. Stopford, 'Trench Warfare' lecture (O)
'XV Corps Operations 1917–18'.
'Operations of 17th Division 21 Aug. 1918 to 11 Nov. 1918'.
Reports on General Staff Conferences
'Why Germany Capitulated on November 11 1918 – a Brief Study Based on Documents in Possession of the French General Staff'

Published sources

(a) Parliamentary papers

Hansard
PP, 1902, Cmd. 803: 'Imperial Yeomanry Organisation and Equipment'
PP, 1904, XXXI, Cmd. 2062 and 2063: 'Minutes of Evidence taken before the Royal Commission on the Militia and Volunteers'
PP, 1903, XL, Cmd. 1789 and 1790, 'Minutes of Evidence taken before the Royal Commission on the War in South Africa'
PP, 1907, XLVI Cmd. 3294: 'Interim Report of the War Office Committee on the provision of Officers (a) for Service with Regular Army in War, and (b) for the Auxiliary Forces'
PP, 1921, XX, Cmd. 1193: General Annual Reports of the British Army (including the Territorial Force) for the period from 1 October 1913 to 30 September 1919
PP, 1922, Cmd. 1734: 'Report of the War Office Committee of Enquiry into "Shell-Shock"'

(b) Journals, magazines, newspapers

Army Review
The Barncroft Magazine
Blackwood's Magazine
Blunderbuss
Cavalry Journal
City of London Rifles Quarterly Journal
Daily Mirror
Daily Sketch
The East Kent Yeoman
Essex Review
The Graphic

The Guardian
The Listening Post
The LRB Record
Journal of the East Surrey Regiment
Journal of the Royal Army Medical Corps
Journal of the Royal United Services Institution
Mufti
Nineteenth Century and After
19th Division RFA Old Comrades Association [*newsletter*]
Observer
Old Comrades Journal, 2/4 The Queen's
Royal Military College Record
Public Secondary Schools Cadet Association Camp Magazine
The Scotsman
Sunday Chronicle
The Times
The Times Literary Supplement
The War Illustrated
The White Rose
Yorkshire Post

(c) Official military manuals and orders

Army Orders
Field Service Regulations, Part 1, Operations 1909. HMSO, reprinted with amendments 1914.
Infantry Training (HMSO, 1914)
Notes for Commanding Officers (Aldershot, 1918)
Notes for Young Officers (HMSO, 1917)
Regimental Standing Orders of His Majesty's Irish Regiment of Footguards (London, 1911)
Regulations for the Officers Training Corps, 1908
Report of the Committee on the Lessons of the Great War
Report on an Inspection of the Military Forces of the Commonwealth of Australia by General Sir I. Hamilton (May, 1914)
Standing Orders of the 2nd Battalion Essex Regiment (Aldershot, 1903)
Statistics of the Military Effort of the British Empire ... 1914–20 (HMSO, 1922)

(d) Correspondence, diaries, memoirs, and autobiographical fiction

Ackerley, J.R. *My Father and Myself*. Harmondsworth: Penguin, 1971. (O)
Ackrell, P.G. *My Life in the Machine Gun Corps*. Ilfracombe: Stockwell, 1966. (R)
Adams, B. *Nothing of Importance*. Stevenage: Strong Oak, 1988. (O)
Aldington, R. *Death of a Hero*. London: Consul, 1965. (R/O)
Andrews, W.L. *Haunting Years*. London: Hutchinson, nd. (R)
Anon., [Bell, D.H.] *A Soldier's Diary of the Great War*. London: Faber, 1929. (R/O)
Anon., [Pilcher, T.D.] *A General's Letters to His Son*. London: Cassell, 1917. (G)
Anon., *George Elton Sedding: the Life and Work of an Artist Soldier*. Letchworth: privately published, n.d. (R)
Ashurst, G. *My Bit: a Lancashire Fusilier at War 1914-18*. Marlborough: Crowood, 1988. (R)

Barbusse, H. *Under Fire*. London: Everyman, 1926. (R)
Bartlett, V. *No Man's Land*. London: Allen & Unwin, 1930. (O)
Beckett, I.F.W. (ed.). *The Army and the Curragh Incident, 1914*. London: Bodley Head, 1986.
Behrend, A. *As from Kemmel Hill*. London: Eyre & Spottiswoode, 1963. (O)
Benstead, C.R. *Retreat*. London: Methuen, 1930. (O)
Bion, W.R. *The Long Week-End 1897-1917*. London: Free Association, 1982. (O)
Bird, A. (ed.). *Honour Satisfied*. Swindon: Bird, 1990. (O)
Blatchford, R. *My Life in the Army*. London: Clarion. 1910. (R)
Blomfield, C.J. *Once an Artist Always an Artist*. London: Page, 1921. (O)
Bloch, M. *Memoirs of War 1914-15*. Cambridge, Cambridge University Press, 1980. (R)
Blunden, E. *Undertones of War*. Harmondsworth: Penguin, 1982. (O)
Bond, B. and Robbins, S. (eds). *Staff Officer*. London: Leo Cooper, 1987. (O)
Boustead, H. *The Wind of Morning*. London: Chatto & Windus, 1974. (R/O)
Bowra, C.M. *Memories 1898–1939*. London: Weidenfeld & Nicolson, 1966. (O)
Bridges, T. *Alarms and Excursions*. London: Longman, 1938. (O/G)
Brown, A. *Destiny*. Bognor: New Horizon, 1979. (R)
Brennan, G. *A Life of One's Own: Childhood and Youth*. Cambridge: CUP, 1979. (O)
Brereton, C.B. *Tales of Three Campaigns*. London: Selwyn & Blount, 1926. (O)
Campbell, P.J. *In the Cannon's Mouth*. London: Hamish Hamilton, 1986. (O)
Campbell, P.M. *Letters from Gallipoli*. Edinburgh: privately published, 1916. (O)
Carrington, C.E. *Soldier from the Wars Returning*. London: Hutchinson, 1965. (R/O)
Carton de Wiart, A. *Happy Odyssey*. London: Cape, 1950) (O/G)
Chapman, G. *A Passionate Prodigality*. London: MacGibbon & Kee, 1965. (O)
Childs, W. *Episodes and Reflections*. London: Cassell, 1930. (G)
Clarkson, D. *Memoirs of a Company Runner*. Edinburgh: Scottish National Institute for War Blinded, 1972. (R)
Cliff, N.D. *To Hell and Back with the Guards*. Brauton: Merlin, 1988. (R)
Coldicott, R. *London Men in Palestine*. London: Arnold, 1919 (O)
Coppard, G. *With a Machine Gun to Cambrai*. London: Macmillan, 1986. (R)
Corbett, A.F. *Service Through Six Reigns*. Norwich: privately published, 1953. (R)
Croney, P. *Soldier's Luck*. Ilfracombe: Stockwell, 1965 (R)
Crozier, F.P. *A Brass Hat in No Man's Land*. London: Cape, 1930 (R/O/G)
Crozier, F.P. *Impressions and Recollections*. London: T. Werner Laurie, 1930 (R/O/G)
Crozier, F.P. *The Men I Killed*. London: Joseph, 1937. (R/O/G)
Crutchley, C.E. *Machine Gunner 1914-19*. London: Purnell, 1975 (R)
Crutchley, C.E. *Shilling a Day Soldier*. Bognor: New Horizon, 1980. (R)
Dalton, H. *With the British Guns in Italy*. London: Methuen, 1919. (O)
Davison, C. *The Burgoyne Diaries*. London: Harmsworth, 1985. (O)
Dearden, H. *Medicine and Duty*. London: Heinemann, 1928. (O)
Devonald-Lewis, R. (ed.). *From the Somme to the Armistice*. London: Kimber, 1986. (O)
Dixon, A. *Tinned Soldier*. London: Cape, 1941. (R)
Douglas of Kirtleside. *Years of Combat*. London: Quality, 1963. (O)
Douie, C. *The Weary Road*. London: Murray, 1931. (O)

Dunham F. (Haigh R.H. and Turner, P.W. eds). *The Long Carry*. Oxford: Pergamon, 1970. (R)
Eberle, V.F. *My Sapper Venture*. London: Pitman, 1973. (R/O)
Eden, A. *Another World*. London: Allen Lane, 1976. (O)
Edmonds, C. [Carrington, C.E.] *A Subaltern's War*. London: Peter Davies, 1929. (R/O)
van Emden, R. (ed.). *Tickled to Death to Go*. Staplehurst: Spellmount, 1996. (R)
'Ex Pte X' [Burrage, A.M.] *War is War*. London: Gollancz, 1930 (R)
Eyre, G. *Somme Harvest*. London: Jarrolds, 1938. (R)
Fay, S. *The War Office at War*. Wakefield, EP, 1973.
Feilding, R. *War Letter to a Wife*. London: Medici Society, 1929. (O)
'F.O.O.' *The Making of a Gunner*. London: Eveleigh Nash, 1916. (O)
Foot, S. *Three Lives – and Now*. London: Heinemann, 1937. (O)
Fraser-Tytler, N. *With Lancashire Lads and Field Guns in France, 1915-18*. Manchester: Heywood, 1922. (O)
French, A. *Gone for a Soldier*. Kineton: Roundwood Press, 1972. (R)
Fussell, P. (ed.). *The Ordeal of Alfred M. Hale*. London: Leo Cooper, 1975) (R)
Gibbons, J. *Roll On, Next War*. London: Muller, 1935. (R)
Gilbert, M. *Winston S. Churchill* Companion Vol. III, 2 parts. London: Heinemann, 1972. (O)
Gillam, J. *Gallipoli Diary*. Stevenage: Strong Oak, 1989. (O)
Gladden, N. *Somme 1916*. London: Kimber, 1974. (R)
Glover, M. *The Fateful Battle Line*. London: Leo Cooper, 1993. (R/O)
Gordon, H. *The Unreturning Army*. London: Dent, 1967. (O)
Gough, H. *The Fifth Army*. London: Hodder, 1931. (G)
Graham, S. *A Private in the Guards*. London: Macmillan, 1919. (R)
Graham, S. *The Challenge of the Dead*. London: Cassell, 1921. (R)
Graham, S. *Part of the Wonderful Scene*. London: Collins, 1964. (R)
Graves, R. *Goodbye to All That*. Harmondsworth: Penguin, 1960.(O)
Grey, F. *Confessions of a Private*. Oxford: Blackwell, 1920. (R)
Griffith, W. *Up to Mametz*. London, Severn House, 1981. (O)
Groom, W.H.A. *Poor Bloody Infantry*. New Malden: 1983. (R)
'GSO' [Fox, F.] *GHQ*. London: Philip Allan, 1920. (O)
Hall, J.N. *Kitchener's Mob*. London: Constable, 1916. (R)
Hanbury-Sparrow, A.A. *The Land-Locked Lake*. London: Barker, 1932. (O)
Hankey. D. *A Student in Arms*. London: Melrose, two vols. 1916–17. (R/O)
Hargrave, J. *The Suvla Bay Landing*. London: Macdonald, 1964. (R)
Harris, L.H. *Signal Venture*. Aldershot: Gale and Polden, 1951. (R)
Harvey, H.E. *Battleline Narratives 1915–18*. London: Brentano's. 1928 (R)
Hatton, S.F. *The Yarn of a Yeoman*. London: Hutchinson, n.d. (R)
Hawke, J. *From Private to Major*. London: Hutchinson, 1938 (R/O)
Hawkings, F. (Taylor, A. ed.). *From Ypres to Cambrai*. Morley: Elmfield, 1974. (R/O)
Haworth, C. *March to Armistice 1918*. London: Kimber, 1968. (R)
Hay, I. *The First Hundred Thousand*. Edinburgh: Blackwood's, 1916. (O)
Hay, I. *Carrying On – after the First Hundred Thousand*. Edinburgh: Blackwood's, 1917. (O)
Herbert, A. *Mons, Anzac and Kut*. London: Hutchinson, n.d. (O)
Hitchcock, F.C. *'Stand To': a Diary of the Trenches*. Norwich: Glidden, 1988. (O)
Hodges, F.J. *Men of 18 in 1918*. Ilfracombe: Stockwell, 1988. (R)

Hooper, W. (ed.) *Letters of C.S. Lewis*. London: Fontana, 1988. (O)
Hope, T.S. *The Winding Road Unfolds*. London: Tandem, 1965. (R)
Horrocks, B. *Corps Commander*. London: Sidgwick & Jackson, 1977. (O)
[Howell, R.] *Philip Howell: a Memoir by his Wife*. London: Allen & Unwin, 1942. (O/G)
Housman, L. (ed.). *War Letters of Fallen Englishmen*. London: Gollancz, 1930.
Hutchison, G.S. *Warrior*. London: Hodder, 1932. (O)
Jacomb, C.E. *Torment*. London: Melrose, 1920. (R)
Jobson, A. *Via Ypres*. London:Westeminster City, 1934. (R)
Jones, D. *In Parenthesis*. London: Faber, 1963. (R)
Jones, P. *War Letters of a Public School Boy*. London: Cassell, 1918. (O)
Joynt, W.D. *Saving the Channel Ports 1918*. North Blackburn: Wren, 1975. (O)
Kernahan, C. *An Author in the Territorials*. London: Pearson, 1908. (O)
Killick, A. *Mutiny!* Brighton: Militant, n.d. (R)
Kingsmill, A.G. *The Silver Badge*. Ilfracombe: Stockwell, 1966. (R)
Latham, B. *A Territorial Soldier's War*. Aldershot: Gale & Polden, 1967. (R/O)
Linklater, E. *The Man on My Back*. London: Macmillan, 1950. (R)
Linklater, E. *Fanfare for a Tin Hat*. London: Macmillan, 1970. (R)
Lloyd, R.A. *A Trooper in the 'Tins'*. London: Hurst and Blackett, 1938. (R)
Lucy, J. *There's a Devil in the Drum*. London: Naval and Military Press, 1992. (R/O)
MacGill, P. *The Amateur Army*. London: Jenkins, 1915. (R)
MacGill, P. *The Red Horizon*. London: Caliban, 1984. (R)
MacGill, P. *The Great Push*. London: Jenkins, 1916. (R)
Mackenzie, C. *Gallipoli Memories*. London: Panther, 1965. (O)
Macmillan, H. *Winds of Change*. London: Macmillan, 1966. (O)
Macready, N. *Annals of an Active Life*. London: Hutchinson, 2 vols, n.d. (G)
Maitland, F.H. *Hussar of the Line*. London: Hurst and Blackett, 1951. (R)
Majdaleny, F. *The Monastery*. London: Corgi, 1957.
Manning, F. *The Middle Parts of Fortune*. London: Granada, 1977. (R/O)
'Mark VII', [Plowman, M.] *A Subaltern on the Somme*. London: Dent, 1928. (O)
Martin, B. *Poor Bloody Infantry*. London: Murray, 1987. (O)
Marks, T.P. *The Laughter Goes from Life*. London: Kimber, 1977. (R)
[Maxwell, C.A. ed.] *Frank Maxwell, Brig.-General VC CSI DSO: a Memoir and Some Letters*. London, Murray, 1921. (O/G)
[Maxwell, C.A.]. *I am Ready*. London: Hazell, Watson, Viney, 1955. (O/G)
Mays, S. *Fall Out the Officers*. London: Eyre & Spottiswoode, 1969 (R)
Mersey, Viscount. *Journals and Memories*. London: Murray, 1952. (O)
Mellersh, H.E.L. *Schoolboy into War*. London: Kimber, 1978. (O)
Metcalfe, T.W. *Memorials of the Military Life*. London: Ivor Nicholson and Watson, 1936. (O)
[Mitchell, J.M.] *The New Army in the Making by an Officer*. London: Kegan Paul …, 1915. (O)
Monash, J. *The Australian Victories in France in 1918*. London, IWM, n.d. (G)
Montague, C.E. *Disenchantment*. London: Chatto and Windus, 1928. (R/O)
Montgomery of Alamein, *Memoirs*. London: Reader's Union, 1960. (O)
Moran, Lord. *The Anatomy of Courage*. London: Constable, 1945. (O)
Morice, J. (ed.). *Six-Bob-A-Day Tourist*. Ringwood, Victoria, Penguin, 1985. (R)
Mosley, O. *My Life*. London: Nelson, 1968. (O)

Mottram, R.H. *Journey to the Western Front*. London: Bell, 1936. (O)
Mottram, R.H., Easton, J., Partridge, E. *Three Personal Accounts of the War*. London: Scolartis, 1929.
Moynihan, M. (ed.). *A Place Called Armageddon*. Newton Abbot: David and Charles, 1974.
Mugge, M.A. *The War Diary of a Square Peg*. London: Routledge, 1920. (R)
Munday, H. *No Heroes, No Cowards*. Milton Keynes: People's Press, 1981. (R)
Nash, T.A.M. (ed.). *The Diary of an Unprofessional Soldier*. Chippenham: Picton, 1991. (R/O)
Nettleton, J. *The Anger of the Guns*. London: Kimber, 1979. (R/O)
Neville, J.E.H. *The War Letters of a Light Infantryman*. London: Sifton Praed, 1930. (O)
Newman, B. *The Cavalry Went Through*. London: Gollancz, 1930. (R)
Nicholson, W.N. *Behind The Lines*. London: Cape, 1939. (O)
O'Riordan, C. *et al. A Martial Medley*. London: Scolartis,1931.
Packer, C. *Return To Salonika*. London: Cassell, 1964. (R)
Parker, E. *Into Battle 1914–18*. London: Longman, 1964. (R/O)
Pearce Smith, K. *Adventures of an Ancient Warrior*. Privately published, 1984. (O)
Prendergast, J. *Prender's Progress*. London: Cassell, 1979. (O)
Quigley, H. *Passchendaele and the Somme*. London: Methuen, 1928. (O)
Raymond, E. *Tell England*. London: Cassell, 1928. (O)
Raymond,E. *The Story of My Days*. London: Quality, 1968. (O)
Rees, R.T. *A Schoolmaster at War*. London:Haycock, n.d. (O)
Richards, F. *Old Soldiers Never Die*. London: Faber, 1965. (R)
Richards, F. *Old Soldier Sahib*. London: Faber, 1965. (R)
Rogerson, S. *Twelve Days*. Norwich, Gliddon, 1988. (O)
Roy, R.H. (ed.). *The Journal of Private Fraser*. Victoria, BC: Sono Nis Press, 1985. (R)
Rule, A. *Students under Arms*. Aberdeen: Aberdeen University Press, 1934. (R/O)
Russell, A. *With the Machine Gun Corps*. London: Dranes, 1923. (R)
Seely, J.E.B. *Adventure*. London: Heinemann, 1931. (G)
Scrivenor, J.B. *Brigade Signals*. Oxford: Blackwell, 1937. (O)
Sheffield, G.D. and Inglis, G.I.S, (eds). *From Vimy Ridge to the Rhine: the Great War Letters of Christopher Stone*. Marlborough: Crowood, 1989. (R/O)
Shephard, E. *A Sergeant-Major's War*. Marlborough: Crowood, 1987.(R/O)
Sherriff, R.C. *Journey's End*. London: Heinemann, 1958. (O)
Sherriff, R.C. *No Leading Lady*. London: Gollancz, 1968. (O)
Slim, W. *Unofficial History*. London: Corgi, 1970. (O)
Smith, A. *Four Years on the Western Front, by A Rifleman*. London: Stamp Exchange, 1987. (R)
de Sola Pinto, V. *The City that Shone*. London: Hutchinson, 1969. (O)
Sparrow, G. and MacBean Ross, J.N. *On Four Fronts with the Royal Naval Division*. London: Hodder, 1918 (O)
Spears, E.L. *The Picnic Basket*. London: Secker, 1967. (O)
Steuart, R.H.J. *March, Kind Comrade*. London: Sheed & Ward, 1931. (O)
Swinton, E. (ed.). *Twenty Years After*. London: Newnes, 3 vols, n.d. but c.1938–39.
Tawney, R.H. *The Attack and Other Papers*. Nottingham: Spokesman, 1981. (R)
Taylor, F.A.J. *The Bottom of the Barrel*. London: Regency, 1979. (R)

Terraine, J. *General Jack's Diary 1914–1918*. London: Eyre & Spottiswoode, 1964. (O/G)
Thomas, A. *A Life Apart*. London: Gollancz, 1968. (O)
Thompson, E. *These Men Thy Friends*. London: Knopf, 1927. (O)
Thorburn, A.D. *Amateur Gunners*. Liverpool: Potter, 1933. (O)
Thornton, R.K.K. (ed.). *Ivor Gurney: War Letters*. London: Hogarth, 1984. (R)
Thurtle, E. *Time's Winged Chariot*. London: Chaterson, 1945. (R)
Tilsley, V.W. *Other Ranks*. London: Cobden-Sanderson, 1931. (R)
Townshend, E. (ed.). *Keeling Letters and Recollections*. London: Allen and Unwin, 1918. (R)
Tucker, J.F. *Johnny Get Your Gun*. London: Kimber, 1978. (R)
Turner P.W. and Haigh, R.H. *Not for Glory*. London: Maxwell, 1969. (R/O)
Vaughan, E.C. *Some Desperate Glory*. London: Macmillan, 1985. (O)
Voigt, F.A. *Combed Out*. London: Cape, 1929. (R)
'Wagger', *Battery Flashes*. London: Murray, 1916. (R/O)
Wheatley, D. *Officer and Temporary Gentleman*. London: Hutchinson, 1978. (O)
Williams, H.R. *The Gallant Company*. Sydney: Angus & Robertson, 1933. (R/O)
Williamson, H. *The Golden Virgin*. London: Macdonald, 1984. (R/O)
Williamson, H. *A Fox under My Cloak*. London: Macdonald, 1985. (R/O)
Wilson, R. *Palestine 1917*. Tunbridge Wells: Costello, 1987. (R/O)
Woods, E.S. (ed.). *Andrew R. Buxton: the Rifle Brigade*. London: Robert Scott, 1918. (O)
Woolley, G.H. *Sometimes a Soldier*. London: Benn, 1963. (R/O)
Wyndham, H. *The Queen's Service*. London: Heinnemann, 1899. (R)
Yeates-Brown, F. *Bengal Lancer*. London: Mott, 1984. (O)
Zilboorg, E. (ed.). *Richard Aldington & H.D.* Bloomington: Indiana University Press, 1992. (R/O)

(e) Other contemporary and near-contemporary material

'A British Officer' [Cairnes, W.E.] *Social Life in the British Army*. London: Long, 1910.
Alderson, E.A.H. *Pink and Scarlet, or Hunting as a School for Soldiering*. London: Hodder, 1913.
Alderson, E.A.H. *Lessons from 100 Notes made in Peace and War*. Aldershot: Gale & Polden, 1908.
'A Lieutenant-Colonel in the British Army', *The British Army*. London: Sampson Low, 1899.
Arthur, G. *Life of Lord Kitchener* II. London: Macmillan, 1920.
Banning, S.T. *Military Law Made Easy*. Aldershot: Gale & Polden, 1904.
'Basilisk', *Talks on Leadership Addressed to Young Officers*. London: Rees, 1941.
Beaver, P. (ed.). *The Wipers Times*. London: Macmillan, 1973.
Buchan, J. (ed.). *The Long Road to Victory*. London: Nelson, 1920.
[Cairnes, D.S.] *The Army and Religion*. London: Macmillan, 1919.
[Cairnes, W.E.]. *The Army from Within, by the Author of 'An Absent Minded War'*. London: Sands, 1901.
Codrington, G.R. *The Territorial Army*. London: Sifton Praed, 1938.
Collet, D.W. *St Dunstan's College, Catford, Roll of Honour for the First World War 1914–19*. London: privately published, 1928.
Dawbarn, C. *Joffre and his Army*. London: Mills and Boon, 1916.

Deane, H.F.W. and Evans, W.A. *Public Schools Year Book*. London: privately published, 1908.
Edmondson, R. *John Bull's Army from Within*. London: Griffiths, 1907.
'Ex-Trooper', *The French Army from Within*. London: Hodder, 1914.
Fuller, J.F.C. *The Army in My Time*. London: Rich and Cowan, 1935.
Germains, V.W. *The Kitchener Armies*. London: Peter Davies, 1930.
Haig, D. *A Rectorial Address Delivered to the Students in the University of St. Andrews, 14th May 1919*. St Andrews: Henderson, 1919.
Henderson G.F.R. (Malcolm, N. ed.). *The Science of War*. London: Longman, 1906.
Henderson, G.F.R. *Stonewall Jackson and the American Civil War*. London: Longman, 1898.
Henty, G.A. *et al. Hazard and Heroism*. Edinburgh, Chambers 1904.
Henty, G.A. *In Time of Peril*. London: Griffith, Farran, 1881.
The Hoghunters Annual II. Bombay: Times of India Press, 1929.
In Memoriam Royal Grammar School, Newcastle-upon-Tyne. Newcastle, n.d.
Jacomb, C.E. *God's Own Country*. London: Max Goschen, 1913.
Kipling, R. *Traffics and Discoveries*. London: Macmillan, 1904.
Kipling, R. *Soldiers Three*. London: Macmillan, 1911.
Kipling, R. *Wee Willie Winkie*. London: Macmillan, 1920.
Kipling, R. *Puck of Pook's Hill*. Harmondsworth: Puffin, 1987.
Kipling, R. *The New Armies in Training*. London: Macmillan, 1915.
Lake, B.C. *Knowledge for War – Every Officer's Handbook for the Front*. London: Harrison, n.d.
MacLean, A.H.H. *Public Schools and the War in South Africa*. London: Stanford, 1902.
MacNutt, F.B. *The Church in the Furnace*. London: Macmillan, 1918.
Mr. Punch's History of the Great War. London: Cassell, 1919.
Mitchell, F. *Tank Warfare*. Stevenage: Spa Books, 1987.
Myers, C.S. *Shell-Shock in France, 1914-18*. Cambridge: Cambridge University Press, 1940.
Newham-Davis, N. *Military Dialogues*. London: Long, n.d.
Nicolai, W. *The German Secret Service*. London: Stanley Paul, 1924.
O'Toole, T. *The Way They Have in the Army*. London: John Lane, 1916.
'Regular', *Customs of the Army: a Guide for Cadets and Young Officers*. London, 1917.
Richards, W. *His Majesty's Territorial Army*. London: Virtue, 6 vols, n.d.
Spurr, J. *Conscript Titch*. London: Chambers, 1917.
Trappman, A.H. *Straight Tips for 'Subs'*. London, 1915.
Taylor, A.W. *How to Organise and Administer a Battalion*. London: Rees, 1915.
Vivian, E.C. *The British Army from Within*. London: Hodder, 1914.
Williams, B. *Raising and Training the New Armies*. London: Constable, 1918.

Secondary sources

(a) Official histories and monographs

Bean, C.E.W. (ed.). *Official History of Australia in the War of 1914–1918*. Sydney: Angus and Robertson, 12 vols, 1921–42.
Edmonds, J.E. (ed.). *Military Operations, France and Belgium*. London: Macmillan, 14 vols, 1922–49.

Edmonds, J.E. *The Occupation of the Rhineland*. London: HMSO, 1987.
MacPhail, A. *Official History of the Canadian Forces in the Great War 1914–19: the Medical Services*. Ottawa, Dept of National Defence, 1925.
Morgan, M.C. *The Second World War 1939–45, Army, Army Welfare*. London: War Office, 1950.
Nicholson, G.W.L. *Canadian Expeditionary Force 1914–1919*. Ottowa, Queen's Printer, 1962.

(b) Unit and formation histories

Anon., *Artillery and Trench Mortar Memories, 32nd Division*. London: Unwin, 1932.
Anon., *The History of the London Rifle Brigade*. London: Constable, 1921.
Anon., *A Short History of the 19th (Western) Division 1914–18*. London: Murray, 1919.
Anon., *Sixteenth, Seventeenth, Eighteenth, Nineteenth Battalions The Manchester Regiment: a Record 1914-1918*. Manchester: Sherrat & Hughes, 1923.
Anon., *66th East Lancashire Division Dinner Club*. Manchester: Falkener, 1923.
Anon., *The 23rd London Regiment 1798–1919*. London: Times, 1936.
Anon., *The War History of the 1/4th Battalion the Loyal North Lancashire Regiment 1914–19*. privately published, 1921.
Atteridge, A.H. *History of the 17th (Northern) Division*. Glasgow: Maclehose, 1929.
Bailey, O.F. and Hollier, O.F. *The Kensingtons: 13th London Regiment*. London: privately published, 1935.
Bolitho, H. *The Galloping Third*. London: Murray, 1963.
Boyes, R. *In Glass Houses*. Colchester: MPSC Assoc., 1988.
Bryant, A. *Jackets of Green*. London: Collins, 1972.
Bryant, P. *Grimsby Chums*. Hull: Humberside Leisure Services, 1990.
Cooper, B. *The Tenth (Irish) Division in Gallipoli*. London: Jenkins, 1918.
Cowper, L. *The King's Own: the Story of a Royal Regiment*, II. Oxford: Oxford University Press, 1939.
Denman, T. *Ireland's Unknown Soldiers*. Dublin: Irish Academic Press, 1992.
Dudley Ward, C. *The Welsh Regiment of Foot Guards 1915–1918*. London: Murray, 1936.
Dunn, J.C. *The War the Infantry Knew*. London: Jane's, 1987.
Ewing, J. *The History of the 9th (Scottish) Division 1914–1919*. London: Murray, 1921.
Fellows, G. and Freeman, B. *Historical Records of the South Nottinghamshire Hussars Yeomanry*. Aldershot: Gale & Polden, 1928.
Gladstone, E.W. *The Shropshire Yeomanry 1795–1945*. Manchester: Whitethorn Press, 1953.
Haig-Brown, A.R. *The OTC and the Great War*. London: Country Life, 1915.
Hall, J. *The Coldstream Guards 1885–1914*. Oxford: Oxford University Press, 1920.
Headlam, C. *History of the Guards Division in the Great War 1915–1918*. 2 vols. London: Murray, 1924.
Henriques, J.Q. *The War History of the 1st Battalion Queen's Westminster Rifles 1914–1918*. London: Medici Society, 1923.
Hurst, G.B. *With the Manchesters in the East*. Manchester: Manchester University Press, 1917.

Keeson, C.A.C. *The History and Records of Queen Victoria's Rifles*. London: Constable, 1923.
Keith-Falconer, A. *The Oxfordshire Hussars in the Great War*. London: Murray, 1927.
Kingham, W.R. *London Gunners*. London: Methuen, 1919.
Jamie, J.P.W. *The 177th Brigade 1914–18*. Leicester: Thornley, 1931.
Jerrold, D. *The Royal Naval Division*. London: Hutchinson, n.d.
Lindsay, J.H. *The London Scottish in the Great War*. London: privately published, 1925.
McGilchrist, A.M. *The Liverpool Scottish 1900–1919*. Liverpool: Young, 1930.
McWilliams, J.L. and Steel, R.J. *The Suicide Battalion*. St Catharines, Ontario: Vanwell, n.d.
Maddox, G. *Liverpool Pals*. Barnsley: Wharncliffe, 1991.
Matthews, E.C. *With the Cornish Territorials on the Western Front*. Cambridge: Spalding, 1921.
Maude, A.H. (ed.). *The 47th (London) Division, 1914–19*. London: Amalgamated Press, 1922.
Miles, W. *The Durham Forces in the Field 1914–18* II. London: Cassell, 1920.
Milner, L. *Leeds Pals*. Barnsley: Wharncliffe, 1991.
Mitchell, G.S. *'Three Cheers for the Derrys'*. Derry: YES! Publications, 1991.
Mitchinson, K.W. *Gentlemen and Officers*. London: IWM, 1995.
Nichols, G.H.F. *The 18th Division in the Great War*. Edinburgh: Blackwood, 1922.
Ponsonby, C. *West Kent (QO) Yeomanry and 10th (Yeomanry) Battalion, the Buffs 1914–19*. London: Melrose, 1920.
Potter, C.H. and Fothergill, A.S.C. *The History of the 2/6th Lancashire Fusiliers*. Rochdale: privately published, 1927.
Priestley, R.E. *Breaking the Hindenburg Line*. London: T. Fisher Unwin, 1919.
Richardson, G. *For King and Country and Scottish Borderers*. Hawick: privately published, 1987.
Riddell, E. and Clayton, M.C. *The Cambridgeshires 1914–18*. Cambridge: Bowkes, 1934.
Rutter, O. (ed.). *The History of the Seventh Service Battalion of the Royal Sussex Regiment 1914–19*. London: Times, 1934.
Sellers, L. *The Hood Battalion*. London: Leo Cooper, 1995.
Sheffield, G.D. *The Redcaps: a History of the Royal Military Police and its Antecedents from the Middle Ages to the Gulf War*. London: Brassey's, 1994.
Scott, A.J.L. *Sixty Squadron R.A.F. 1916–1919*. London: Greenhill, 1990.
Stewart, P.F. *The History of the XII Royal Lancers (Prince of Wales)*. London: Oxford University Press, 1950.
Stone, C. *A History of the 22nd (Service) Battalion Royal Fusiliers (Kensington)*. London: privately published, 1923.
Strachan, H. *History of the Cambridge University Officers Training Corps*. Tunbridge Wells: Midas, 1976.
Tempest, E.V. *History of 6th Battalion West Yorkshire Regiment*. Bradford: Percy Lund, 1921.
Turner, W. *Pals: the 11th (Service) Battalion (Accrington) East Lancashire Regiment*. Barnsley: Wharncliffe, n.d.
Verdin, R. *The Cheshire (Earl of Chester's) Yeomanry*. Birkenhead: privately published, 1971.

Verney, P. *The Micks*. London: Pan, 1973.
Wallace, E. *Kitchener's Army and the Territorial Forces*. London: Newnes, n.d.
Wallis Grain, H.W. *16th (Public Schools) (S) Battalion Middlesex Regiment and the Great War*. London: privately published, 1935.
Wilson, S.J. *The Seventh Manchesters*. Manchester: Manchester University Press, 1920.
Wurtzburg, C.F. *The History of the 2/6 (Rifle) Battalion 'The King's' (Liverpool Regiment) 1914–19*. Aldershot: Gale & Polden, 1920.
Younghusband, G.J. *The Story of the Guides*. London: Macmillan, 1908.

(c) Other books

Adair, J. *Developing Leaders*. Maidenhead: Talbot Adair, 1988.
Adams, R.J.Q. (ed.). *The Great War, 1914–18*. College Station, TX: Texas A&M, 1990.
Allison, W. and Fairley, J. *The Monocled Mutineer*. London: Quartet, 1978.
Andreski, S. *Military Organisation and Society*. London: Routledge & Kegan Paul, 1968.
Andrews, E.M. *The Anzac Illusion*. Cambridge: Cambridge University Press, 1993.
Arhrenfeldt, R.H. *Psychiatry in the British Army in the Second World War*. London: Routledge & Kegan Paul, 1958.
Arnold, G. *Held Fast for England*. London: Hamish Hamilton, 1979.
Ascoli, D. *The Mons Star*. London: Harrap, 1981.
Ashworth, T. *Trench Warfare 1914–1918*. London: Macmillan, 1980.
Audoin-Rouzeau, S. *Men at War 1914–1918*. Oxford: Berg, 1992.
Babington, A. *For the Sake of Example*. London: Leo Cooper, 1983.
Babington, A. *The Devil to Pay*. London: Leo Cooper, 1991.
Barclay, C.N. *Armistice 1918*. London: Dent, 1968.
Barnett, C. *The Collapse of British Power*. Gloucester: Sutton, 1984.
Bartlett, F.C. *Psychology and the Soldier*. Cambridge: Cambridge University Press, 1927.
Bartov, O. *Hitler's Army*. New York: Oxford University Press, 1991.
Baynes, J. *Morale*. London: Cassell, 1967.
Baynes, J. *The Forgotten Victor*. London: Brassey's, 1989.
Beckett, I.F.W. *Riflemen Form*. Aldershot: Ogilby Trust, 1982.
Beckett, I.F.W. *The Amateur Military Tradition*. Manchester: Manchester University Press, 1992.
Beckett, I.F.W. and Simpson, K. (eds). *A Nation In Arms*. Manchester: Manchester University Press. 1985.
Benyon, H. *Working For Ford*. Harmondsworth: Penguin, 1984.
Berton, P. *Vimy*. Markham, Ontario: Penguin, 1987.
Bidwell, S. and Graham, D. *Fire-Power*. London: Allen & Unwin, 1982.
Bond, B. *The Victorian Army and the Staff College 1854–1914*. London: Eyre Methuen, 1972.
Bond, B. *British Military Policy between the Two World Wars*. Oxford: Clarendon, 1980.
Bond, B. *War and Society in Europe, 1870–1970*. London: Fontana, 1984.
Bond, B. (ed.). *The First World War and British Military History*. Oxford: Clarendon, 1991.

Bonham-Carter, V. *Soldier True*. London: Muller, 1963.

Bourke, J. *Dismembering the Male: Men's Bodies, Britain and the Great War*. London: Reaktion, 1996.

Bourne, J.M. *Britain and the Great War 1914–1918*. London: Arnold, 1989.

Bracco, R.M. *Merchants of Hope*. Oxford: Berg, 1993.

Brophy, J. and Partridge, E. *The Long Trail*. London: Deutsch, 1965.

Brown, M. *The Imperial War Museum Book of the First World War*. London: Sidgwick & Jackson, 1991.

Brown, M. *Tommy Goes to War*. London: Dent, 1978.

Brugger, S. *Australians and Egypt 1914–1919*. Carlton, Victoria: Melbourne University Press, 1980.

Bryant, P. *Grimsby Chums*. Hull: Humberside Libraries & Arts, 1990.

Burness, P. *The Nek*. Kenthurst, NSW: Kangaroo Press, 1996.

Calder, A. *The People's War*. London: Panther, 1971.

Carew, T. *Wipers*. London: Hamish Hamilton 1972.

Carpenter, H. *J.R.R. Tolkien: a Biography*. London: Unwin Paperbacks, 1978.

Carrington, C.E. *Rudyard Kipling: His Life and Work*. Harmondsworth: Penguin, 1970.

Cassar, G.H. *Kitchener: Architect of Victory*. London: Kimber, 1977.

Cecil, H. *The Flower of Battle*. London: Secker, 1995.

Cecil, H. and Liddle, P. *Facing Armageddon*. London: Leo Cooper, 1996.

Clayton, A. *The British Empire as a Superpower*. Athens, GA: University of Georgia Press, 1986.

von Clausewitz, C. (Howard, M. and Paret, P. eds). *On War*. Princeton, NJ: Princeton University Press, 1976.

Constantine, S., M.W. Kirby and M.B. Rose, *The First World War in British History*. London: Edward Arnold, 1995.

Copp, T. and McAndrew, B. *Battle Exhaustion*. Montreal: McGill-Queens University Press, 1990.

van Creveld, M. *Fighting Power*. London: Arms & Armour, 1983.

Cunningham, H. *The Volunteer Force*. London: Croom Helm, 1975.

Dakers, C. *The Countryside at War 1914–1918*. London: Constable, 1987.

Dallas, G, and Gill, D. *The Unknown Army*. London: Verso, 1985.

Dennis, P. *The Territorial Army 1907–1940*. Woodbridge: Boydell, 1987.

Duffy, C. *Russia's Military Way to the West*. London: Routledge & Kegan Paul, 1981.

Duffy, C. *Military Experience in the Age of Reason*. London: Routledge & Kegan Paul, 1987.

Duffy, C. *Red Storm on the Reich*. New York: Atheneum, 1991.

Ellis, J. *Eye Deep in Hell*. London: Fontana, 1977.

English, J. *The Canadian Army and the Normandy Campaign*. New York: Praeger, 1991.

Essame, H. *The Battle for Europe, 1918*. London: Batsford, 1972.

Farwell, B. *For Queen and Country*. London: Allen Lane, 1983.

Ferro, M. *The Great War 1914–1918*. London: Routledge, 1973.

Fernadez-Armesto, F. *Millennium*. New York: Scribner, 1995.

Firkins, P. *The Australians in Nine Wars*. London: Pan, 1973.

Fletcher, D. *Tanks and Trenches*. Stroud: Sutton, 1994.

Fraser, D. *And We Shall Shock Them*. London: Sceptre, 1988.

Fuller, J. *Troop Morale and Popular Culture in the British and Dominion Armies 1914–1918*. Oxford: Clarendon, 1991.
Fussell, P. *The Great War and Modern Memory*. London: Oxford University Press, 1975.
Fyfe, A.J. *Understanding the First World War: Illusions and Realities*. New York: Peter Long, 1988.
Gabriel, R. *The Painful Field*. New York: Greenwood, 1989.
Gammage, B. *The Broken Years*. Ringwood, Victoria: Penguin, 1975.
Gathorne-Hardy, J. *The Public School Phenomenon*. London: Hodder, 1977.
Gilbert, *The Challenge of War: Winston S. Churchill 1914–1916*. London: Minerva, 1990.
Girouard, M. *The Return to Camelot: Chivalry and the English Gentleman*. London: Yale University Press, 1981.
Gregory, A. *The Silence of Memory*. Oxford: Berg, 1994.
Grey, J. *A Military History of Australia*. Cambridge: Cambridge University Press, 1990.
Grieves, K. *Sir Eric Geddes*. Manchester: Manchester University Press, 1989.
Griffith, P. (ed.). *British Fighting Methods in the Great War*. London: Cass, 1996.
DeGroot, G. *Blighty*. London: Longman, 1996.
Hamilton, N. *Monty: the Making of a General*. London: Hamish Hamilton, 1981.
Handy, C.B. *Understanding Organisations*. Harmondsworth: Penguin, 1985.
Harries-Jenkins, G. *The Army in Victorian Society*. London: Routledge & Kegan Paul, 1977.
Harris, J. *Private Lives, Public Spirit: Britain 1870–1914*. London: Penguin, 1994.
Harris, K. *Attlee*. London: Weidenfeld & Nicolson, 1982.
Harris, S.J. *Canadian Brass*. Toronto: University of Toronto Press, 1988.
Heathcote, T.A. *The Indian Army*. Newton Abbot: David and Charles, 1974.
Heathcote, T.A. *The Military in British India, 1600–1947*. Manchester: Manchester University Press, 1995.
Henderson, D.M. *Highland Soldier*. Edinburgh: Donald, 1989.
Henderson, W.D. *Cohesion: the Human Element in Combat*. National Defense University: Washington, DC, 1985.
Hobsbawm, E.J. *Industry and Empire*. Harmondsworth: Penguin, 1968.
Holmes, R. *The Little Field Marshal*. London: Cape, 1981.
Holmes, R. *Firing Line*. London: Cape, 1985.
Holt, R. *Sport and the British*. Oxford: Clarendon, 1989.
Horn, P. *Rural Life in England in the First World War*. New York: St. Martin's Press, 1984.
Horne, A. *The Price of Glory*. Harmondsworth: Penguin, 1964.
Horne, A. *Macmillan 1894–1956*. London: Macmillan, 1988.
Horne, J. *The Best of Good Fellows*. London: J. Horne Publishers, 1995.
Hyatt, A.M.J. *General Sir Arthur Currie*. Toronto: Toronto University Press, 1987.
Howkins, A. *Poor Labouring Men*. London: Routledge & Kegan Paul, 1985.
Hynes, S. *A War Imagined*. London: Bodley Head, 1990.
Ireland, W. *The Story of Stokey Lewis VC*. Haverford West: privately published, n.d.
James, E.A. *British Regiments, 1914–18*. London: Sampson, 1978.
James, J. *The Paladins*. London: Macdonald, 1990.
James, L. *Mutiny*. London: Buchan and Enright, 1987.

James, R.R. *Anthony Eden*. London: Weidenfeld & Nicolson, 1986.
Jeffery, K. *The British Army and the Crisis of Empire*. Manchester: Manchester University Press, 1984.
Joyce, P. *Work, Society and Politics*. London: Methuen, 1980.
Keegan, J. *The Face of Battle*. Harmondsworth: Penguin, 1978.
Kellett, A. *Combat Motivation*. Boston, MA: Kluwer-Nijhoff, 1982.
Kingston, B. *The Oxford History of Australia, III, 1860–1900*. Melbourne: Oxford University Press, 1988.
Kiernan, V.G. *European Empires from Conquest to Collapse 1815–1960*. London: Fontana, 1982.
Klein, H. (ed.). *The First World War in Fiction*. London: Macmillan, 1978.
Laffin, J. *Digger*. London: Cassell, 1959.
Lamb, D. *Mutinies*. Oxford and London: Solidarity, nd.
Leed, E.J. *No Man's Land*. Cambridge: Cambridge University Press, 1979.
Lewin, R. *Slim the Standardbearer*. London: Pan, 1978.
Liddle, P.H. *Home Fires and Foreign Fields*. London: Brassey's, 1985.
Liddle, P.H. *The Soldier's War 1914–1918*. London: Blandford, 1988.
Livingstone, R.W. (ed.). *The Legacy of Greece*. Oxford, Oxford University Press, 1921.
Lowerson, J. *Sport and the English Middle Classes 1870–1914*. Manchester: Manchester University Press, 1993.
Luvaas, J. *The Education of an Army*. London: Cassell, 1964.
McCarthy, D. *Gallipoli to the Somme*. London: Leo Cooper, 1983.
McIntosh, P. *Physical Education in England since 1800*. London: Bell, 1952
MacKenzie, J.M. *Propaganda and Empire*. Manchester: Manchester University Press, 1984.
McKenzie, R. and Silver, A. *Angels in Marble*. London: Heinemann, 1968.
MacKenzie, S.P. *Politics and Military Morale*. Oxford: Clarendon, 1992.
McKibben, R. *The Ideologies of Class*. Oxford: Clarendon, 1994.
McLaine, I. *Ministry of Morale*. London: Allen & Unwin, 1979.
McMichael, S.R. *An Historical Perspective on Light Infantry*. Ft Leavenworth, KS: Combat Studies Institute, 1987).
Marwick, A. *The Deluge*. Harmondsworth: Penguin, 1967.
Mason, P. *A Matter of Honour*. London: Cape, 1974.
Mason, P. *The English Gentleman*. London: Pimlico, 1993.
Maurice, F. *The Life of Lord Rawlinson of Trent*. London: Cassell, 1928.
Meacham, S. *A Life Apart*. London: Thames & Hudson, 1977.
Middlebrook, M. *The First Day on the Somme*. London: Allen Lane, 1971.
Middlebrook, M. *The Kaiser's Battle*. London: Allen Lane, 1978.
Millett, A.R. and Murray, W. *Military Effectiveness*, vol. I. London: Unwin Hyman, 1988.
Moran, Lord. *The Anatomy of Courage*. London: Constable, 1945.
Morton, D. *When Your Number's Up*. Toronto, Random House, 1993.
Moorhouse, G. *Hell's Foundations*. London: Sceptre, 1992.
Newby, H. *Country Life*. London: Cardinal, 1987.
Nicholson, N. *Alex*. London: Weidenfeld & Nicolson, 1973.
Ogilvie, R.M. *Latin and Greek*. London: Routledge, 1964.
Omissi, D. *The Sepoy and the Raj*. London: Macmillan, 1996.
Orwell, S, and Angus, I. (eds). *The Collected Essays, Journalism and Letters of*

George Orwell. 4 vols. Harmondsworth: Penguin, 1970.
Panichas, G.A. *Promise of Greatness*. London: Cassell, 1968.
Parker, P. *The Old Lie*. London: Constable, 1987.
Pedroncini, G. *Les Mutineries de 1917*. Paris: Presses Universitaires de France, 1967.
Perry, F.W. *The Commonwealth Armies*. Manchester: Manchester University Press, 1988.
Price, R. *An Imperial War and the British Working Class*. London: Routledge & Kegan Paul, 1972.
Porch, D. *The March to the Marne*. Cambridge: Cambridge University Press, 1981.
Powell, G. *Plumer: the Soldier's General*. London: Leo Cooper, 1990.
Pugsley, C. *Gallipoli: the New Zealand Story*. Auckland: Hodder, 1984.
Pugsley, C. *On the Fringe of Hell*. Auckland: Hodder, 1991.
Putkowski, J. and Sykes, J. *Shot at Dawn*. Barnsley: Wharncliffe, 1989.
Quinn, T. *Tales of the Old Soldiers*. Stroud: Sutton, 1993.
Reader, W.J. *'At Duty's Call'*. Manchester: Manchester University Press, 1988.
Richards, J. *Visions of Yesterday*. London: Routledge & Kegan Paul, 1973.
Roberts, D. *Paternalism in Early Victorian England*. New Brunswick, NJ: Rutgers University Press, 1979.
Roberts, R. *The Classic Slum*. Harmondsworth: Penguin, 1973.
Roebuck, J. *The Making of Modern English Society from 1850*. London: Routledge & Kegan Paul, 1973.
Ross, J. *The Myth of the Digger*. Sydney: Hale & Iremonger, 1985.
Rothstein, A. *The Soldiers' Strikes of 1919*. London: Journeyman, 1985.
Serle, G. *John Monash*. Carlton, Victoria: Melbourne University Press, 1982.
Sillars, S. *Art and Survival in First World War Britain*. New York: St. Martin's Press, 1987.
Simkins, P. *Kitchener's Army*. Manchester: Manchester University Press, 1988.
Simon, B. and I. Bradley, *The Victorian Public School*. Dublin: Gill & Macmillan, 1975.
Simpson, A. *Hot Blood and Cold Steel*. London: Tom Donovan, 1993.
Skelly, A.R. *The Victorian Army at Home*. London: Croom Helm, 1977.
Smith, L.V. *Between Mutiny and Obedience*. Princeton: Princeton University Press, 1994.
Snelling, S. *VCs of the First World War – Gallipoli*. Sutton: Stroud, 1995.
Spears, E.L. *Two Men Who Saved France*. London: Eyre & Spottiswoode, 1966.
Spiers, E.M. *The Army and Society 1815–1914*. London: Longman, 1980.
Spiers, E.M. *Haldane: an Army Reformer*. Edinburgh: Edinburgh University Press, 1980.
Spiers, E.M. *The Late Victorian Army 1868–1902*. Manchester: Manchester University Press, 1992.
Springhall, J. *Youth, Society and Empire*. London: Croom Helm, 1977.
Stephen, M. *The Price of Pity*. London: Leo Cooper, 1996.
Steppler, G.A. *Britons, To Arms!* Stroud: Sutton, 1992.
Stevenson, J. *Popular Disturbances in England 1700–1870*. London: Longman, 1979.
Stevenson, J. *British Society 1914–45*. Harmondsworth: Penguin, 1984.
Stouffer, S.A. *et al., The American Soldier*. New York: Science Editions, two vols, 1965.
Thompson, A. *Anzac Memories*. Melbourne, Oxford University Press, 1994.

Thompson, F.M.L. *English Landed Society in the Nineteenth Century*. London: Routledge & Kegan Paul,1971.
Thompson, F.M.L. *The Rise of Respectable Society*. London: Fontana, 1988.
Thompson, P. *The Edwardians*. London: Paladin, 1977.
Travers, T. *The Killing Ground*. London: Unwin Hyman, 1987.
Travers, T. *How the War Was Won*. London: Routledge, 1992.
Waites, B. *A Class Society at War*. Leamington Spa: Berg, 1987.
Wagner, G. *The Chocolate Conscience*. London: Chatto & Windus, 1987.
Walvin, J. *Leisure and Society*. London: Longman, 1978.
Walvin, J. *A Child's World*. Harmondsworth: London, 1982.
Ward, S.R. *The War Generation*. Port Washington, NY: 1975.
Webb, B. *Edmund Blunden: a Biography*. New Haven: Yale University Press, 1990.
Williams, J. *Byng of Vimy*. London: Leo Cooper, 1983.
Williamson, D.G. *The British in Germany 1918–1930*. Oxford: Clarendon, 1991.
Wilson, T. *The Myriad Faces of War*. Cambridge: Polity, 1986.
Winter, D. *Death's Men*. Harmondsworth: Penguin, 1979.
Winter, D. *Haig's Command*. London: Viking, 1991.
Winter, D. *25 April 1915: the Inevitable Tragedy*. St Lucia: University of Queensland Press, 1994.
Winter, J.M. *The Great War and the British People*. London: Macmillan, 1987.
Winter, J.M. *Sites of Memory, Sites of Mourning*. Cambridge, Cambridge University Press, 1995.

(d) Theses

Badsey, S.D. 'Fire and the Sword: the British Army and the *Arme Blanche* Controversy, 1871–1921'. PhD, University of Cambridge, 1981.
Barr, N.J.A. 'Service not Self – the British Legion 1921–1939, PhD, University of St Andrews, 1994.
Bet-El, I.R. 'Experience into Identity: the Writings of British Conscript Soldiers, 1916–1918'. PhD, University of London, 1991.
Crang, J.A. 'A Social History of the British Army 1939–45'. PhD, University of Edinburgh, 1992.
Leese, P.J. 'A Social and Cultural History of Shellshock, with Particular Reference to Experience of British Soldiers during and after the Great War'. PhD, Open University, 1989.
Losinger, I.D. 'Officer–Man Relations in the Canadian Expeditionary Force, 1914–1919'. MA, Carleton University, 1990.
Muenger, E.A. 'The British Army in Ireland, 1886–1914'. PhD, University of Michigan, 1981.
Morris, P.M. 'Leeds and the Amateur Military Tradition: the Leeds Rifles and its Antecedents, 1815–1918'. PhD, University of Leeds, 1983.
Savill, L.E.M. 'The Development of Military Training in Schools in the British Isles'. MA, University of London, 1937.
Sheffield, G.D. 'The Effect of War Service on the 22nd Battalion Royal Fusiliers (Kensington) 1914–18, with Special Reference to Morale, Discipline and the Officer–Man Relationship'. MA, University of Leeds, 1984.
Spillan, G.F. 'Manpower Problems in the British Army 1918–1939: the Balancing of Resources and Commitments'. DPhil, University of Oxford, 1985.

Wahlert, G. 'Provost: Friend or Foe?: the Development of an Australian Provost Service 1914–1945'. MA, University of New South Wales, 1996.

Wilson, J.B. 'Morale and Discipline in the British Expeditionary Force, 1914–18'. MA, University of New Brunswick, 1978.

(e) Chapters, articles and unpublished papers

Andrews, E.M. 'Bean and Bullecourt: Weaknesses and Strengths of the Official History of Australia in the First World War', *RIHM*, 72 (1990).

Barlow, S. 'The Diffusion of "Rugby" Football in the Industrial Context of Rochdale, 1868–90: a Conflict of Ethical Values' *International Journal of the History of Sport*, X, (1993).

Barnett, C. 'A Military Historian's View of the Great War', *Essays by Divers Hands, Being the Transactions of the Royal Society of Literature* XXXVI (1970).

Baynes, J. 'The Officer–Other Rank Relationship in the British Army in the First World War', *Quarterly Review* (Oct. 1966).

Beckett, I.F.W. 'The Problem of Military Discipline in the Volunteer Force, 1859–1899', *JSAHR*, 56, (1978).

Beckett, I.F.W. 'The Nation in Arms', in Beckett and Simpson, *Nation*.

Beckett, I.F.W. 'The Territorial Force', in Beckett and Simpson, *Nation*.

Beckett, I.F.W. 'The Amateur Military Tradition in Britain', *W & S*, 4 (1986).

Beckett, I.F.W. 'The British Army 1914–18: the Illusion of Change', in Turner, J. (ed.) *Britain and the First World War*. London: Unwin Hyman, 1988.

Beckett, I.F.W. 'The Real Unknown Army: British Conscripts 1916–1919', *GW*, 2, (1989).

Bessell, R. and Englander, D. 'Up from the Trenches: Some Recent Writing on the Soldiers of the Great War', *European Studies Review*, 11 (1981).

Best, G. 'Militarism and the Victorian Public School', in Simon and Bradley, *Victorian Public School*.

Bogacz, T. 'War Neurosis and Cultural Change in England, 1914–22: the Work of the War Office Committee of Enquiry into 'Shell-Shock', *JCH* 24 (1989).

Bourne, J.M. 'The British Working Man in Arms', in Cecil and Liddle, *Facing Armageddon*. London: Leo Cooper, 1996.

Bourne, J.M. 'British Generals in the First World War' in Sheffield G.D. (ed.), *Leadership and Command: the Anglo-American Military Experience since 1861*. London: Brassey's, 1997.

Brett-James, A. 'Some Aspects of British and French Morale on the Western Front, 1914–1918', unpublished paper, 1978.

Bryant, C., 'Obedience and Power: Changing Power Relations in a Habsburg Army Regiment, 1914–15' unpublished paper, 1996.

Carrington, C.E. 'Some Soldiers', in Panichas, *Promise of Greatness*.

Carrington, C.E., Review of *The Great War and Modern Memory* in *Newsletter of the Friends of Amherst College*, Winter, 1976.

Carrington, C.E. 'Kitchener's Army: the Somme and After', *JRUSI*, 1 (1978).

Cecil, H. 'Henry Williamson: Witness of the Great War', in Sewell, B. (ed.), *Henry Williamson: the Man, the Writings*. Padstow: Tabb House, 1980.

Cecil, H. 'The Literary Legacy of the War: the Post-war British War Novel, a Select Bibliography', in Liddle, *Home Fires*.

Chell, R.A. 'My First Command – Service with an Essex Battalion', *Essex Countryside*, Dec. 1972.

Danchev, A. '"Bunking" and Debunking: the Controversies of the 1960s', in Bond, *First World War.*
Davies, R. 'Donald Hankey, a Student in Arms' *ST!* 47, Sept. 1996.
Dean, E.T. '"We Will All Be Lost and Destroyed" Post-Traumatic Stress Disorder and the Civil War', *Civil War History* XXXVII (1991).
Dewey, P.E. 'Military Recruiting and the British Labour Force during the First World War', *HJ* 27, (1984).
Denman, T. 'The Catholic Irish Soldier in the First World War: the "Racial Environment"', *IHS* XXVII, 108 (1991).
Diehl, J.M. 'Germany: Veterans' Politics – Under Three Flags', in Ward, *War Generation.*
Diest, W. 'The Military Collapse of the German Empire: the Reality of the Stab-in-the-Back Myth', *WIH* 3, (1996).
Edwardes, M. '"Oh to meet an Army man": Kipling and the Soldiers', in Gross, J. (ed.), *Rudyard Kipling: the Man, his Work and his World.* London: Weidenfeld & Nicolson, 1972.
Englander, D. 'The French Soldier, 1914–18', *French History*, 1 (1987).
Englander, D. 'Soldiering and Identity: Reflections On the Great War,' *WIH*, I
Englander, D. and Osborne, J. 'Jack, Tommy, and Henry Dubb: the Armed Forces and the Working Class', *HJ*, 21 (1978).
Gooch, J. 'The Armed Services' in Constantine *et al.*, *The First World War in British History.*
Greenhut, J. 'Sahib and Sepoy: an Inquiry into the Relationship between the British Officers and Native Soldiers of the British Indian Army', *MA* XLVIII (1984)
Grieves, K.R. 'Making Sense of the Great War: Regimental Histories, 1918–23', *JSAHR*, LXIX (1991).
Grieves, K.R. '"Lowther's Lambs": Rural Paternalism and Voluntary Recruitment in the First World War', *Rural History*, 4 (1993).
Grieves, K. 'C.E. Montague and the Making of *Disenchantment* 1914–21', *WIH*, 4 (1997).
Griffith, L.W. 'The Pattern of One Man's Remembering', in Panichas, *Promise of Greatness.*
Hanbury-Sparrow, A.A. 'Discipline or Enthusiasm?', *British Army Review*, 20 (1965).
Hargreaves, R. 'Promotion from the Ranks', *Army Quarterly*, LXXXVI (1963).
Hartley, L.P. 'Three Wars', in Panichas, *Promise of Greatness.*
Howard, M. 'Men against Fire: the Doctrine of the Offensive in 1914', in P. Paret (ed.), *The Makers of Modern Strategy.* Oxford: Clarendon, 1986.
Hurt, J.S. 'Drill, Discipline and the Elementary School Ethos' in McCann, P. (ed.), *Popular Education and Socialisation in the Nineteenth Century.* London: Methuen, 1977.
Inglis, K.S. 'Anzac and the Australian Military Tradition', *RIHM*, 72 (1990).
Jeffery, K. 'The Post-War Army', in Beckett and Simpson, *Nation.*
Joll, J. '1914: The Unspoken Assumptions' in Koch, H. (ed.), *The Origins of the First World War.* London: 1972.
Kennedy, P. 'Britain in the First World War', in Millett and Murray, *Military Effectiveness* I.
King, T. 'The Influence of the Greek Heroic Tradition on the Victorian and

Edwardian Military Ethos', unpublished paper, 1991.
de Lee, N. 'Oral History and British Soldiers' Experience of Battle in the Second World War' in Addison, P. and Calder, A. *Time to Kill*. London: Pimlico, 1996.
Little, R.W. 'Buddy Relations and Combat Performance', in M. Janowitz (ed.), *The New Military*. New York: Norton, 1969.
MacKenzie, S.P. 'Morale and the Cause: the Campaign to Shape the Outlook of Soldiers of the British Expeditionary Force, 1914–18', *Canadian Journal of History*, XXV (1990).
McKibbin, R. 'Why Was There no Marxism in Great Britain?', *English Historical Review*, 99 (1984).
Mackinlay, A. 'The Pardons Campaign', *WFA Bulletin*, 39 (June 1994).
Mangin, J.A. 'Athleticism: a Case Study of the Evolution of an Educational Ideology', in Simon and Bradley, *Victorian Public School*.
Melling, J. '"Non-Commissioned Officers": British Employers and their Supervisory Officers, 1880–1920', *Social History*, 5 (1980).
Messerschmidt, M. 'German Military Law in the Second World War', in Deist, W. (ed.), *The German Military in the Age of Total War*. Leamington Spa: 1985.
Mitchinson, K.W. 'The Reconstruction of 169 Brigade: July–October 1916', *ST!*, 29 (1990).
Mordike, J. 'The Story of Anzac: a New Approach', *JAWM*, 16 (1990).
Morris, J. 'Richard Aldington and *Death of a Hero* – or Life of an Anti-Hero?', in Klein, *The First World War in Fiction*.
Morton, D. 'The Canadian Military Experience in the First World War, 1914–18', in Adams, *Great War*.
Nasson, W. 'Tommy Atkins in South Africa', in Warwick, P. (ed.), *The South African War*. London, 1980.
Newby, H. 'The Deferential Dialectic', *Comparative Studies in Society and History*, 17 (1975).
Otley, C.B. 'Militarism and Militarization in the Public Schools, 1900–1972', *BJS*, 29 (1978).
Paget, D. 'Remembrance Play: *Oh! What a Lovely War* and History', in Tony Howard and J. Stokes, *Acts of War: the Representation of Military Conflict on the British Stage and Television since 1945*. Aldershot: Scolar, 1996.
Peacock, A.J. 'Crucifixion No. 2', *Gunfire*, 4 (1986).
Peacock, A.J. 'A Rendezvous with Death', *Gunfire*, 5 (1986).
Petter, M. '"Temporary Gentlemen" in the Aftermath of the Great War: Rank, Status and the Ex-Officer Problem', *HJ*, 37 (1994).
Philips, G. 'The Social Impact' in Constantine *et al.*, *First World War*.
Pollard, S. 'Factory Discipline in the Industrial Revolution', *Economic History Review*, XVI (1963).
Porch, D. 'The French Army in the First World War', in Millett and Murray, *Military Effectiveness* I.
Prior, R. and Wilson, T. 'Paul Fussell at War', *WIH*, 1 (1994).
Putkowski, J. 'Toplis, Etaples & "The Monocled Mutineer"', *ST!*, 18 (1986).
Putkowski, J. 'British Army Mutinies in World War One', unpublished paper.
Putkowski, J. 'The Mutiny of 1st Bn. Connaught Rangers', unpublished paper.
Putkowski, J. 'A2 and the "Reds in Khaki"', *Lobster*, 27 (1994).
Putkowski, J. 'A2' *Lobster*, 28 (1994).

Razzell, P.E. 'The Social Origins of British Officers in the Indian and British Home Army', *BJS*, 14 (1963) 248-60.

[Scott, P.T. ed.]. 'A Dugout in War and Before', serialised in *GW*, from 1 (May 1989) to 3 (Nov. 1990).

[Scott, P.T. ed.]. 'Mesopotamian Diary: With the 5th Buffs along the Tigris 1915–16', serialised in *GW* from 1 (May 1989) to 3 (Nov. 1990).

Scott, P.T. 'Law and Orders: Discipline and Morale in the British Armies in France, 1917', in Liddle, P.H. *Passchendaele in Perspective*. London: Leo Cooper, 1997.

Seaton, R.W. 'Deterioration of Military Work Groups under Deprivation Stress', in Janowitz, *New Military*.

Sheffield, G.D. 'The Effect of the Great War on Class Relations in Britain: the Career of Major Christopher Stone DSO MC', *W&S*, 7 (1989).

Sheffield, G.D. '"Disillusionment" and Other Myths of British Army Morale in the First World War', unpublished paper, 1992.

Sheffield, G.D. 'The Operational Role of British Military Police on the Western Front, 1914–18', in Griffith, *British Fighting Methods*.

Sheffield, G.D. *The Morale of the British Army on the Western Front, 1914–1918*. Occasional Paper 2, Institute for the Study of War and Society, De Montfort University Bedford, 1995.

Sheffield, G.D. '"Oh! What a Futile War": Representations of the Western Front in the Modern British Media and Popular Culture', in I. Stewart and S. Carruthers, *War, Culture and the Media*. Trowbridge: Flicks Books, 1996.

Shepherd, B. 'Shell-Shock on the Somme', *JRUSI*, 141 (1996).

Sherriff, R.C. 'The English Public Schools in the War', in Panichas, *Promise of Greatness*.

Shills, E.A. and Janowitz, M. 'Cohesion and Disintegration in the Wehrmacht', Lerner, D. (ed.), *Propaganda in War and Crisis*. New York: Arno, 1972.

Simkins, P. 'The Four Armies 1914–18' in Chandler, D. and Beckett, I. *The Oxford History of the British Army*. Oxford: Oxford University Press, 1994.

Simkins, P. 'The War Experience of a Typical Kitchener Division: the 18th Division, 1914–18', in Cecil and Liddle, *Facing Armageddon*.

Simkins, P. 'Co-Stars or Supporting Cast? British Divisions in the "Hundred Days", 1918', in Griffith, *British Fighting Methods*.

Simpson, D.M. 'Morale and Sexual Morality among British troops in the First World War', unpublished paper, 1996.

Simpson, K. 'Capper and the Offensive Spirit', *JRUSI*, 118 (1973).

Simpson, K. 'The Officers', in Beckett and Simpson, *Nation*.

Simpson, K. 'The British Soldier on the Western Front', in Liddle, *Home Fires*.

Simpson, K. 'Dr. James Dunn and Shell-Shock', in Cecil and Liddle, *Facing Armageddon*.

Sly, J.S. 'The Men of 1914', *ST!*, 35 (1992).

Smith, C.N. 'The Very Plain Song of It: Frederic Manning, Her Privates We', in Klein, *The First World War in Fiction*.

Smith, L.V. 'The Disciplinary Dilemma of French Military Justice, September 1914–April 1917: the Case of 5e Division d'Infanterie', *Journal of Military History*, 55 (1991).

Smith, L.V. 'War and "Politics": the French Army Mutinies of 1917', *WIH*, 2 (1995).

Smith, M. 'The War and British Culture', in Constantine, S., Kirby, M.W. and Rose, M.B. *The First World War in British History*. London: Arnold, 1995.
de Sola Pinto, V. My First War', in Panichas, *Promise of Greatness*.
Spiers, E.M. 'The British Cavalry 1902–14', *JSAHR*, LVII (1979).
Spiers, E.M. 'The Regular Army', in Beckett and Simpson, *Nation*.
Spiers, E.M. 'The Scottish Soldier at War' in Cecil and Liddle, *Facing Armageddon*.
Summers, A. 'Edwardian Militarism' in Samuel, R. *Patriotism: the Making and Unmaking of the British National Identity*, vol. I. London: Routledge: 1989.
Teagarden, E.M. 'Lord Haldane and the Origins of the Officer Training Corps', *JSAHR*, XLV (1967).
Thompson, E.P. 'Time, Work Discipline and Industrial Capitalism', *Past and Present*, 38 (1967).
Travers, T.H.E. 'From Surafend to Gough: Charles Bean, James Edmonds, and the Making of the Australian Official History', *JAWM*, 27 (1995).
Various authors, 'Condemned: Courage and Cowardice, *JRUSI*, 143 (1998).
Veitch, C. 'Play up! Play up! and Win the War': Football, the Nation and the First World War 1914–15', *JCH* I, 20 (1985).
Waites, B. 'The Government of the Home Front and the "Moral Economy" of the Working Class', in Liddle, *Home Fires*.
Ward, S.R. 'Great Britain: Land Fit for Heroes Lost', in Ward, *War Generation*.
Waugh, A. 'A Light Rain Falling', in Panichas, *Promise of Greatness*.
Whitehead, I. 'Not a Doctor's Work? The Role of the British Regimental Medical Officer in the Field', in Cecil and Liddle, *Facing Armageddon*.
Williams, D. 'An Artilleryman's War 1914–19', *ST!*, 29 (1990).
Worthington, I. 'Antecedent Education and Officer Recruitment': the Origins and Early Development of the Public School–Army Relationship', *Military Affairs*, XLI (1977).
Wyatt, R.J. 'The Major "Minor Horror of War"', *ST!*, 23 (1988).

Index